TELEPHONY & TELEG

TELEPHONY AND TELEGRAPHY

AN INTRODUCTION TO INSTRUMENTS AND SWITCHING SYSTEMS
(ELECTROMECHANICAL AND REED-ELECTRONIC)

by
SYDNEY F. SMITH
B.Sc.(Eng.), C.Eng., M.I.E.E.

THIRD EDITION

OXFORD UNIVERSITY PRESS
1978

Oxford University Press, Walton Street, Oxford OX2 6DP

OXFORD LONDON GLASGOW NEW YORK
TORONTO MELBOURNE WELLINGTON CAPE TOWN
IBADAN NAIROBI DAR ES SALAAM LUSAKA ADDIS ABABA
KUALA LUMPUR SINGAPORE JAKARTA HONG KONG TOKYO
DELHI BOMBAY CALCUTTA MADRAS KARACHI

Casebound ISBN 0 19 859358 9
Paperback ISBN 0 19 859357 0

First edition 1969
Reprinted 1972
Second edition 1974
Third edition 1978

Filmset by BAS Printers Limited, Over Wallop, Hampshire and printed in Great Britain by J. W. Arrowsmith Ltd., Bristol.

CONTENTS

Preface Page xi

Acknowledgements xii

How to use this book xiii

Chapter 1. TELEPHONES AND TELEPRINTERS 1

Part 1. Introduction 1

1.1.1 What is telecommunication? 1
1.1.2 Topics discussed in this first chapter 1

Part 2. General Principles 1

1.2 Telephone and telegraph networks 1
1.2.1 What are telephony and telegraphy? 1
1.2.2 Exchanges and switching 3
1.2.3 General description of the public telephone network 3
1.2.4 General description of the public telegraph network 4

1.3 Principles of speech apparatus 5
1.3.1 The carbon granule microphone (transmitter) 5
1.3.2 The simple telephone receiver 5
1.3.3 The magneto bell (ringer) 6
1.3.4 Simple induction coil 6
1.3.5. Anti-sidetone induction coil 9
1.3.6 Local battery and central battery operation 10

1.4 Principles of subscribers' signalling apparatus 11
1.4.1 Contacts and springset operation 11
1.4.2 The dial 12
1.4.3 Start–stop telegraph instruments (teleprinters) 13
1.4.4 Teleprinter transmitters 14
1.4.5 Automatic transmitters 15
1.4.6 Teleprinter receivers 18

Part 3. Typical Practical Apparatus 18

1.5 Telephone instruments 18
1.5.1 Auto. and C.B. telephone (B.P.O. type 332) 18
1.5.2 Auto. telephone (B.P.O. type 706) 19

1.6 The trigger dial 21

Part 4. Summary 1.7.1—1.7.3 25

Part 5. Self-teaching Notes 1.8.1—1.8.5 27

Chapter 2. RELAYS 29

Part 1. Introduction 29

2.1.1 Definition of a relay 29
2.1.2 Uses of relays 29

2.1.3 Topics discussed in this chapter 30

Part 2. Principles of Relay Construction and Operation 30

2.2 Telephone relays 30

2.2.1 Magnetic circuit 30

2.2.2 Contacts 31

2.2.3 Operating and releasing characteristics 33

2.2.4 Operate and release times 34

2.2.5 Slugged relays 37

2.2.6 Circuit methods of controlling relay timings 38

2.2.7 Isthmus armatures 38

2.2.8 Nickel-iron sleeves 40

2.2.9 Polarization of telephone relays 40

2.3 Telegraph relays 42

2.3.1 General features 42

2.3.2 Polarization of telegraph relays 43

2.4 Relay circuits and diagrams 44

Part 3. Typical Practical Relays 46

2.5 Typical telephone relays 46

2.5.1 British post office 3000 type 46

2.5.2 British post office relays types 12 and 600 48

2.5.3 High speed relays 48

2.6 Typical telegraph relays 49

2.6.1 Polarized telegraph relay, type 299 etc. 49

2.6.2 The Carpenter relay (P.O. polarized relay 2B) 51

2.7 Special purpose relays 52

2.7.1 Thermal relays 52

2.7.2 Dry reed relay 53

Part 4. Summary 2.8.1—2.8.3 54

Part 5. Self-teaching Notes 2.9.1—2.9.4 56

Chapter 3. SIGNALLING 58

Part 1. Introduction 58

3.1.1 Why is signalling necessary? 58

3.1.2 Topics discussed in this chapter 58

Part 2. Signalling Principles 58

3.2 Signals required 58

3.2.1 Subscribers' lines 58

3.2.2 Junctions between exchanges 59

3.2.3 Telegraph signals 59

3.3 Types of signal 59

3.3.1 Single-wire signals 60

3.3.2 A.C. and d.c. signalling 61

3.3.3 Loop signalling 62

3.3.4 Loop-disconnect pulsing 63

3.3.5 Single-current signals 63

3.3.6 Double-current signals 65

3.3.7 Tones 65

3.3.8 Differential signalling 66

3.3.9 Voice frequency (v.f.) signalling 67

3.4 Telegraph signalling codes 67

3.4.1 Morse code 67
3.4.2 Five-unit code 68

Part 3. Typical Signalling Systems 69

3.5 Subscribers' line signalling 69
3.5.1 Dial telephone system 69
3.5.2 Multi-frequency (m.f.) signalling telephone 71
3.5.3 Automatic Telex subscriber's line 72

3.6 Telephone exchange junction signalling 74
3.6.1 Loop signalling junctions 74
3.6.2 Signalling system AC9 75

Part 4. Summary 3.7.1—3.7.2 77

Part 5. Self-teaching Notes 3.8.1—3.8.2 78

Chapter 4. SWITCHING SYSTEMS 80

Part 1. Introduction 80
4.1.1 Manual exchanges 80
4.1.2 Automatic exchanges 80
4.1.3 Topics discussed in this chapter 80

Part 2. Switching System Principles 80

4.2 Manual exchanges 80
4.2.1 General facilities 80
4.2.2 Co-ordinate switching 81
4.2.3 Plug-and-cord switching 82
4.2.4 Auto-manual and trunk switchboards 84

4.3 Automatic exchanges 85
4.3.1 Advantages of automatic switching 85
4.3.2 Step-by-step operation 86
4.3.3 Strowger selectors 86
4.3.4 Strowger trunking principles 88
4.3.5 Common control 92
4.3.6 The crossbar switch 92
4.3.7 Use of crossbar switches 94
4.3.8 Blocking 97
4.3.9 Common control of matrix switches 98

Part 3. Typical Practical Systems 99

4.4 Strowger 99
4.4.1 Non-director 99
4.4.2 Director 100

4.5 Crossbar 101
4.5.1 Crossbar switch mechanisms 101
4.5.2 British Post Office system TXK 1 (Plessey 5005) 104
4.5.3 British Post Office system TXK 3 (ITT Pentaconta) 105

4.6 Reed electronic 107
4.6.1 British Post Office system TXE 2 (Plessey Pentex) 107
4.6.2 British Post Office system TXE 4 (STC) 108

Part 4. Summary 4.7.1—4.7.2 111

Part 5. Self-teaching Notes 4.8.1—4.8.4 114

Chapter 5. STROWGER SYSTEM—SELECTOR MECHANISMS 115

Part 1. Introduction 115

5.1.1 Why Strowger? 115
5.1.2 What is a selector mechanism? 115
5.1.3 Topics discussed in this chapter 115

Part 2. Principles of Strowger Selectors 115

5.2 Uniselector principles 115
5.2.1 Uniselector mechanisms 115
5.2.2 Uniselector banks and wipers 117

5.3 Two-motion selector principles 120
5.3.1 Vertical and rotary stepping mechanisms 120
5.3.2 Interrupter contacts 122
5.3.3 Wipers and banks (two-motion selectors) 123
5.3.4 Mechanically operated springsets 125

Part 3. Typical Practical Switching Apparatus 126

5.4 Typical selectors 126
5.4.1 25-point uniselector (B.P.O. types 2 and 3) 126
5.4.2 Pre-2000 type two-motion selector 128
5.4.3 2000 type two-motion selector 129
5.4.4 4000 type two-motion selector 132

5.5 Selector racks 135

Part 4. Summary 5.6 137

Part 5. Self-teaching Notes 5.7.1—5.7.4 138

Chapter 6. STROWGER SYSTEM—SELECTOR CIRCUITS 140

Part 1. Introduction 140

6.1.1 Relay control of selectors 140
6.1.2 Topics discussed in this chapter 140

Part 2. Circuit Elements 141

6.2 Line signalling 141

6.3 Pulse controlled stepping circuits 143
6.3.1 *B* relay 143
6.3.2 *CD* relay 145
6.3.3 Selector vertical stepping circuit 145
6.3.4 Selector vertical and rotary stepping circuit 147

6.4 *P*-wire and rotary hunting circuits 149
6.4.1 Use of the 'private' wire (*P*-wire) 149
6.4.2 Uniselector absence-of-earth testing 150
6.4.3 Group selector absence-of-earth testing 152
6.4.4 Absence-of-earth dual (200 outlet) testing 154
6.4.5 Battery testing 155
6.4.6 Final selector outlet testing 158
6.4.7 *P*-wire metering 159

6.5 Selector release circuits 160

Part 3. Typical Complete Circuits 161

6.6 Telephone exchange selector circuits 161
6.6.1 Subscribers' uniselector line circuit 161
6.6.2 Group selector (absence-of-earth testing) 163

6.6.3 Final selector 165
Part 4. Summary 6.7.1—6.7.2 168
Part 5. Self-teaching Notes 6.8.1—6.8.4 171

Chapter 7. TRAFFIC AND TRUNKING 173
Part 1. Introduction 173
7.1.1 What is traffic? 173
7.1.2 Trunking 173
7.1.3 Topics discussed in this chapter 173
Part 2. Theory of Traffic and Grading 173
7.2 The nature of traffic 173
7.2.1 Units of measurement 173
7.2.2 Traffic variations 176
7.2.3 The erlang and 'probability' 178
7.3 Effect of traffic on plant provision 180
7.3.1 Grade of service 180
7.3.2 Traffic efficiency of large groups 181
7.3.3 Traffic data for new exchanges 182
7.4 Principles of grading 183
7.4.1 Full availability and limited availability 183
7.4.2 Interconnecting 185
7.4.3 Grading 186
7.5 The design of gradings 188
7.5.1 Number of graded groups 188
7.5.2 'Best' gradings 189
7.5.3 Four-group gradings (algebraic method) 191
7.5.4 Six-group gradings (algebraic method) 192
7.5.5 Larger gradings (transposition method) 193
Part 3. Typical Practical Grading Arrangements 194
7.6 Grading terminals 194
7.6.1 Trunk distribution frame (T.D.F.) 195
7.6.2 Grading on uniselector rack 197
7.6.3 Grading on group selector racks 197
7.7 Grading charts 199
Part 4. Summary 7.8.1—7.8.2 200
Part 5. Self-teaching Notes 7.9.1—7.9.6 203

Chapter 8. DISTRIBUTION FRAMES 205
Part 1. Introduction 205
8.1.1 Definition of a distribution frame 205
8.1.2 Topics discussed in this chapter 205
Part 2. Facilities Provided by Distribution Frames 205
8.2 Exchange cabling 205
8.3 Method of jumpering 208
8.4 Protection and testing 209
Part 3. Construction and Apparatus of Typical Frames 210
8.5 Main distribution frame 210
8.6 Intermediate distribution frame 215
8.7 Jumpers 217

Part 4. Summary 8.8 — 217

Part 5. Self-teaching Notes 8.9.1—8.9.3 — 219

Chapter 9. EXCHANGE POWER PLANT — 221

Part 1. Introduction — 221

9.1.1 Power requirements — 221

9.1.2 Topics discussed in this chapter — 221

Part 2. Principles of Exchange Power Supplies — 221

9.2 Sources of energy and power — 221

9.2.1 Units — 221

9.2.2 Prime movers — 223

9.2.3 Mains supplies and a.c. machines — 223

9.2.4 Direct current machines — 226

9.2.5 Mains rectifiers — 228

9.3 Storage batteries — 232

9.3.1 Secondary cells — 232

9.3.2 Charge-discharge and float operation — 234

9.4 Power switchboard — 234

9.4.1 General — 234

9.4.2 Switchboard apparatus for float systems — 237

Part 3. Typical Practical Exchange Power Plants — 241

9.5 Telephone exchange power plants — 241

9.5.1 Charge discharge system — 241

9.5.2 Partial float (parallel battery) system — 242

9.5.3 Full float system — 244

Part 4. Summary 9.6.1—9.6.2 — 244

Part 5. Self-teaching Notes 9.7.1—9.7.4 — 247

Appendix 1 International Morse code — 249

2 International five-unit teleprinter code — 250

3 Effect of polarization — 251

4 Maximum power transfer (matching) — 253

5 Circuit diagram symbols — 255

6 Bibliography — 256

ANSWERS (to questions in self-teaching notes) — 257

INDEX — 273

PREFACE TO THE THIRD EDITION

The earlier editions were written to meet the needs of those studying for the *City and Guilds of London Institute* examination in *Telephony and Telegraphy A*. With the advent of the *Technician Education Council (TEC)*, technical education in the UK is now based on programmes designed by individual colleges or groups of colleges and comprising some college-devised units and some TEC standard units or local adaptations of these standard units

The need for this third edition arises from these changes. It is still, however, primarily a students' book intended for those embarking on a career in the industry. As well as trainee technicians following TEC programmes, this includes engineers transferring from related fields and newly qualified graduates, both of whom will need to know something of the established techniques in the specialized fields of telephony and telegraphy.

Chapter 4 has been substantially revised to include outline descriptions of the principal systems currently in service in the UK, Strowger director and non-director, crossbar TXK 1 and TXK 3, and reed-electronic TXE 2 and TXE 4. For consistency, some material on Strowger formerly in Chapter 5 now appears in Chapter 4.

In addition to all the material for the TEC standard unit *Telephone Switching Systems II*, this third edition retains those parts of the original text necessary to present a balanced view of the subject for the general reader. The telephone switching parts of TEC standard unit *Telecommunications Systems I* are also included to meet the needs of TEC level II students having exemption from level I. Much of the original text will also be of value to students taking TEC *Telephone Switching Systems III*.

Joyden's Wood S.F.S.
1977

ACKNOWLEDGEMENTS

My immediate source for the first edition was my own set of lecture notes. Compiled originally mainly from the sources listed in the bibliography (Appendix 6), they were continuously revised from my frequent contact with many people actively engaged in the industry, in particular those with Associated Electrical Industries Ltd, The General Electric Company Ltd and the British Post Office. I thank all these people for their generous assistance in this, especially Mr N. C. Smart and Mr T. R. Hockney (GEC); my former colleagues Mr K. D. H. Beck, Mr R. W. Hunter, Mr A. E. Lee, Mr R. Luff and Mr W. H. Philpott (AEI); and the staffs of New Cross, Lee Green, Greenwich and Duncan telephone exchanges in London.

The method of presentation of the material, including the provision of 'self-teaching' notes, owes much to the preliminary discussions involving Mr J. C. Cartwright and Mr B. Winch of South East London Technical College and members of staff of Oxford University Press. I am also indebted to Mr B. Winch for much helpful criticism of the manuscript.

My thanks are due also to the City and Guilds of London Institute for permission to reproduce questions from past examination papers in *Telephony and Telegraphy A*. The answers and the comments on the answers to these questions are my own, however, and no responsibility for their accuracy or otherwise can be attributed to the Institute.

Thanks are due also to my employers, Standard Telephones & Cables Ltd, for permission to publish the third edition and for the use of the photograph on the cover.

S.F.S.

Some advice on how to use this book

1 It is unlikely that all readers will want the same things from this book. It is certainly no thriller so probably no one will start at page 1 and read steadily on to the end!

The following notes are written with the intention of helping you to get just what you want from these pages with as little trial and error or fruitless searching as possible.

2 **An introduction to telephony or telegraphy** for those who already have some training or experience in some related field is most easily obtained by selective reading. To facilitate this, each chapter is divided into parts as follows:

Part 1—Introduction
2—Principles
3—Practical apparatus
4—Summary
5—Self-teaching notes and exercises.

If, for example, you require principles of the subject without details of practical apparatus, you would read part 2 of each chapter only.

3 **Reading to learn** does *not* mean scanning the text and glancing casually at the odd diagram here and there nor does it mean reading solidly through at a constant pace from beginning to end.

You should give each passage the consideration it deserves. Pass quickly over introductions and things you already know, read more than once anything you find particularly difficult, refer back to previous paragraphs where necessary and study each diagram and worked example carefully to make sure you see its connexion with what is said in the text.

You may also find it helps if you make brief notes of anything which strikes you as important as you go along.

4 **A complete course** is provided here, for those who need it, covering the old City and Guilds of London Institute examination in *Telephony and Telegraphy A* as well as the new Technician Education Council (TEC) programme in *Telephone Switching Systems II* together with the telephone switching parts of TEC *Telecommunications Systems I.*

It is not intended that students should obtain TEC certificates without proper tutorial supervision. If however you are obliged to be absent temporarily from a college course you have been attending or to miss part of the course by changing colleges in mid-session, the *self-teaching notes* at the end of each chapter enable you to use this book as a substitute. Remember, however, that they are *self*-teaching notes: a book cannot teach—you have to do it yourself! This means doing the test questions conscientiously:
Answer the short questions in writing, not in your head.
Answer the longer (examination type) questions fully, as if taking an examination or doing a class exercise but do not spend more than half an hour on a question. When checking your answers, you will have to decide whether what you have written means

the same as the answer given. If in doubt, re-read the appropriate part of the text as indicated on the answer page.

5 Revision is something you should do at convenient intervals during any course and again at the end. Do not leave it until the last minute.

The *summary* at the end of each chapter should be used for revision. Look through it and if anything in it does not appear to make sense to you, it probably means you have forgotten it, so look it up.

Next, glance through the questions in the self-teaching notes, checking the answers to any questions you are not sure of, and if necessary refer to the paragraph of the text indicated on the answer page.

6 Reference to specific points can be made in two ways. Generally, for small items, the best way is to consult the *index*.

For more general topics, or items you cannot find in the index, use the *contents* list at the front of the book. The subject matter of the book has been arranged as far as possible in a logical manner so that similar items are grouped together in the same part of the book.

7 The TEC programme content is subject to variations from college to college to meet the needs of local industry. Students are advised to consult their tutors if they need guidance on which topics to study in depth.

1 · TELEPHONES AND TELEPRINTERS

For notes on how to use this book, see page xiii
For self-teaching notes for this chapter, see page 27

PART 1 INTRODUCTION

1.1.1 What is telecommunication? You know what 'communication' is, of course. When you know something which you want someone else to know, you tell them. That is communication—the exchange of information or the passing of messages. It includes things like talking and writing and is clearly an important part of our lives.

Sometimes, however, information must be conveyed over long distances almost at once if it is to be of any use. A distress signal from a sinking ship or a telephone call to arrange a meeting with a friend are two different examples of this. For these, we need the special forms of communication known as *telecommunication*.

The ancient Greek word 'tele' means 'far', so the word 'telecommunication' means literally 'communication over long distances'.

There are, of course, many ways of communicating over long distances, e.g. smoke signals, beacons, drums and semaphore. For engineering purposes, though, we usually take telecommunication to include only the electrical systems of long distance communication. Smoke signal technology has little in common with the telephone!

We can therefore define telecommunication more fully as *sending information by electrical means over distances which can be greater than the normal range of the senses.*

The two forms of telecommunication dealt with in this book are:

(*a*) *telephony* (from the Greek word 'phone', sound) which enables speech to be conveyed beyond the range of the human voice and hearing, and
(*b*) *telegraphy* (from the Greek word 'grapho', I write) which enables written messages to be sent similar distances much faster than a letter can be carried.

1.1.2 In this first chapter, you will learn exactly what telephony and telegraphy are and how and where each is used. We shall then see how the devices work which are used for sending and receiving messages in each of these two systems with practical examples of each.

PART 2 GENERAL PRINCIPLES

1.2 Telephone and telegraph networks

1.2.1 What are telephony and telegraphy? In *telephony*, messages are sent and received in spoken form using instruments called telephones. The sound waves comprising speech are used to control the transmission of electrical signals in the form of alternating currents at the same frequencies as the sound waves and these are used to reproduce the original sounds at the receiving end. The essential parts of a simple telephone circuit are illustrated in Fig. 1.1a. The operation of these component parts is described later (**1.3.1** and **1.3.2**).

The telephone is easy to use without special training. This means that people making calls can converse directly with each other but there is not normally any permanent record of messages sent or received and both parties must be available for the call at the same time.

A sound wave travels through the air as a rapid succession of changes in the pressure of the air at each point as it passes. Each small pressure increase is followed by a decrease of pressure; this cycle is repeated continuously. The number of complete cycles per second is called the *frequency* of the sound wave.

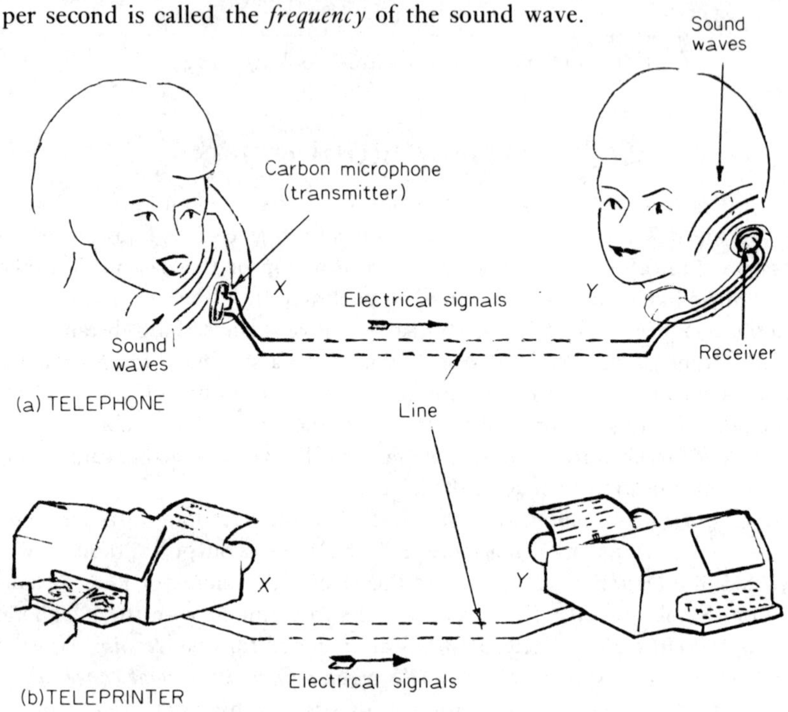

Fig. 1.1 Telephony and telegraphy. In the telephone system (a), the words spoken at *X* are reproduced at *Y*. In the teleprinter telegraph system (b), the message typed at *X* is printed at *Y* (1.2.1).

The unit of frequency is the hertz (Hz).
A frequency of 1 Hz is 1 cycle per second.

Speech consists of a complex combination of such sound waves at a number of different frequencies from about 50 Hz to 10 000 Hz. To provide for the transmission of this complete range of frequencies would be too expensive so telephone systems are designed in practice for a frequency range of 300 Hz to 3 400 Hz (i.e. a *bandwidth* of 3 100 Hz). This is adequate for the words to be understood and for the talker's voice to be recognizable.

In *telegraphy*, messages are sent and received letter by letter as in writing, each letter being represented by its own unique combination of electrical signals according to a pre-arranged code.

One of the simpler forms of telegraphy uses the *Morse code* in which each letter is represented by a combination of long and short pulses of current known as dashes and dots (see Appendix 1). These pulses can be produced by a hand-operated switch called a *Morse key* and can be used to operate a sounder at the receiving end. The sounds so produced have to be interpreted by a skilled operator who writes the message down.

In modern public telegraph systems, messages are typed on machines called *teleprinters* which have keyboards similar to typewriters. The machine sends the appropriate signals over the line to control another teleprinter at the receiving end which automatically prints the message on paper (see Fig. 1.1b).

To use a Morse key or to make efficient use of a teleprinter requires skill and so trained operators are employed to work them. A permanent record of messages is available, however, and messages can be received automatically so that they can be sent at a time convenient to the sender even if the called subscriber or receiving operator is absent.

Telegraph signals are simpler than speech and can therefore use a narrower bandwidth (approx. 100 Hz) than telephone signals. The telegraph signals convey only the words of a message but the more complicated speech currents of a telephone system enable the listener to recognize the talker and the expression or intonation of his voice.

1.2.2 Exchanges and switching. In the simple examples shown in Fig. 1.1, the user of either the telephone or telegraph instrument (at *X*) can only make a call to one other place (*Y*). There are, indeed, some cases where this is all that is required but the most useful thing about modern telephone and telegraph systems is that any subscriber can call any other subscriber connected to the same system. This does not mean that every telephone, for example, must have its own separate line to every other telephone. That would need millions of wires connected to each instrument which is obviously impossible. In practice, each instrument is connected by its own line to a central office or *exchange*.

In some cases a number of instruments may be connected to one line so that any message sent by one can be received by them all.

A *telegraph* network may operate either as a *point-to-point* system or as a *switched* system. In point-to-point operation, messages are recorded either by human operators or by automatic machines (**1.4.5**) at the exchanges and then re-transmitted to the called subscriber via other exchanges as required. In the switched method of operation, subscribers' lines are connected to each other at the exchanges for each call as required so that messages can be sent directly from one subscriber to another.

The advantage of switching over point-to-point operation is that it requires fewer operating staff at the exchanges, provides a faster service and enables any queries concerning a message to be dealt with immediately. For these reasons switching is now used at all exchanges in the public telegraph network in Great Britain.

In the case of a public *telephone* service, point-to-point operation would mean having an operator repeat all conversations which would be quite unacceptable to subscribers. Point-to-point operation cannot be used therefore and all public telephone systems include switching facilities at the exchanges.

Both telephone and telegraph switching can be either *manual* or *automatic* (auto.). In *manual* switching systems, connexions are set up through the exchanges by human operators as requested verbally by the calling subscribers. In *automatic* switching systems, the connexions are set up by automatic apparatus under the control of the calling subscriber's dial or similar signalling device.

1.2.3 General description of the public telephone network. Telephone instruments are installed at private subscribers' premises or made available to the general public at call offices (paystations) with coin collecting boxes (C.C.B.). Each of these installations is connected by a pair of wires called the *subscriber's line* to a *local* telephone exchange which typically serves up to about 10 000 lines. In some cases more than one subscriber's instrument may be connected to the same line which is then described as a *party line*. In some countries a rural party line may serve as many as twenty subscribers but in the U.K. the only party lines are those serving two subscribers per line on a system known as *shared service*.

Each local exchange provides the means of connecting any of its subscribers' lines to each other. The local exchange is also connected to other similar exchanges in the same area (e.g. the same town) by further lines (pairs of wires) called *junctions*. For calls outside this area, junctions are provided to *trunk* exchanges which provide the means of connecting calls over the long distance routes (trunk lines) to subscribers in other areas anywhere in the world. These long distance circuits may use cables (lines) or they may use radio for transmission between the exchanges.

There are too many trunk exchanges for each to have its own route to all others so two or more ranks of exchange of different status are provided. In the UK, for example, there are three ranks of trunk exchange as shown in Fig. 1.2.

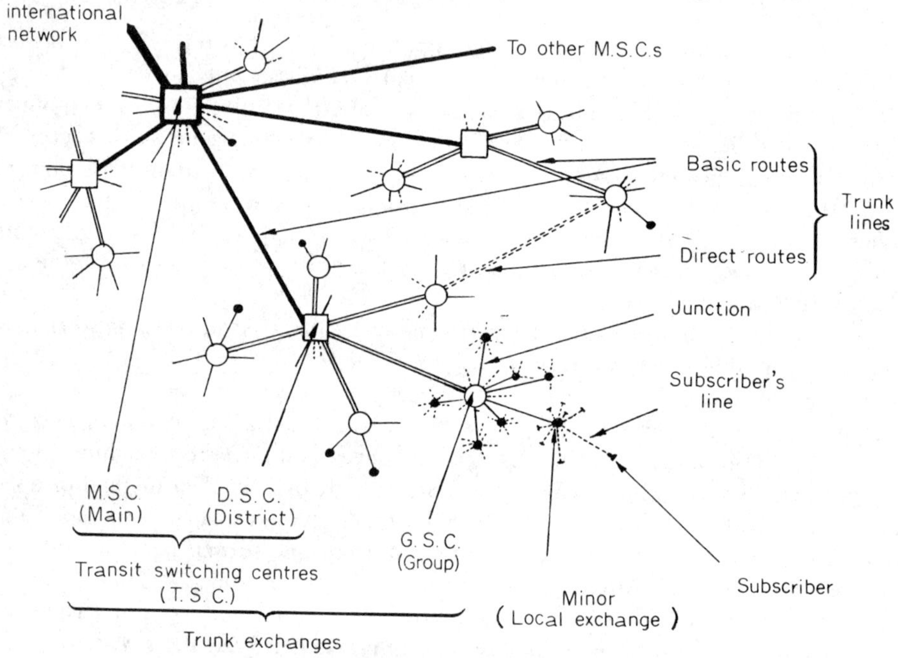

Fig. 1.2 Typical telephone exchange network (1.2.3).

For every exchange, there is an optimum or ideal position, not necessarily at the centre of the area, at which the line plant required would be a minimum. Each local exchange is built on the nearest suitable site to this position. When siting a trunk exchange, the optimum position for it is determined and the trunk exchange is then usually installed in the same building as the nearest existing local exchange to that position. For this reason, it can be seen in Fig. 1.2 that the trunk exchanges are shown as serving the subscribers in their immediate area in the capacity of a local exchange.

1.2.4 General description of the public telegraph network. The *Telex* system in the U.K. consists of teleprinters at subscribers' premises interconnected over a network of lines and automatic exchanges similar to but separate from those serving telephone subscribers. Post Office operated teleprinter offices connected to the Telex network are used for transmitting messages on behalf of members of the public If these messages are handed in at a post office in writing, they are called *telegrams* or, if they are dictated verbally from a private telephone or from a public telephone call box, they are called *phonograms*. Sub post offices do not usually handle enough telegram traffic to justify the cost of a teleprinter so their telegrams are dictated to and from a Head post office. This is known as *telephone–telegram* working.

1.3. Principles of speech apparatus

1.3.1 The carbon-granule microphone (transmitter). The carbon microphone illustrated in Fig. 1.3 works by causing variations in the direct current (d.c.) supplied by a battery. It does not generate an alternating electromotive force (e.m.f.) of its own as some other types of microphones do.

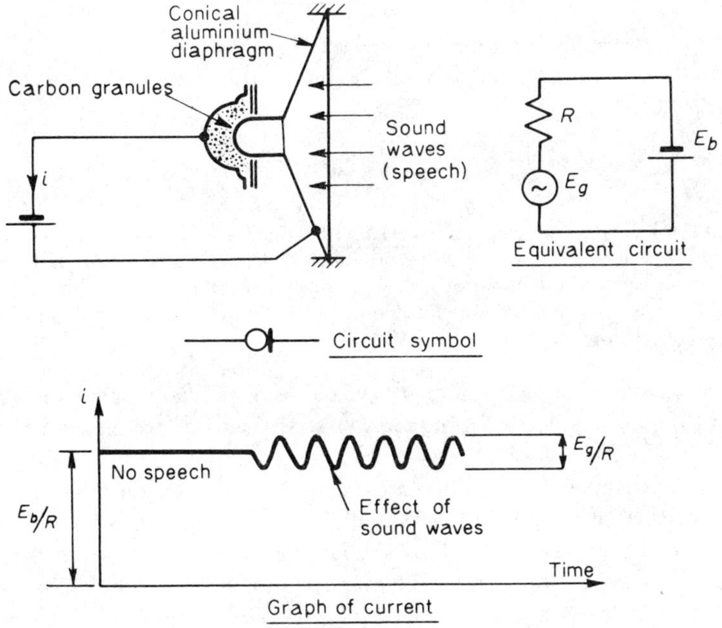

Fig. 1.3 Principle of carbon microphone. The sound waves cause the diaphragm to vibrate producing resistance variations in the carbon granules and thus fluctuations in the current i (1.3.1).

Sound waves (**1.2.1**) cause the flexible cone-shaped diaphragm to vibrate, alternately increasing and decreasing the pressure it exerts on the carbon granules. Increasing the pressure on the granules brings them into closer contact and this reduces the electrical resistance between them so that the current increases. Decreasing the pressure increases the electrical resistance and reduces the current. The direct current fluctuates in this way at the same frequency as the original sound waves.

The effect is exactly the same as connecting an alternating current (a.c.) generator at the speech frequency in series with the battery so that the current consists of both a.c. and d.c. components. The a.c. component is called the *speech current* and the d.c. component is called the *microphone feed current* (transmitter feed) or the *polarizing current*.

1.3.2 The simple telephone receiver. The receiver is shown in Fig. 1.4 and consists of a flexible iron diaphragm in the form of a disk positioned in the field of an electromagnet polarized by a permanent magnet.

Alternating speech currents from the line flow in the electromagnet coils and produce an alternating magnetic flux. On one half-cycle of the alternating current, the flux produced by the electromagnet is in the same direction as the flux from the permanent magnet so increasing the pull on the diaphragm and bending it more towards the magnet poles. On the next half-cycle, current in the reverse direction reduces the pull on the diaphragm allowing it to recover. The diaphragm thus flexes

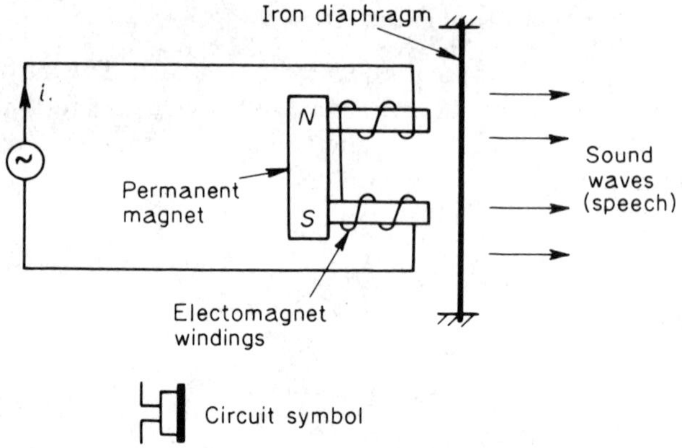

Fig. 1.4 Principle of telephone receiver. Alternating speech current *i* in the electromagnet windings produces a magnetic flux aiding and then opposing that of the permanent magnet on alternate half-cycles. This varies the pull on the diaphragm which thus vibrates to produce sound waves (1.3.2).

once in each direction for each complete cycle of the speech current causing pressure variations in the air beside it to produce sound at the same frequency as the speech current.

If the permanent magnet were not provided, the electromagnet would *pull* on the diaphragm for current in either direction, allowing it to recover as the current passed through zero twice in each cycle. This would produce sound at twice the frequency of the original speech. The permanent magnet also increases the sensitivity of the receiver so that it gives a greater sound output (see Appendix 3). Direct current should not be connected to the receiver because it would change the value of the polarizing flux and could de-magnetize the permanent magnet.

Modern instruments have an improved version of the receiver known as a rocking armature receiver (**1.5.3** and Fig. 1.19).

1.3.3 The magneto bell (ringer) is a polarized device in which alternate half-cycles of a low frequency a.c. ringing supply cause a centrally pivoted armature to rock first to one side and then to the other side as shown in Fig. 1.5. The frequency of a.c. ringing current is typically $16\frac{2}{3}$ Hz in countries such as Great Britain using a mains frequency of 50 Hz and is 20 Hz in countries such as the U.S.A. where the mains frequency is 60 Hz. Modern practice is to use a higher frequency, typically 23·5 Hz or 25 Hz. The ringing voltage is usually 75 V.

The coils are arranged so that current in one direction adds to the pull of the permanent magnet at one end of the armature while opposing it at the other end. Current in the other direction reverses this effect, rocking the armature the other way. The result is that the striker attached to the armature hits each of the two gongs once for each cycle of the ringing current.

The use of a polarizing permanent magnet in a bell increases its sensitivity and prevents frequency doubling in much the same way as in a receiver (see Appendix 3).

An alternative to the use of a bell is *tone calling* in which the ringing current from the line is used to power an electronic oscillator tuned to about 400 Hz. The oscillator output drives a transducer essentially the same as a telephone receiver (**1.3.2**) to produce a sound output at the oscillator frequency modulated by the ringing frequency.

1.3.4 Simple induction coil. The electrical power of speech currents is very small so as much as possible of this power must be transferred from the microphone to the line at the transmitting end, and then from the line to the receiver at the receiving end.

To see how the maximum transfer of power can be achieved, consider first a simple d.c. circuit comprising a battery in series with an external resistance (the load). Calculation of the power delivered to different values of load resistance will show that *maximum power is transferred when the load resistance is equal to the internal resistance of the battery*. In an a.c. circuit, the ratio of voltage to current is called the *impedance* (Z) and depends on the inductance and capacitance of the circuit as well as its resistance. Another effect produced by the inductance and capacitance is that the voltage and current waveforms are not in step; there is a *phase difference* between them. *For maximum a.c. power transfer, the generator and load impedances must be equal in magnitude* (i.e. in numerical value) and should give equal but opposite phase differences. When these conditions are met, the generator and load are said to be *matched* (see Appendix 4).

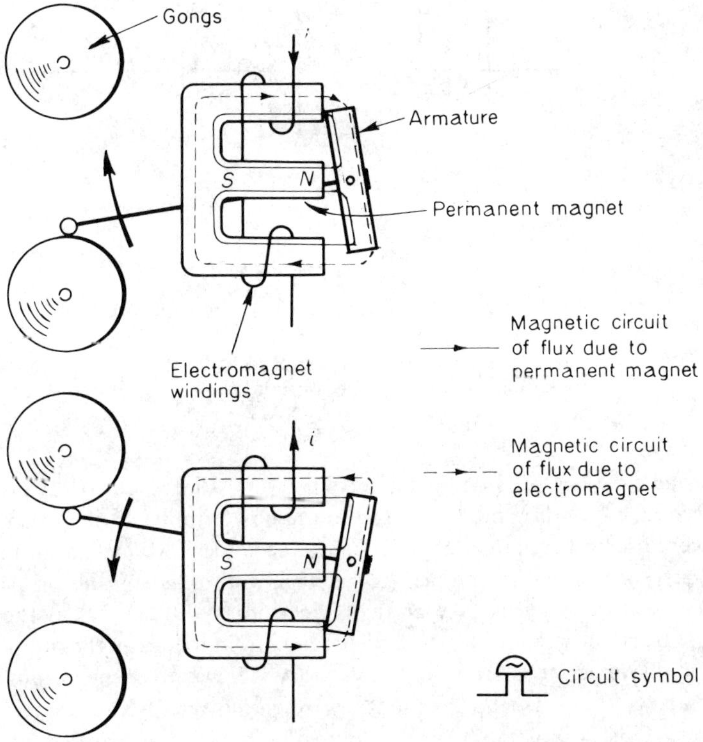

Fig. 1.5 Principle of magneto bell. The current i is an a.c. ringing supply. The directions indicated in the two diagrams are for alternate half-cycles (1.3.3).

For maximum speech power transfer, the impedances of the microphone, receiver and line should all be matched. Practical limitations on physical size and d.c. resistance make it impossible for all three to be made with the same impedance. An *induction coil* using the transformer principle can, however, be used to make them all present the correct effective impedance to each other.

A transformer or induction coil is made up of two (or more) windings consisting of several turns of wire linked by a magnetic circuit such as an iron core. A change of current in one winding produces a change in its magnetic flux which is linked with both windings and induces the same e.m.f. in each turn of both.

When a.c. flows in one winding it continuously induces an alternating e.m.f. in every turn because, by its very nature, an alternating current is a continuously changing current. A winding connected to an a.c. supply is called a *primary* winding and a winding from which the output is taken is called a *secondary* winding. The total induced e.m.f. in each winding is proportional to the number of turns in that winding. The primary e.m.f. is the same as the supply voltage so that, by choosing suitable numbers of turns in the windings, the supply voltage at the primary winding can be transformed to any required output voltage at the secondary winding. This is illustrated in Fig. 1.6.

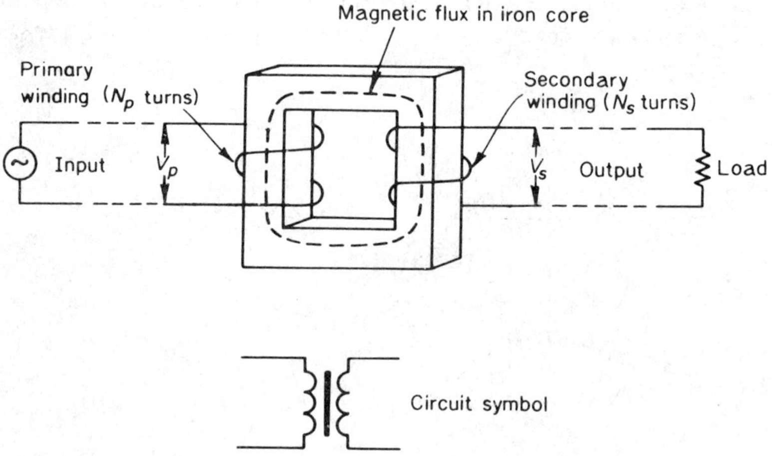

Fig. 1.6 Transformer principle. An a.c. input of Vp volts at the primary winding of Np turns produces an a.c. output of Vs volts at the secondary winding of Ns turns where $\frac{Vp}{Vs}=\frac{Np}{Ns}$ the turns ratio (1.3.4).

Transformers do not provide any gain of power. All the electrical power output from the secondary winding must come from the power input to the primary winding. If the power loss is negligible, therefore, the secondary voltage multiplied by the secondary current is equal to the primary voltage multiplied by the primary current.

If the secondary voltage is greater than the primary voltage, then the secondary current must be smaller than the primary current and in this case, the input impedance (i.e. the ratio of voltage to current) of the primary is lower than the output impedance of the secondary. If the secondary voltage were the lower, its impedance would also be the lower. This is illustrated by the following simple numerical example:

Example: Consider a transformer having a primary winding of 400 turns and a secondary winding of 200 turns. If an a.c. source of 10 V is connected at the primary winding, the secondary voltage will be 5 V.

If a load of resistance 1 000 Ω is connected to the secondary winding, the current in the load will be 5 mA and the power delivered to the load will be 25 mW (5 V × 5 mA).

This power must come from the original a.c. source which is at 10 V and therefore to deliver 25 mW we must supply a current of 2·5 mA. This is the same current which would be taken by a load of 4 000 Ω connected directly to the a.c. source without the use of a transformer.

The load of 1 000 Ω thus has a *reflected impedance* of 4 000 Ω at the input to the primary winding of the transformer.

If the two windings of a transformer are connected in series, the induced e.m.f. across the two is proportional to the total number of turns as shown in Fig. 1.7. A given number of turns can thus provide a much larger transformation and a given transformation can be provided by fewer turns by this means. Transformers of this type are known as *auto-transformers* and this is the form used in telephone instrument circuits, except in those circuits where a d.c. connexion between primary and secondary windings is to be avoided.

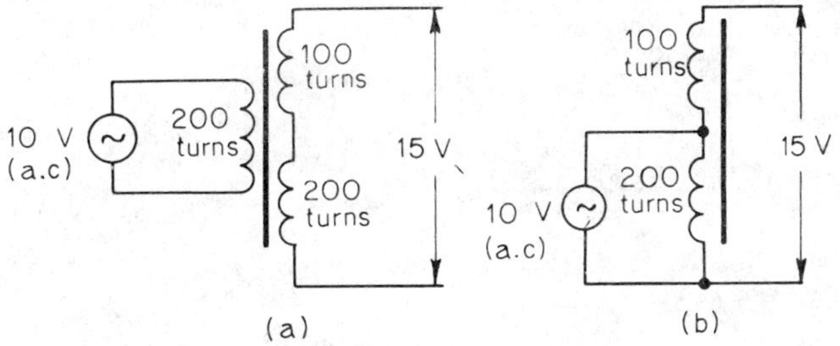

Fig. 1.7 Auto-transformer. The auto-transformer at (b) uses only 300 turns to provide the same transformation as the transformer (a) using 500 turns (1.3.4).

A typical induction coil circuit matching both microphone and receiver to the line is shown in Fig. 1.17. The battery shown in this circuit provides the necessary direct current for the operation of the carbon microphone and the induction coil prevents this current from flowing in the receiver.

1.3.5 Anti-sidetone induction coil. Sidetone is the reproduction in the receiver of sounds picked up by the microphone of the same telephone instrument. It occurs because the instrument has its microphone and receiver connected to one two-wire line. An example is the circuit of Fig. 1.17. Microphone output current in the 400 turn winding induces an e.m.f. in the 1 000 turn winding so that speech current flows in the receiver. There would be no sidetone if a four-wire connexion were used between subscribers' instruments to give separate transmit and receive pairs.

The most comfortable conditions for the user of a telephone have been found to be when he hears his own voice in the receiver at about the same loudness as he would hear it in the form of direct sound waves in the air in normal conversation. Complete absence of sidetone is bad because the telephone then appears to be 'dead'. Too much sidetone is also bad for the following reasons:

(*a*) the talker lowers his voice because he hears it too loudly,
(*b*) the talker's ears fatigue, reducing subsequent listening ability and
(*c*) the interfering effect of local room noise at the listener's telephone is increased.

One commonly used means of reducing sidetone employs the *balanced transformer* or *hybrid coil* principle in the induction coil which is then known as an *anti-sidetone induction coil* or *A.S.T.I.C.* A simple form of this is the three winding transformer shown in Fig. 1.8, the operation of which is as follows:

On transmitted speech, the microphone output speech current divides into two parts, i_1 via winding P_1 to the line (Z_L), and i_2 via P_2 to the *balance network* (Z_B). These currents induce opposing e.m.f.s in winding S. If Z_B is made equal to Z_L, the microphone current divides equally between them so that $i_1 = i_2$.

If P_1 and P_2 have the same number of turns as each other, the effect of both currents is equal, but as they are in opposite directions their combined effect is to produce no resultant e.m.f. or current in winding S.

When receiving speech currents, the current in winding P_1 induces an e.m.f. in winding P_2 equal to the potential across P_1 itself. If the turns on windings P_1 and S are chosen so that the reflected impedance of the receiver in winding P_1 equals the impedance of the microphone, the line voltage V_L will be divided equally between P_1 and the microphone. The voltage across P_2 thus equals the voltage across the microphone, giving zero volts across Z_B and therefore no current in it. The current in P_1 induces an e.m.f. in S and this provides the speech current required to drive the receiver.

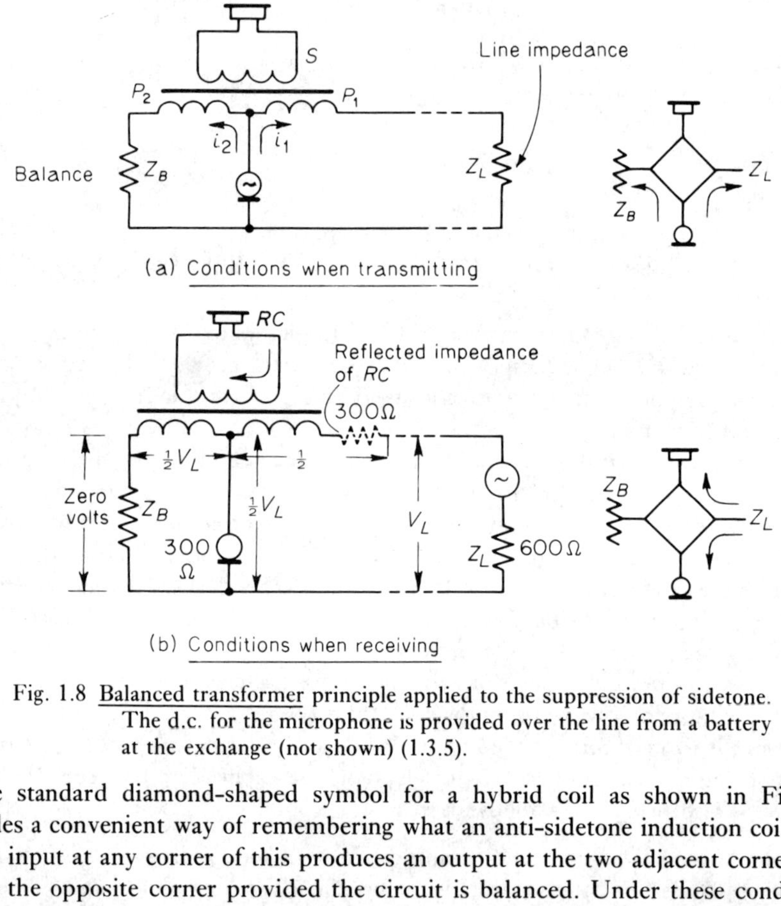

Fig. 1.8 Balanced transformer principle applied to the suppression of sidetone. The d.c. for the microphone is provided over the line from a battery at the exchange (not shown) (1.3.5).

The standard diamond-shaped symbol for a hybrid coil as shown in Fig. 1.8 provides a convenient way of remembering what an anti-sidetone induction coil does. Power input at any corner of this produces an output at the two adjacent corners but not at the opposite corner provided the circuit is balanced. Under these conditions, half the received power is wasted in the microphone impedance, and half the transmitted power in the balance network. The total impedance the instrument presents to the line consists of the microphone and the reflected impedance of the receiver in series. The impedance of modern receivers and microphones is such that this provides adequate matching without the need for additional impedance transformation.

Most practical circuits use auto-transformer type anti-sidetone induction coils as illustrated later in this chapter (**1.5.1** etc.).

1.3.6 Local battery and central battery operation. Carbon microphone feed, or polarizing current (**1.3.1**) can be provided either by a *local battery* at the subscriber's premises or by a *central battery* at the exchange (also called *common battery*).

The disadvantages of local battery operation are the cost of providing the battery at every subscriber's premises, the cost of providing maintenance and replacement services for these batteries and the variations in microphone efficiency which occur due to changes in the condition of the batteries. The advantage of the system is that the microphone feed current is not affected by the resistance of the line and the over-all energy consumption is less than with a central battery. The provision of more sophisticated facilities than a simple telephone circuit (e.g. some extension plans, data terminals, etc.) is often impossible if the central battery is the only source of power, and a local mains power unit or battery is then necessary.

The circuit shown in Fig. 1.8 receives its microphone feed current from the central battery at the exchange. Each line engaged on a call is connected to the exchange battery through coils having the property of inductance which gives them a high impedance to a.c. speech currents. This prevents overhearing or interference between different calls by effectively blocking any flow of speech current from one connexion (e.g. subs *A* and *B* in Fig. 1.9) to another (e.g. subs *C* and *D*) via the battery leads. The required connexion between the two parties on each call can be made via capacitors which have a low impedance to a.c. speech currents. This arrangement is known as a *transmission bridge* and provides separate microphone feeds to the two lines.

If the microphone circuit is fed via the operating coils of suitable switching apparatus (e.g. relays) at the exchange, this apparatus can be controlled by contacts in the line connexions at the subscriber's instrument. Typical contacts used in this way are the pulsing contacts of the dial described in **1.4.2**.

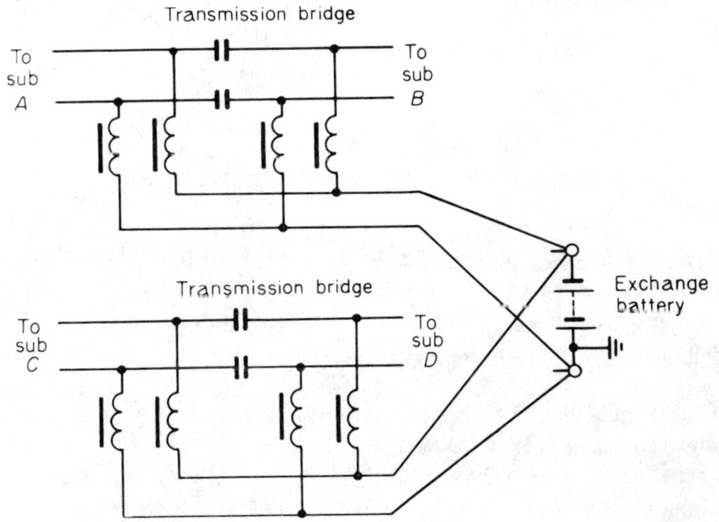

Fig. 1.9 Transmission bridge. This illustrates the condition when calls are established between subscribers *A* and *B* and between *C* and *D* (1.3.6).

1.4. Principles of subscribers' signalling apparatus

1.4.1 Contacts and springset operation. For sending telegraph signals or controlling exchange switching apparatus, it is necessary to use *contacts* to 'make' or 'break' an electric circuit (see also **1.3.6**). Materials such as silver, platinum, and palladium are used for the contacts because these materials provide a low contact resistance and are not liable to the formation of high resistance oxides on the contact surface like the rust which would form, for example, if iron were used. Most contact materials are relatively expensive, so the contacts are made small and are fixed to flat springs made of materials such as nickel-silver or phosphor-bronze alloys.

Movement of the contacts to open (break) or close (make) the circuit is achieved by bending the springs which are separated at their fixed ends by flat insulators of plastic materials. A complete assembly of contact springs and insulators is called a *springset*. A typical springset of the type used on relays is shown in Figs 2.4 and 2.17 in Chapter 2.

The operation of the springset is often controlled by the movement of part of a mechanism such as a teleprinter, a selector switch or the receiver cradle of a telephone instrument. The movement to be utilized may not be in the same direction or over the same distance as that in which the contact spring is to be moved, and in these cases, levers are used to convert the motion. This is illustrated by the gravity switch contacts in Fig. 1.10 which are opened when the handset is placed on its rest (cradle) on a telephone instrument.

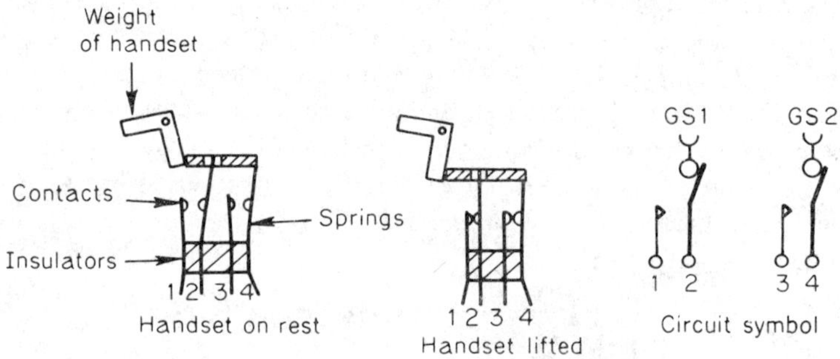

Fig. 1.10 Gravity switch contacts as used on telephone instruments. This is an example of the use of a lever to change the direction of the controlling force, in this case the weight of the handset (1.4.1).

In some other cases, a contact or a lever is required to operate repeatedly. A convenient device for this purpose is the *cam*. One form of this is a rotating disk with projections placed so that a lever or a contact spring resting on it rises and falls regularly as the cam rotates. Another form is a cylinder with a groove cut in its surface along a path which guides a lever regularly to left or right along the rotating cylinder.

Mechanical devices basically similar to those used for operating contacts are also used for controlling the printing mechanisms of teleprinters.

If something such as a cam is to be rotated at some times and not at others, a *clutch* mechanism may be used to couple the cam to a continuously rotating motor shaft as required. One form of clutch used in teleprinters is illustrated in Fig. 1.12. In this, the shaft and *ratchet wheel* rotate continuously and a lever called the *pawl abutment* rests against a *pawl* which is attached to the cam. In this condition, the cam is at rest.

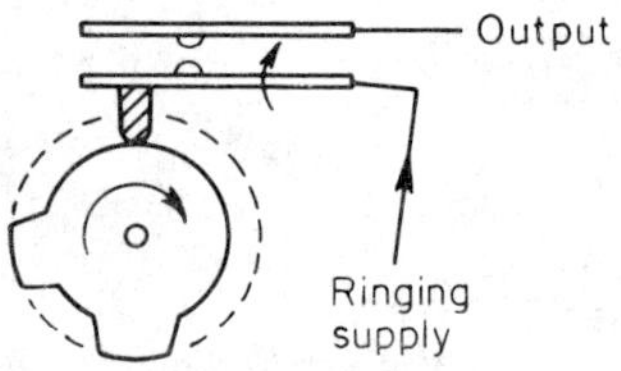

Fig. 1.11 Typical cam. This disk-shaped cam is used to operate a contact to provide interrupted ringing (1.4.1).

When the cam is required to rotate, the pawl abutment is moved away from the pawl which is thus pushed into engagement with the ratchet wheel.

To stop the cam, the pawl abutment is restored to its original position and when the pawl comes round to its starting point, it hits the pawl abutment and is lifted clear of the ratchet.

With this type of clutch, the cam always comes to rest in the same position.

1.4.2 The dial (Fig. 1.20) is a mechanical device which is used in both the telephone system and the Telex system to provide pulses to control the selecting mechanisms at an automatic exchange. The method of using the dial, known as *dialling*, is to place the finger in the hole corresponding to the *digit* (figure) to be sent and to pull the *finger plate* round in a clockwise direction until the rotation is arrested by the *finger stop*. This action winds up a *mainspring* similar to that in a clock and this restores the finger plate to normal when the finger is withdrawn.

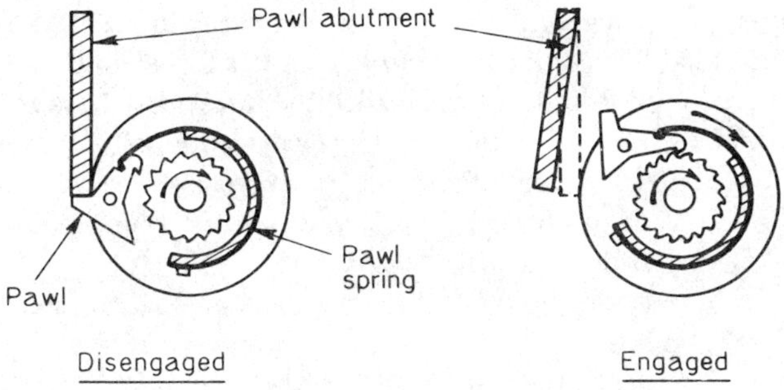

Fig. 1.12 Clutch as used on teleprinter receiving and transmitting cams (1.4.1).

The required pulses are produced during this return rotation, the number of pulses depending on the digit dialled. In the usual form of the dial, one pulse is sent when digit '1' is dialled, two pulses for digit '2' and so on up to ten pulses for digit '0'. An exception is the dial used in New Zealand which has the figures marked on the number ring in a different order from that used in the UK. The New Zealand dial thus sends ten pulses for the digit '0', nine pulses for digit '1', and so on to one pulse for digit '9'.

The dial pulses are produced by the cam operated *pulsing* contacts which may be used, for example, to interrupt the central battery line current at a telephone instrument to provide a train of 'break' pulses (**1.6**).

Contacts used for disconnecting the current flowing in the coils of magnetic devices such as relays are liable to damage due to sparking. The spark is the visible effect of the heat produced by the current which the back e.m.f. of the coil inductance maintains for a short time after the contact has opened. One way of reducing this effect is to connect a *spark quench* in the form of a capacitor in series with a resistor across the contact to provide an alternative path for the 'spark' currents as in Fig. 1.17c.

In addition to the pulsing contacts, another set of contacts is provided which operate when the finger plate is moved off-normal. These *off-normal* contacts are operated by a cam provided for the purpose and can be used for a variety of functions, a typical application being to short circuit the subscriber's telephone receiver during dialling. This prevents the dial pulses being heard as uncomfortably loud 'clicks' by the subscriber. Some types of push-button telephone are arranged to produce the same pulses as a dial so that they can interwork with existing exchange equipment. Large scale integrated circuits (l.s.i.) in the instrument provide memory, control, and counting circuits to store each operation of a digit button and generate the appropriate number of pulses.

1.4.3 Start-stop telegraph instruments (teleprinters). A teleprinter can be considered as consisting of two distinct parts, a transmitter and a receiver. During the transmission of a message, the transmitter at one teleprinter is connected to the receiver of another teleprinter by a line as in Fig. 1.1. By using suitable transmitting and receiving sets they could of course be connected together via a radio link. In most types of teleprinter, the transmitter and receiver are mechanical devices powered by a small electric motor and both the transmitting mechanism and the receiving mechanism are at rest except when a signal is being sent or received. A modern development is the use of electronic devices such as transistors in place of some of the moving parts to perform some of the signalling functions. The motor is then used only to drive the printing mechanism of the receiver. Only the mechanical type of teleprinter will be described here.

The transmitter of a teleprinter is an automatic signalling device using the *five-unit code* and controlled by a keyboard having a similar layout to that of a typewriter.

When a key is depressed, it causes a cam called the *transmitting cam* to rotate through one revolution. During this revolution, a signal consisting of a combination of five pulses, each of either positive potential (called spacing) or negative potential (called marking) are connected to the line (**3.4.2**). There is a different combination of these pulses for each key (see Appendix 2).

Each signal of five pulses is preceded by a start pulse (spacing) and followed by a stop pulse (marking).

The receiver of the teleprinter at the other end of the line selects and prints the character corresponding to the particular combination of negative and positive pulses in the received signal. To do this the receiver must perform a series of operations in step with the transmitter under the control of a *receiving cam* which completes one revolution in the same time as the transmitting cam.

The speed of both the transmitting cam and the receiving cam is controlled to within $\frac{1}{2}\%$ of a standard speed. Even this small possible difference between them could result in the transmitter and receiver getting out of step over a period of time if both cams were continuously rotating.

To ensure that the transmitting and receiving cams remain in step, a system known as *start–stop telegraphy* is used. When any key is depressed, the transmitting cam starts to rotate and a start pulse is transmitted. This start pulse causes the receiving cam at the receiving end to start its revolution. The transmitting cam completes one revolution and stops again. At the end of the revolution of the transmitting cam, a stop signal is transmitted during which the receiving cam completes its revolution and is brought to rest until the next start pulse is received.

Both cams are thus in phase at the beginning and end of each complete signal. The stop signal is maintained until the commencement of the next start signal, and it can be regarded as an enforced rest period or pause between characters to allow one machine to catch up with the other.

1.4.4 Teleprinter transmitters. In some types of teleprinter, the combination of pulses to be sent is set up on a group of five contacts as shown in Fig. 1.13. These contacts are operated in different combinations depending on which key is depressed on the keyboard. If the letter ‘A’ is to be transmitted, for example, contacts 3, 4 and 5 would be switched to ‘marking’ and contacts 1 and 2 to ‘spacing’. A distributor then rotates to connect each contact to the line in turn. This distributor may be either a rotating arm driven over fixed segments by the teleprinter motor or it may consist of a set of switching devices such as relays or transistors operated in turn by electrical means.

In one type of teleprinter widely used in the U.K., the pulse combination to be sent is

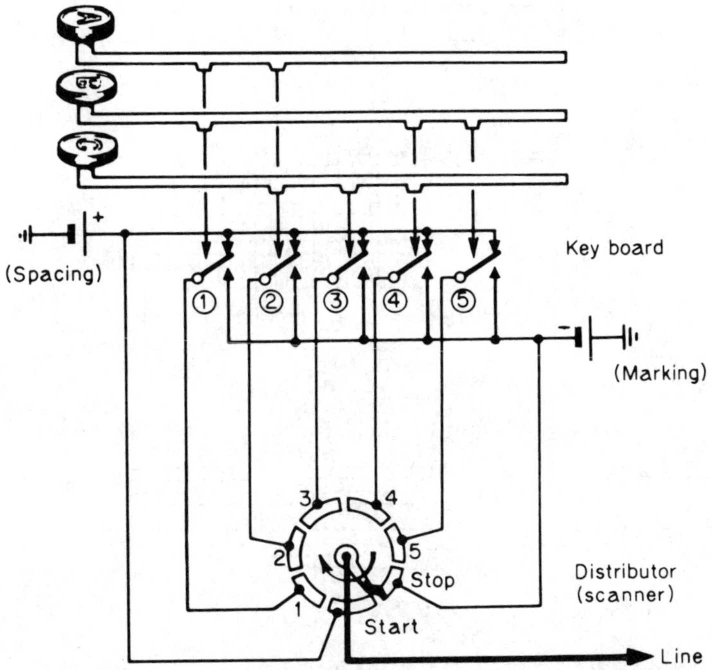

Fig. 1.13 Principle of teleprinter transmitter ('distributor' type). The distributor may be a rotating arm as shown or an electronic 'scanning' circuit performing the same function (1.4.4).

set up mechanically by positioning *five combination bars*. The equivalent function to that of the distributor of Fig. 1.13 is performed by a set of cams operating *selecting levers* one after another. These levers control the transmission of the pulses by the *transmitting contact tongue* which moves between two fixed contacts connected to potentials of 80 V positive and negative respectively.

The block diagram of Fig. 1.14 shows the main parts of the transmitting mechanism which control the contact tongue.

For each key on the keyboard, there is a *keybar* lying across the combination bars which have projections arranged to give a different pattern or combination opposite each keybar. When a key is depressed, its keybar engages these projections so that some combination bars move to the right while others remain stationary. For each key there is thus a particular setting of the combination bars corresponding to the combination of marking and spacing pulses to be transmitted for that letter. Combination bars for those pulses required to be marking are shifted to the right and those for spacing remain stationary.

Each selecting lever is normally held away from the contact operating lever by its own cam. These cams all rotate together and each in turn permits the movement of its associated selecting lever for the duration of the appropriate pulse. If the pulse is to be 'marking' (negative), the combination bar has been moved and so the selecting lever turns about its pivot to move the contact tongue to the left. If a 'spacing' (positive) pulse is required, the selecting lever is prevented from moving by its associated combination bar.

1.4.5 Automatic transmitters. When sending a message, a teleprinter operator will pause between letters for a variety of reasons such as reading the next word to be typed. Such irregularities in operator's typing result in wasted line time when

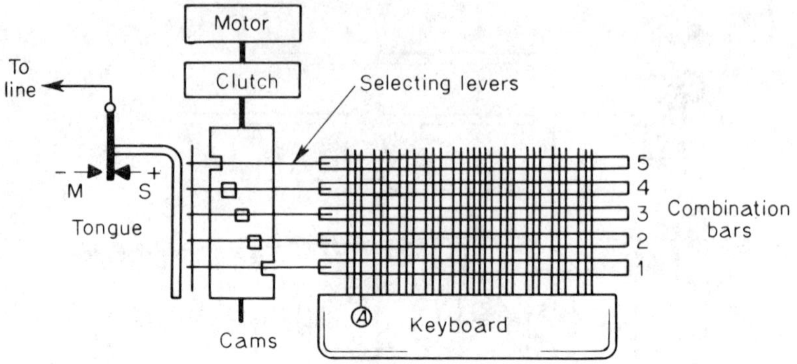

(a) Block diagram (plan view)

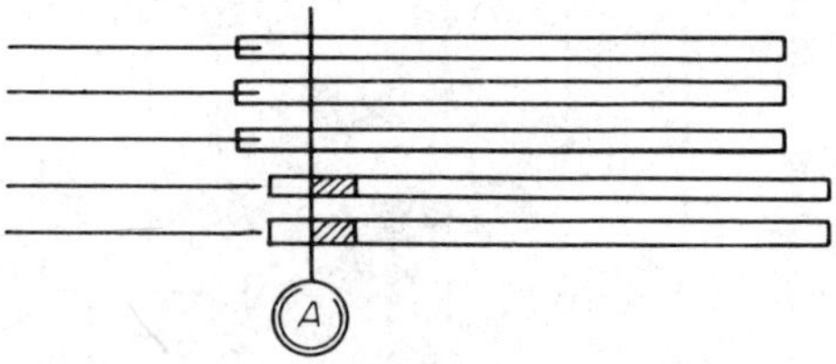

(b) Typical displacement of combination bars (key 'A' depressed)

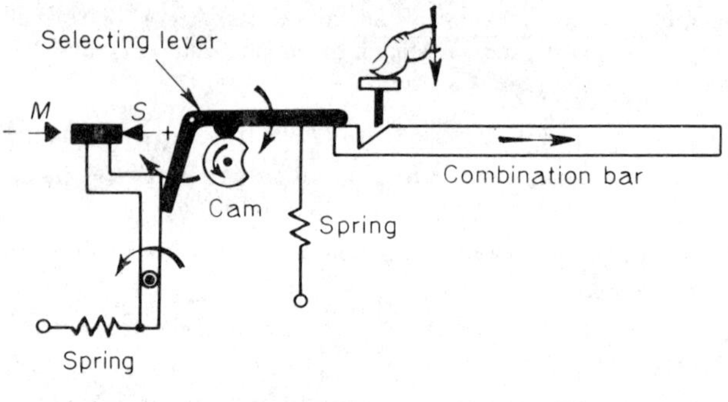

(c) Front view. Arrows indicate direction of movement when transmitting a 'marking' pulse

Fig. 1.14 Principle of teleprinter transmitter (mechanical type) (1.4.4).

teleprinters are directly connected to line as in Fig. 1.1. More effective use of expensive lines is obtained by using automatic transmitters operated by *paper tape* previously punched with holes representing the signals to be transmitted. The tape, which is

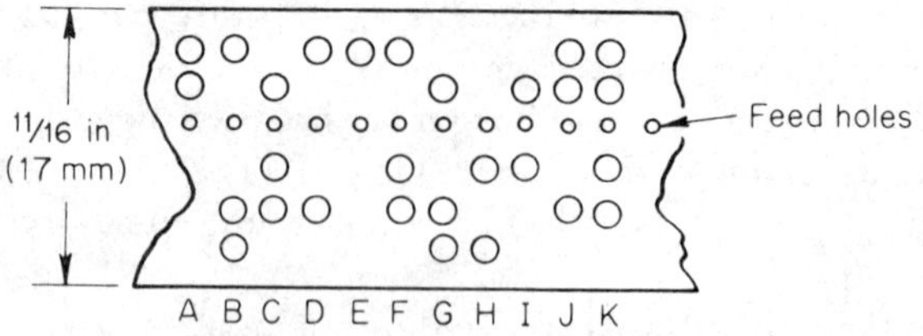

Fig. 1.15 Perforated tape used for controlling automatic transmitters using five-unit code (1.4.5).

illustrated in Fig. 1.15, is prepared on a machine called a keyboard *perforator* having a keyboard similar to that of a teleprinter.

Another application of automatic transmitters is the *torn-tape* method of operation in which incoming messages at a telegraph message-switching centre are received on a *printing re-perforator*. This punches out the message and prints it on the same tape. When the message is complete, the tape is torn off by an operator who reads the address on the tape and conveys it to an automatic transmitter on the required route for re-transmission when a line becomes available. This enables the most efficient use to be made of expensive lines (e.g. international cable routes).

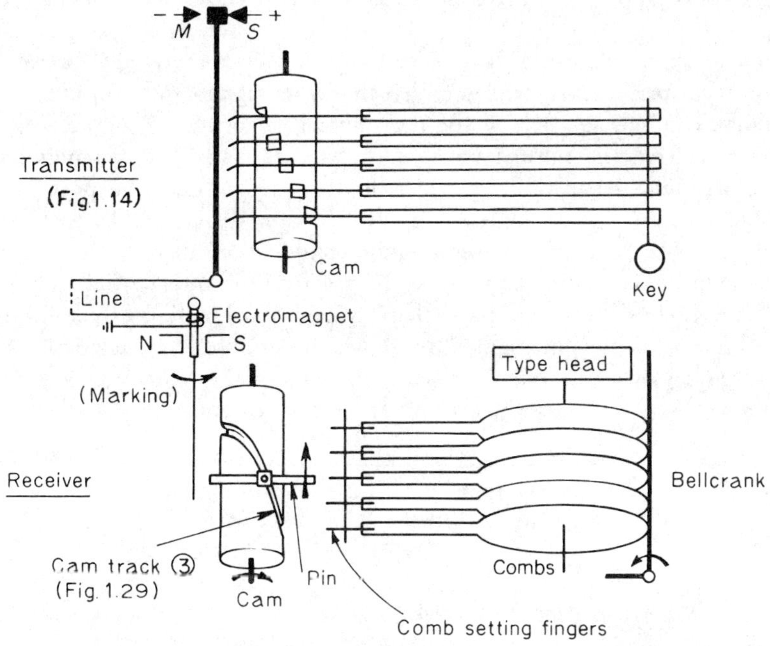

(a) Block diagram showing equivalent parts of transmitting and receiving mechanisms (plan view)

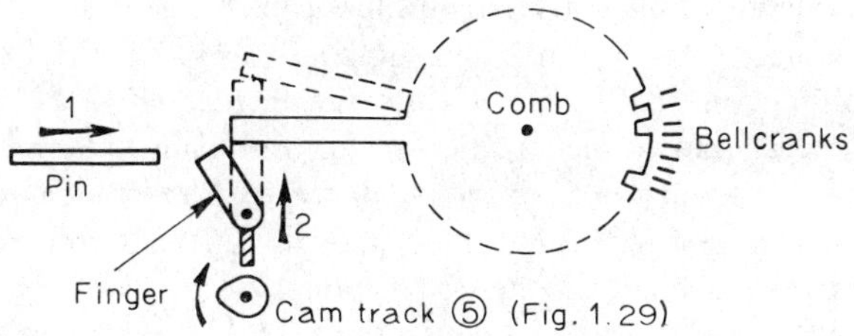

(b) Rear view. Arrows indicate direction of movement which takes place in two stages (shown 1 and 2) on a 'marking' pulse

Fig. 1.16 Principle of teleprinter receiver (1.4.6).

A modern development of this technique is *automatic telegraph message switching*, in which, for example, a computer may be used to store and forward messages automatically.

1.4.6 Teleprinter receivers. The receiving mechanism performs a series of operations in some ways similar to those performed by the transmitter but in the reverse order. The main parts of both transmitter and receiver of a mechanical type of teleprinter commonly used in the UK are shown in block diagram form in Fig. 1.16.

When a start pulse is received, the receiving cam makes one revolution and causes a *finger setting pin* to move past each of five *comb setting fingers* in turn. The pin is made to strike any finger with which it is aligned when a *marking* pulse is received by the *electromagnet* connected to the line. This sets the comb-setting fingers of the receiving teleprinter into a similar combination to the selecting levers at the transmitting teleprinter.

For each of these comb-setting fingers there is a circular *receiving comb* having slots around the periphery arranged in a similar manner to the projections on the corresponding combination bars at the transmitting teleprinter. Each comb is either rotated slightly from the normal position or remains stationary depending on the position of its comb-setting finger, which in turn depends on the polarity of the received pulse.

For each possible character (i.e. signal pulse combination) there is a *bellcrank* lying across the slots in the combs like the keybars across the transmitting combination bars. The comb slots are so arranged that for each combination of comb positions there is just one set of slots in alignment into which the associated bellcrank can drop. For any given set of signal pulses, a particular one of the bellcranks will move to select the appropriate *typebar* which prints the required character.

Now turn to page 27 and test yourself on questions 1–15 there before reading on.

PART 3 TYPICAL PRACTICAL APPARATUS

1.5 Telephone instruments

1.5.1 Auto. and C.B. telephone, B.P.O. type 332. This is the older type of instrument, the circuit of which is shown in Fig. 1.17. With the dial connected as shown, it is suitable for connexion to an automatic exchange (**1.2.2**). The signalling and switching principles of this type of exchange are described in **3.5.1** and Chapters 4, 5 and 6.

The induction coil *IC* is an auto-transformer (Figs 1.7 and 1.17b) incorporating the balanced transformer principle for reducing sidetone illustrated in Fig. 1.8. To explain its operation, it helps if we first rearrange it to form the equivalent circuit shown in Fig. 1.17b. This only involves altering the order in which P_1, S and P_2 are connected. As these are connected in series, the order in which they are connected does not affect the operation. The reduction of sidetone is then achieved as follows:

> Speech current from the microphone divides between windings P_1 and P_2. The combined effect of these two windings is to induce an e.m.f. in the third winding S. The turns ratios are designed so that this induced e.m.f. is nearly the same as the voltage drop across the balance resistor Z_B produced by speech current flowing through it.
>
> The sidetone speech potential is the difference between the induced e.m.f. in S and the potential across Z_B and is, therefore, very small.

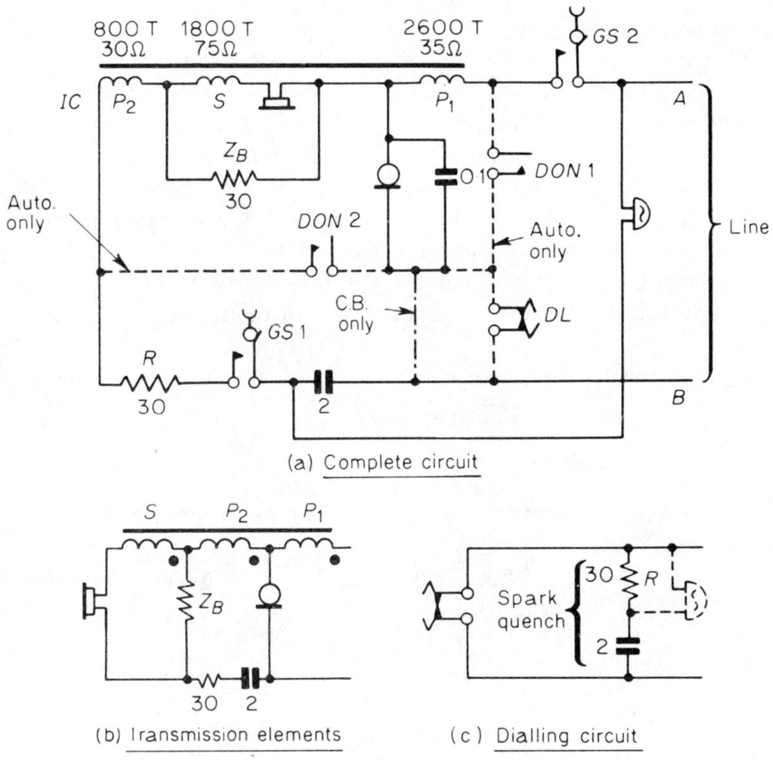

Fig. 1.17 Auto and C.B. telephone (B.P.O. type 332). When used as a C.B. manual telephone, the dial is omitted. *DL* is the dial pulsing contact. *DON*1 and *DON*2 are the dial off-normal contacts (1.5.1).

To prevent current being taken from the exchange battery when the telephone is not in use, gravity switch contacts (*GS*) are connected in series with the line and a 2 μF capacitor is connected in series with the bell. This same capacitor also prevents d.c. from flowing in the receiver during use (**1.3.2**).

Long overhead telephone lines may act like receiving aerials for some radio frequencies and the carbon microphone, having a non-linear voltage/current characteristic, can act as a detector, resulting in broadcast radio programmes being heard in the receiver. The 0·1 μF capacitor by-passes the microphone to currents at radio frequencies to prevent this.

The dial pulse springs are connected in series with the line. One off-normal contact provides a low resistance circuit through the telephone during dialling to obtain the best performance from the exchange apparatus and also prevents dial pulses from being heard in the receiver (**1.4.2**). The other off-normal contact connects a *spark quench* across the dial pulsing contact consisting of the 2 μF capacitor in series with a 30 Ω resistor (wound on the induction coil). This resistor also shunts the bell to prevent it from responding to the dial pulses.

The same instrument is suitable for connexion to a C.B. (central battery) manual exchange. The dial and its associated wiring is then omitted as shown by the broken lines in Fig. 1.17a.

1.5.2 Auto. telephone, B.P.O. type 706. This is the instrument shown in Fig. 1.19. It is a later design than the type 332 (see **1.5.1**) and incorporates a rocking armature receiver, the principle of which is illustrated in Fig. 1.18.

The rocking armature receiver is more sensitive (gives a greater sound output for a given electrical input) than the simple receiver illustrated earlier in Fig. 1.4. This is achieved by using a magnetic circuit like that of the magneto bell so that the alternating flux produced by speech current in the electromagnet does not pass through the permanent magnet. The 'permanence' of the permanent magnet depends on its ability to resist changes in its magnetism. It therefore presents a high reluctance to the alternating flux, reducing the sensitivity of the electromagnet in the simple receiver in which it forms part of the magnetic circuit.

By using an armature separate from the diaphragm, the latter is constructed to give the best acoustic performance without regard to the magnetic properties. This

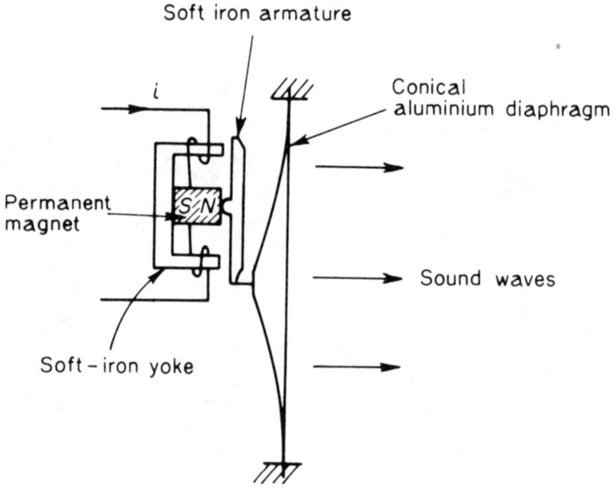

Fig. 1.18 Rocking armature receiver. Compare this with the simple telephone receiver in Fig. 1.4 and the magnetic circuit of the magneto bell in Fig. 1.5 (1.5.2).

results in a more faithful reproduction of speech. The armature, of course, is designed to provide the required magnetic properties without the need to compromise on the acoustic requirements. This results in a lower reluctance and increased sensitivity.

The induction coil (*IC*) and balance network (Z_B) are different from those of the older instrument (see **1.5.1**) to give the greater sidetone reduction necessary with the more sensitive receiver, which would otherwise make the sidetone sound louder.

The turns ratios on the induction coil are such that, on received speech, less than half the received power goes to the receiver, so that not all of the improved sensitivity of the receiver is actually used. The effect of this on transmitted speech, however, is that less than half of the microphone output goes to the balance network. This arrangement thus takes advantage of the more sensitive receiver to give improved transmission as well as reception.

A *regulator* is frequently used to reduce the sensitivity on short lines. The d.c. from the exchange battery is used to bias rectifiers in parallel with the microphone and receiver. The shorter the line, the greater is the direct current and the lower is the resistance of the rectifiers which shunt the microphone and receiver, making them less sensitive (see Fig. 1.19c).

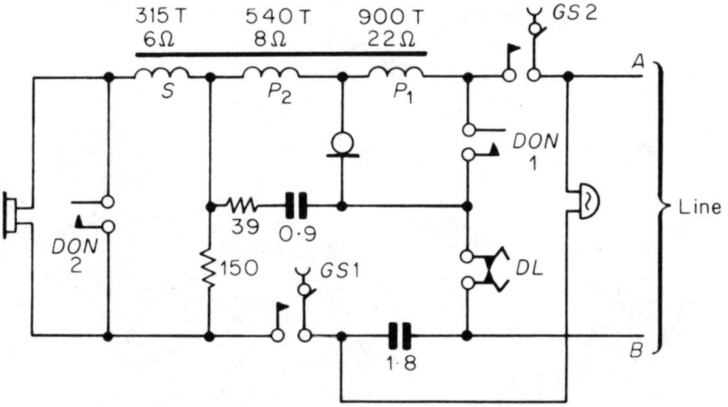

(a) Complete circuit (without regulator)

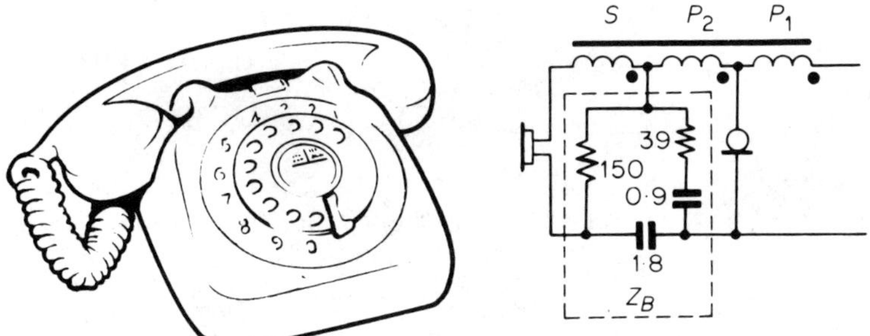

(b) Transmission elements

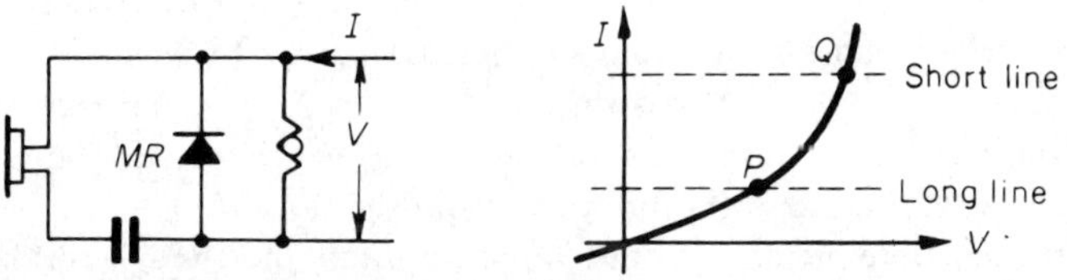

(c) Principle of regulator. The rectifier *MR* has a lower resistance *V/I* at *Q* than at *P* (see also Fig. 2.14)

Fig. 1.19 Auto telephone (B.P.O. type 706) (1.5.2).

1.6 The trigger dial

This type of dial is the one in current use in the U.K. and gets its name from the type of mechanism used to generate the pulses. The general appearance and main components are illustrated in Fig. 1.20. The finger plate and casing of the dial illustrated are made of coloured (or black) plastics and the mechanism and its mounting plate are of metal. Another version has a finger plate and casing of transparent plastics without a separate mounting plate for the mechanism which has many of its moving parts made of hard wearing plastics materials such as nylon.

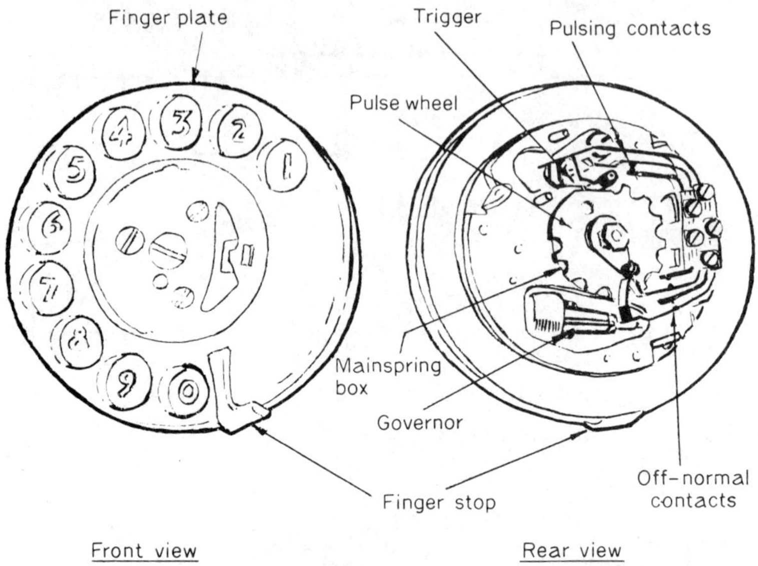

Fig. 1.20 Typical telephone dial (trigger dial) (1.6).

Rotating the finger plate in a clockwise direction winds the mainspring to provide a restoring force when the finger plate is released. The required pulses are produced during this return (counter clockwise) rotation as follows:

Pulsing (Figs 1.21 and 1.22). The pulse wheel rotates with the finger plate and the trigger engages with the teeth of the wheel. When the dial is being wound, the pulse wheel carries the trigger away from the pulsing springs so that no pulses are produced.

On the return rotation, the trigger is carried back towards the springset, and the insulated roller pushes on the pulsing contact spring to open the contact once for each tooth. The pulse wheel has one tooth for each finger hole on the finger plate, so that the number of teeth passing under the trigger is the same as the number dialled and so therefore is the number of pulses produced. If the dial has a nylon trigger, the separate insulated roller is not required and the trigger operates directly on the contact spring.

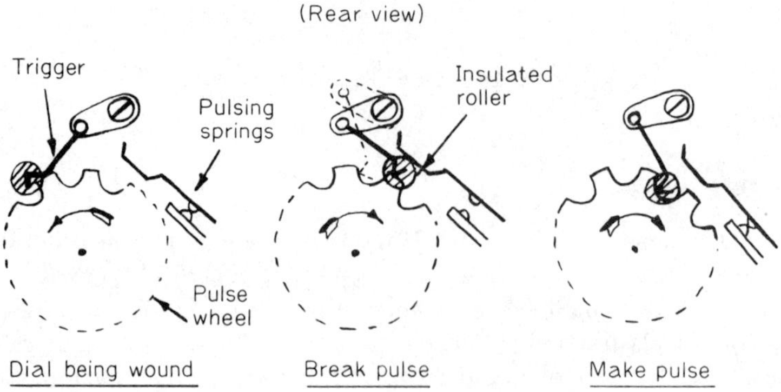

Fig. 1.21 Generation of pulses by trigger dial (1.6).

Lost motion period (Fig. 1.21). The movement of the trigger from left to right in the diagram gives a delay of about 230 ms before the start of pulsing. This is the *lost motion period* which, together with the time taken to rotate the finger plate initially, ensures that a minimum pause of some 400 ms occurs between digits. This *inter digital pause* (*I.D.P.*) allows the automatic apparatus at the exchange to complete each stage of selection before receiving the next digit.

Speed control (*governor*) (Fig. 1.22). At the end of a call, replacing the handset opens the gravity switch contacts to release the connexion. Opening the dial pulsing contacts too long would have the same effect. The dial pulses must therefore not be too long. The selectors at the exchange need sufficient time to operate on each pulse, so the pulses must not be too short either. To ensure that all pulses are of the correct duration, the return rotation of the finger plate is controlled at a uniform speed by the governor shown in Fig. 1.22. This consists of two wings carrying weights which are forced outwards by centrifugal force when the governor is rotated. The wings are adjusted so that when the dial mechanism reaches the required speed, the weights rub on the inside of the brass cup. The friction between the weights and the cup prevents any further increase in speed. The governor is

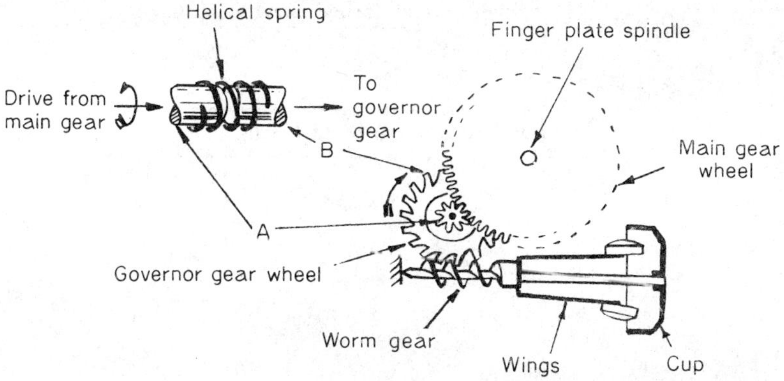

Fig. 1.22 Governor. Rotation of spindle *A* as shown tightens the helical spring on to *B* so that both spindles rotate in the same direction. Rotation of *A* in the opposite direction slackens the spring so that *B* does not rotate with it and the governor is ineffective (1.6).

normally set to give a pulsing speed of between 9 pulses per second (p.p.s.) and 11 p.p.s. (i.e. approximately 10 p.p.s.) but may be allowed to vary within the range 8 p.p.s. to 12 p.p.s. in service before being re-adjusted.

The governor gear wheel has a helical spring clutch which unwinds to give a free wheel action when the dial is being wound. The governor thus offers no resistance to the (clockwise) rotation of the finger plate to the finger stop. On the return rotation, the clutch spring winds up, gripping the spindle to drive the governor via the worm drive.

Off-normal contacts (Fig. 1.23). When the finger plate is in the rest position, the off normal contacts are held open by a cam attached to the main spindle. When the finger plate is rotated, the cam rotates with it, allowing the contacts to operate under the tension of their own springs. If the pulse wheel is made of nylon, the off-normal contact cam consists of a projection moulded as part of the wheel.

The shape of the teeth on the pulse wheel of the standard dial is such that at a speed of 10 p.p.s., the period for which the pulsing contacts are opened on each pulse is $66\frac{2}{3}$ ms, and they are then closed again for $33\frac{1}{3}$ ms. Each complete pulse of 100 ms thus consists of a 'break' of $66\frac{2}{3}$ ms and a 'make' of $33\frac{1}{3}$ ms and is said to have a ratio of $66\frac{2}{3}\%$ break (or $33\frac{1}{3}\%$ make).

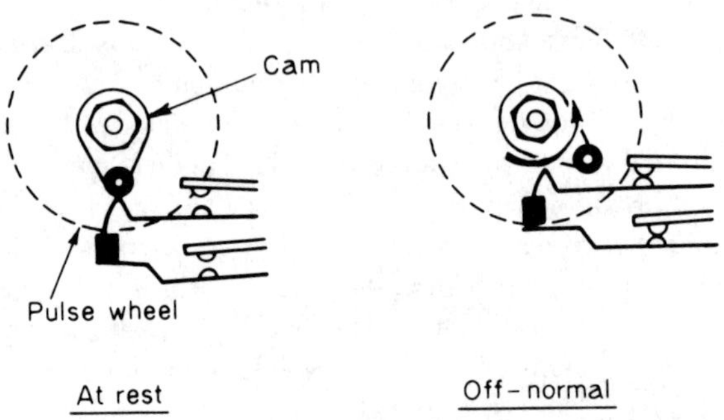

Fig. 1.23 Dial 'off-normal' spring operation (1.6).

Now turn to page 28 and read **1.8.5**

PART 4 SUMMARY

1.7.1 Telephone and telegraph networks

Telephone: easy to use, personal conversation, recognition of talker's voice.
Telegraph: written record, automatic reception, narrow frequency band.

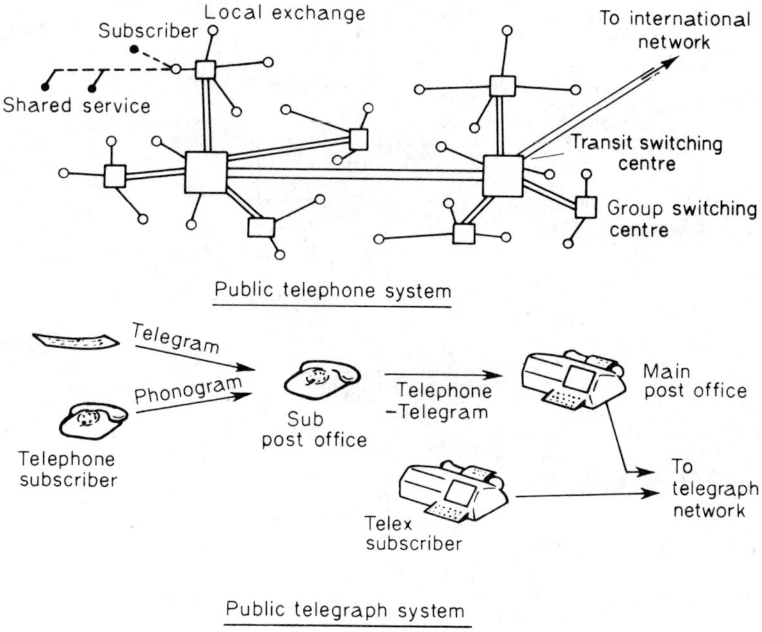

Fig. 1.24 Public networks – summary.

1.7.2 Telephone instruments

Carbon microphone requires d.c. for operation.
Receiver is polarized to prevent frequency doubling and should not be connected to d.c.
Sidetone: Talker lowers voice, hears less effectively.
Listener hears own room noise louder.
Induction coil matches instrument to line, reduces sidetone.
Dial: 8 p.p.s. to 12 p.p.s., $66\frac{2}{3}\%$ break, lost motion period 230 ms for I.D.P. of 400 ms.

1.7.3 Teleprinters

Transmitter: Keybar, combination bar, cam.
Start pulse: po*S*itive, *S*pacing.
Selecting levers, tongue.
Stop pulse: marking, negative.
Receiver: Start pulse, electromagnet, cam.
Finger setting pin, comb setting fingers, receiving combs, bellcrank.
Stop pulse, cams come to rest.
Type hammer prints during next cycle.

Turn to page 28 and try the rest of the questions there before leaving this chapter.

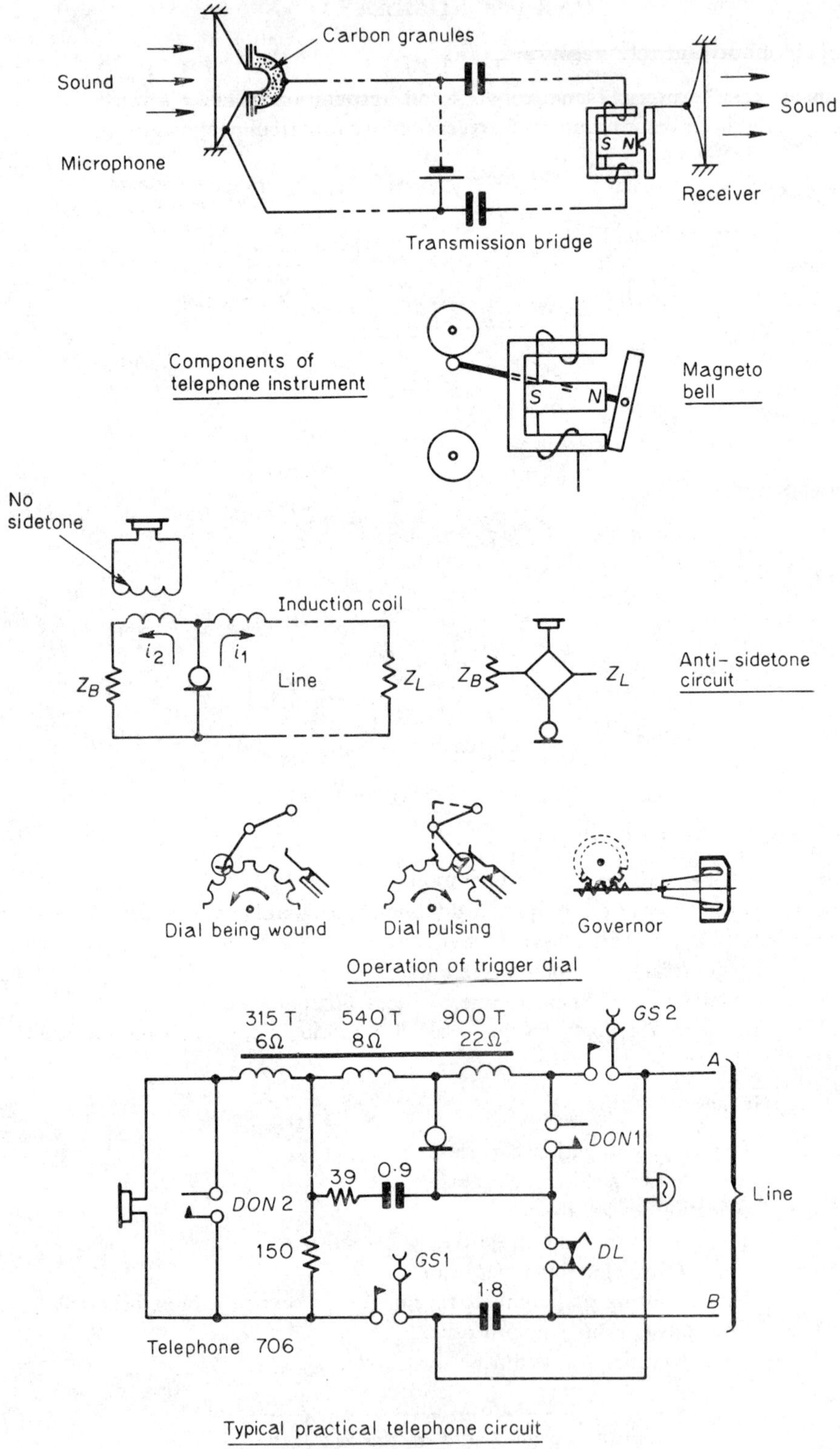

Fig. 1.25 Telephones—summary.

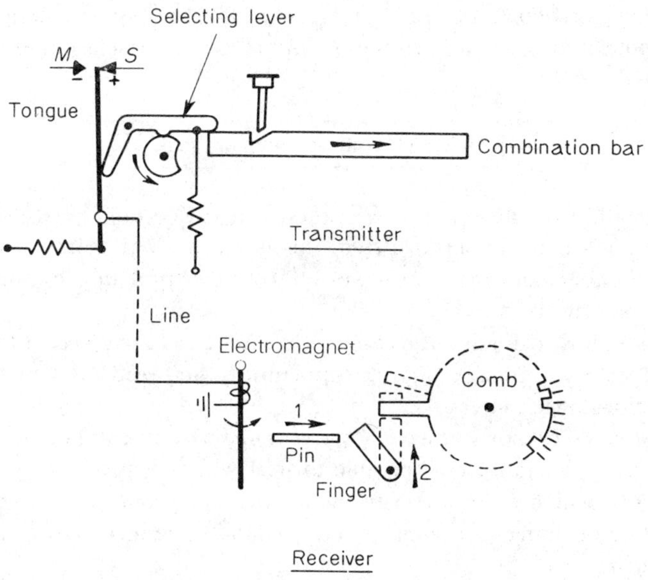

Fig. 1.26 Teleprinters—summary.

PART 5 SELF-TEACHING NOTES

1.8.1 First of all, read the notes entitled 'How to use this book' which you will find on page xiii.

1.8.2 Now read Parts 1 and 2 of Chapter 1, that is **1.1.1** to **1.4.6** inclusive. Take it slowly, think about it and study the diagrams carefully.

1.8.3 Having read **1.1.1** to **1.4.6**, try the following fifteen questions. Don't do this in your head, take a piece of paper and write the answers down for all fifteen questions. Only a few words are necessary for most of the answers.

Q.1 State one advantage of telegraphy compared with telephony.
Q.2 State one advantage of telephony compared with telegraphy.
Q.3 What is 'shared service'?
Q.4 What is the difference between a *telegram* and a *phonogram*?
Q.5 What is 'speech current'?
Q.6 Why is a permanent magnet provided in the simple telephone receiver?
Q.7 What is the condition for maximum power transfer in a d.c. circuit?
Q.8 What component is provided in a telephone instrument to achieve maximum power transfer to and from the line?
Q.9 State one undesirable effect of excessive sidetone.
Q.10 Draw a simple diagram to illustrate the method used to reduce sidetone.
Q.11 What is the first pulse transmitted when a key is depressed on a teleprinter?
Q.12 What prevents the *transmitting cam* from operating the *selecting levers* for those pulses which are to have a positive potential?
Q.13 What is the purpose of the stop pulse in teleprinter operation?
Q.14 What component does the *receiving cam* position next to each *comb setting finger* in turn so that it can operate the finger if a marking pulse is being received at that instant?

Q.15 Having positioned the receiving combs in the required combination, what component drops into the comb slots in order to select the character to be printed?

Now check your answers on page 257

1.8.4 You should have answered most of these fifteen questions correctly.
If you scored less than six out of fifteen, this is much lower than you need if you are to be able to understand the rest of the subject. You probably scanned through the chapter far too quickly.

If you scored less than ten out of fifteen (but more than five), see if the notes on the answer page help you to sort out your difficulties, then read **1.1.1** to **1.4.6** again and try the questions once more.

If you scored ten or more, or have now overcome whatever difficulties prevented you from doing so, you should now be able to deal with a longer question. Write a full answer to Q.16 which follows. Write the answer out properly as if you were taking an examination or doing an exercise in class taking no longer than 20 minutes to half an hour over it.

Q.16 Describe the essential similarities and differences between a telephone and a telegraph service and indicate the main fields in which each service is used.
In a telegraph system, what advantages accrue from the use of a switching system compared with point-to-point working? (C & G 1962)

Now check your answer against the points listed on page 258

Look up anything that you missed in your answer and then read the rest of Chapter 1, starting at **1.5**. This deals with practical applications of the principles dealt with in the earlier part of the chapter and can therefore be covered more quickly.

1.8.5 The whole of Chapter 1 should have been studied before attempting the following questions. A few days should preferably have passed since reading Parts 1, 2 and 3 (**1.1.1** to **1.6**) but a few minutes spent on revision from the summary (Part 4) will be found useful immediately before writing out your answer.

Treat these questions in the same way as Q.16.

Check the answer to each question on page 258 before answering the next one.

Q.17 Draw the circuit of a central battery telephone instrument incorporating an anti-sidetone induction coil. Explain with the aid of a simplified circuit how the induction coil functions on transmitted and received speech. What is gained by reducing sidetone? (C & G 1959)

Q.18 Explain with sketches how
(*a*) the speed of a dial is controlled
and (*b*) pulses are generated by a dial.
Give typical limiting values for dial speed and explain why these limits are necessary. (C & G 1962)

Q.19 Explain with the aid of simple sketches how the positioning of the combination bars in a teleprinter is used to control the transmission of five pulses to signal the required character.

2 · RELAYS

For self-teaching notes, see page 56.

PART 1 INTRODUCTION

2.1.1 A relay is an electrically operated switch in which a current flowing in a coil, or winding, is made to operate one or more contacts which can be used to open or close other electrical circuits.

2.1.2 Relays are used for such purposes as repeating signals, switching connexions through an exchange, or for controlling the operations of other switching devices (e.g. the crossbar switches described in Chapter 4 and the selectors described in Chapter 5).

The need for relays to repeat telegraph signals is shown in Fig. 2.1. This was the earliest application of relays and it is from this 'relaying' of signals from one section of the line to the next that we get the name 'relay'. The signal is carried by a fresh current for each stage of its journey just as the baton is carried by a fresh runner on each stage of the course in a relay race on the sports field.

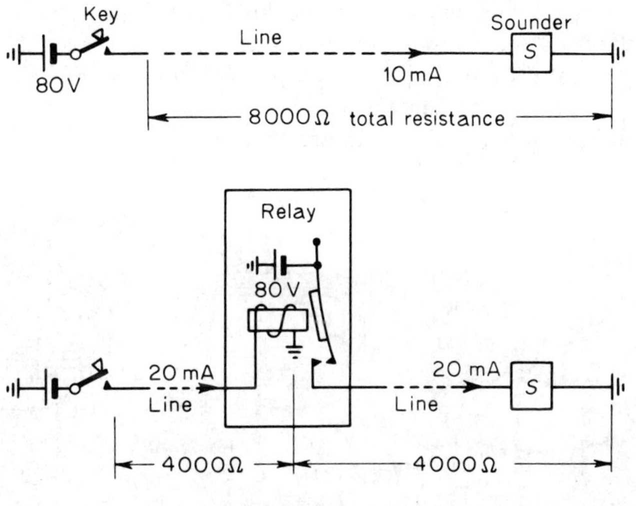

Fig. 2.1 Signal repeating relay (2.1.2).

The relays used in telecommunications are of two main types, *telegraph relays* and *telephone relays*. There are also other types of relay used for such purposes as railway signalling which do not concern us here, and some special-purpose relays used in telecommunications such as reed relays and thermal relays.

The telegraph relays are sensitive (operate on small currents), usually polarized (operation depends on the direction in which the current flows in the coil) and are used mainly for repeating signal pulses.

The telephone relays are less sensitive and are usually non-polarized but allow for variations on the basic design to suit a wide variety of different applications.

In modern practice, telephone relays and some types of telegraph relays are used in both telephony and telegraphy. Relays are perhaps the most widely used of all switching devices in telephony and telegraphy and many examples of their use will be found in every chapter of this book.

2.1.3 In this chapter, you will first learn about the basic features common to all relays and the more important of the variations for special purposes. In the last part of the chapter the construction and characteristics of typical practical relays are described.

PART 2
PRINCIPLES OF RELAY CONSTRUCTION AND OPERATION

2.2 Telephone relays

2.2.1 Magnetic circuit. The essentials of a simple single-contact relay are shown in Fig. 2.1. In this illustration, it is being used to repeat telegraph signals but it will serve equally well to illustrate the principle of the telephone relay.

In this relay, current in the electromagnet coil causes the armature to be magnetically attracted to the core. The magnetic force is referred to as the *tractive effort*. The movement of the armature bends a spring to close the contact, the relay is then said to be *operated*.

When the current is subsequently disconnected, this same spring straightens itself, opens the contact and provides a force to restore the armature to normal. The force is described as the *spring load* and the relay is said to be *released*.

The closed path occupied by the magnetic flux is called the *magnetic circuit* and, in this simple example, is mainly in the air space surrounding the relay. This is not a very efficient arrangement because air has a lower *permeability* than iron, (i.e. more current is required in the coil to maintain a given flux in air than is required to maintain the same flux in an iron circuit of the same size).

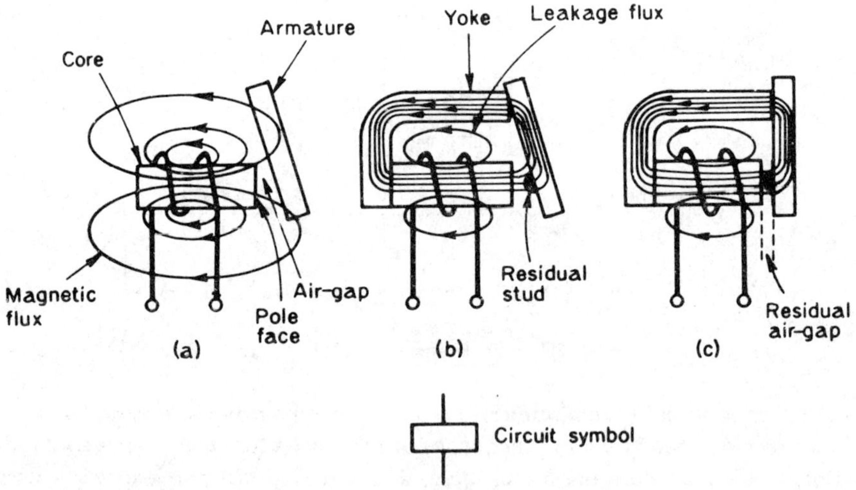

Fig. 2.2 Relay magnetic circuit showing the effect of the yoke and the residual stud. Compare this with Figs 2.17 and 2.27 (2.2.1).

To improve the efficiency of the relay, it is usual to provide an iron *yoke* to complete the iron circuit as in Fig. 2.2.

When the relay is unoperated and the armature is in its normal 'rest' position, there is a gap between the armature and the pole face to allow room for the armature to move. This is called the *armature air-gap*. When the coil is energized and the relay operates, it is necessary to allow some of this air-gap to remain. The air-gap remaining when the relay is operated is called the *residual air-gap*.

The distance the armature is able to move is of course the difference between these two air-gaps and is known as the *armature travel.*

The reason for the residual air-gap is that when the current in the coil is switched off, the iron retains a small amount of its magnetism. This could exert sufficient pull on the armature to prevent the relay from releasing, or at least allow it to release only very slowly, if the armature were touching the pole face. Air, however, has a lower permeability than iron, so quite a small air-gap reduces the flux which can be maintained by this residual magnetism sufficiently to allow the relay to release reliably.

To provide a residual air-gap, some form of stop is necessary to limit the travel of the armature. The most commonly used method is to attach to the armature a *residual stud* (or 'pin') of a non-magnetic material such as brass. Other methods include brass screws attached to the frame or the coil bobbin, plastic adhesive tape on the armature or a thin coat of varnish on the armature or the pole face.

2.2.2 Contacts. When used as a signal repeater as in Fig. 2.1, only one contact unit is needed on the relay, but for switching a speech current path through an exchange, at least two units are necessary, one for each wire. For controlling the functions of a selector mechanism it is often necessary to use relays with more than two contact units. These can be arranged as:

(*a*) *make* contacts which are normally open and which close when the relay operates,
(*b*) *break* contacts which are normally closed and which open when the relay operates,
(*c*) *change-over* contacts in which the break action opens before the make action closes, and
(*d*) *make-before-break* (*K-action*) change-over contacts in which the break action does not open until *after* the make action has closed.

The arrangement of the contact springs to provide these actions are shown in Fig. 2.3 together with their circuit symbols.

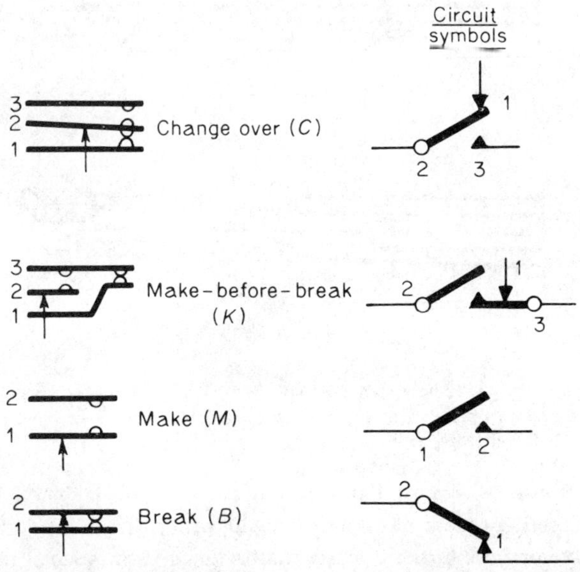

Fig. 2.3 Relay contact actions. The arrows on the lever springs indicate the position and direction of the force exerted by the armature (2.2.2).

To operate a number of contact units from the same armature, it is necessary to link the contacts to each other mechanically without connecting them to one another electrically. Two basic systems are in common use in a variety of forms. These are illustrated in Fig. 2.4. One uses brass pins attached to alternate moving springs bearing on insulated studs on the other moving springs and on the armature. The other system uses 'combs' made of an insulating material such as plastic. These combs are positioned in slots or holes in the springs and the springs rest on the steps in the combs.

The movement of the armature causes the 'lever' or moving springs to make or break contact with the 'buffer' or fixed springs. It is clearly necessary to provide some means of holding the fixed springs in position so that the moving springs shall be able to press firmly against them. This pressure is referred to as *contact pressure* and is necessary to ensure a good low resistance connexion at the contact. In some types of relay, the fixed springs are made thicker and therefore stiffer than the moving springs. Other types of relay use a *buffer block* made usually of a hard ceramic material. The buffer block has steps moulded or cut into it on which the springs rest as shown in Fig. 2.5.

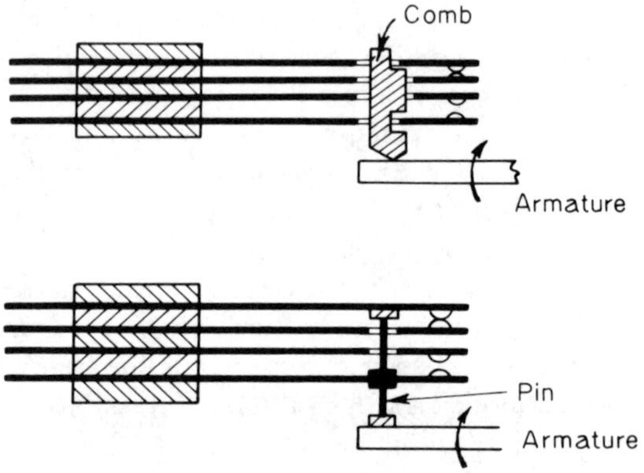

Fig. 2.4 Relay springset operation showing a make and a break contact on a comb type and a pin type springset (2.2.2).

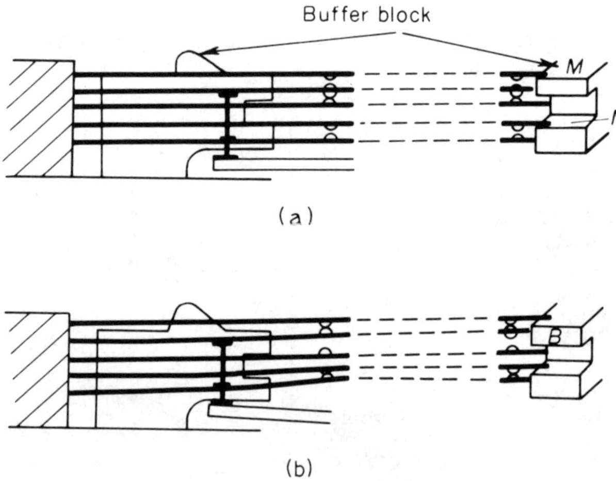

Fig. 2.5 Buffer block. The make springs *M* rest on the buffer block with the relay unoperated at (a) and the break springs *B* rest on the buffer block with the relay operated at (b) (2.2.2).

The contact springs consist of flat strips of some material such as nickel silver carrying dome shaped contacts of silver, a metal which has a low contact resistance and does not easily corrode. Other contact materials used for special purposes include platinum, palladium, tungsten and p.g.s., an alloy of platinum, gold and silver. Some types of relays have wire springs plated with a suitable contact material.

To reduce the possibility of contact corrosion or particles of dust blocking the passage of current through a contact, most types of relay are provided with two contacts on each spring, the end of the spring being forked so that the two sets of contacts can each complete their operation independently. The shape of a typical spring of this sort is shown in Fig. 2.6.

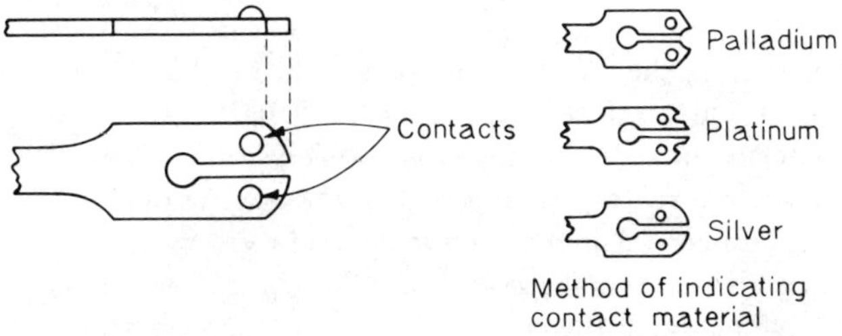

Fig. 2.6 Twin contacts showing the forked end of a contact spring fitted with twin contacts (2.2.2).

Some forms of corrosion produce a high resistance film on the contact surface. Such films can be broken down by an applied potential in excess of about 5 V and a satisfactory conducting path can then be maintained by a current of 1 mA or more. In most applications these conditions are met by the inherent properties of the circuit, e.g. a contact operating a 2000 Ω relay from a 50 V supply.

Sometimes a contact may be used to switch only a low voltage a.c. such as speech current in series with a capacitor. The initial closure of the contact usually breaks down the film by charging or discharging the capacitor through the contact. The film soon re-forms because the a.c. falls to zero twice in every cycle and may even have too low a peak value to maintain conduction. The cure is to employ *contact wetting*. This consists of adding a direct current of 1 mA or more so that the total current never falls to zero. Resistors typically of 22 kΩ or 47 kΩ connected to the exchange 50 V supply are commonly used for this purpose.

2.2.3 Operating and releasing characteristics. A relay operates when the magnetic force of attraction acting on the armature exceeds the restoring force applied by the springs. It releases when the magnetic attraction falls to a value which is less than the load. To determine the conditions necessary for operating or releasing a relay, we must see what factors affect these two forces.

The force applied by the springs depends on the number of contacts on the relay, the type of contact action (make, break, etc.) and on the thickness of the springs.

The magnetic force on the armature (tractive effort) is proportional to the square of the magnetic flux density and to the area of the pole face (see Appendix 3).

Magnetic flux results from current in the coil setting up a magnetomotive force (m.m.f.). Every turn contributes to it, so m.m.f. depends upon the number of turns and upon the current. The SI unit of m.m.f. is the ampere-turn, so m.m.f. can be calculated by multiplying the current by the number of turns.

The flux produced by a given m.m.f. depends upon the 'reluctance' of the complete magnetic circuit:

$$\text{Magnetic flux} = \frac{\text{ampere-turns}}{\text{reluctance}}, \text{ i.e. } \quad = \frac{F}{S}$$

$$\left(\text{This can be compared with Ohm's law, } I = \frac{V}{R}\right)$$

Reluctance is a measure of the opposition offered by a magnetic circuit to a magnetic flux, and is given by:

$$\text{Reluctance} = \frac{\text{length}}{\text{permeability} \times \text{cross-sectional area}}, \text{ i.e. } S = \frac{L}{\mu A}.$$

$$\left(\text{This also can be compared with resistance} = \frac{\text{length}}{\text{conductivity} \times \text{cross-sectional area}}\right)$$

Therefore, increasing the current, the number of turns or the area of the pole face will increase the tractive effort. Increasing the length of the air-gap, using lower permeability materials or reducing the cross-sectional area of any part of the magnetic circuit will increase the reluctance and reduce the tractive effort.

The minimum ampere-turns required to operate a given relay is called the *operating ampere-turns* of the relay and the corresponding current is the *operate current*. Once operated, the value to which the current must be reduced in order to release the relay is called the *release current*. Because the air-gap is less (lower reluctance, more flux) when the armature is operated than when it is released, the release current of the relay is less than the operate current. A *hold* current is also specified in the test data for a relay sometimes. This is higher than the *release* current to allow for manufacturing tolerances, all samples being expected to remain operated on a current of at least the hold value and none to remain operated on the hold current or less.

Anything which either increases the magnetic force acting on the armature or reduces the spring load with the relay unoperated will reduce the operate current value. Increasing the spring load or reducing the magnetic effect will, of course, increase the operate current.

Once the relay is already operated, any factor which increases the magnetic force or which reduces the spring load will lower the release current value. Increasing the spring load or reducing the magnetic effect will raise the release current value.

It follows that if, for example, a relay is required to have a low value of operate current, it should have few springs, short travel, low reluctance, a large number of turns and the contact springs should have the minimum thickness. Such a relay would also have a low value of release current.

If a relay were required to have a high value of release current, it would need as many springs as possible of maximum thickness, long residual stud, few turns and high reluctance.

2.2.4 Operate and release times. When current starts to flow in a relay coil, a certain time elapses before the contacts of the relay open or close. This period of time is known as the operate time or *operate lag* of the relay. When the current in the coil is disconnected, there is a *release lag* before the contacts restore.

There are two main causes of these lags. The operation of the relay involves the physical movement of components such as the armature, springset comb etc. and these all have some mass, however small, which has to be accelerated in order to produce movement. Before this movement can even start, the coil current has to rise to the operate value for the relay, and this does not occur instantaneously: it is delayed by the inductance of the coil.

The effect of inductance in delaying the current may be explained as a reaction between the magnetic field and the current which is producing it. As the current starts to flow, the increase in current produces an increasing magnetic flux which is linked with the coil. Since this flux is changing in value, it induces an e.m.f. in the coil and by Lenz's law the direction of this e.m.f. is such as to oppose the increase in the current. The effect of disconnecting the current to release the relay is to cause the flux to collapse, and in doing so, it tends to keep the current flowing if the circuit arrangements are such as to provide a path for the current at this time.

Another way of looking at the same thing is that the magnetic field has a quantity of energy stored in it. This energy is provided in the first place by the current and takes some time to build up. On release, the energy has to be dissipated and this again takes time. Where no circuit path exists in which the coil current can be maintained by the collapsing magnetic field, the eddy currents induced by this field in the iron core or other nearby metallic components serve to dissipate the stored energy.

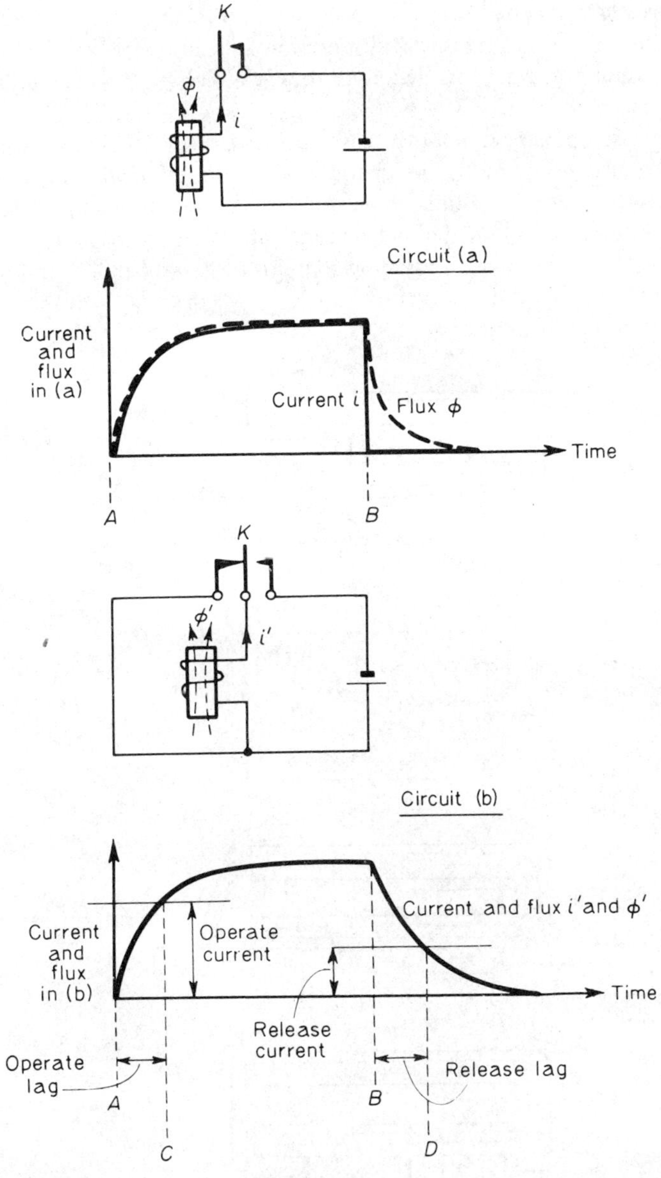

Fig. 2.7 Current and flux in relay coils. In (a) there is no circuit for the maintenance of the coil current during release as there is in (b). The coils are energized in the period of time between A and B and the contacts are operated between C and D (2.2.4).

The inductance of the coil depends on both the number of turns and the reluctance of the magnetic circuit. The more turns the coil has, the higher will be the inductance; the greater the reluctance, however, the less will be the inductance.

If a relay is required to operate quickly, (i.e. it is to have a short operate lag), the mass of the moving parts and therefore their inertia should be as small as possible. The inductance, or more precisely the ratio of inductance to resistance (i.e. the time constant), should be low and the current in the coil under working circuit conditions should be large compared with the relay operate current: a condition which can be achieved by having a low coil resistance and a low value of operate current. The means of obtaining a low operate current are described in **2.2.3**.

The conditions for a short release lag are low inertia, short time constant (L/R) and minimum circuit current. This generally means a high resistance winding and a high value of release current (**2.2.3**).

For both fast operate and fast release features, eddy currents must be kept to a minimum because they produce magnetic flux in a direction which opposes the change occurring in the main relay flux. This can be achieved by using plastic rather than metal cheeks (end plates) on the coil bobbin and using a core material having a high resistivity such as nickel-iron. This alloy contains 36% Ni, 64% Fe and has a resistance ten times that of soft-iron.

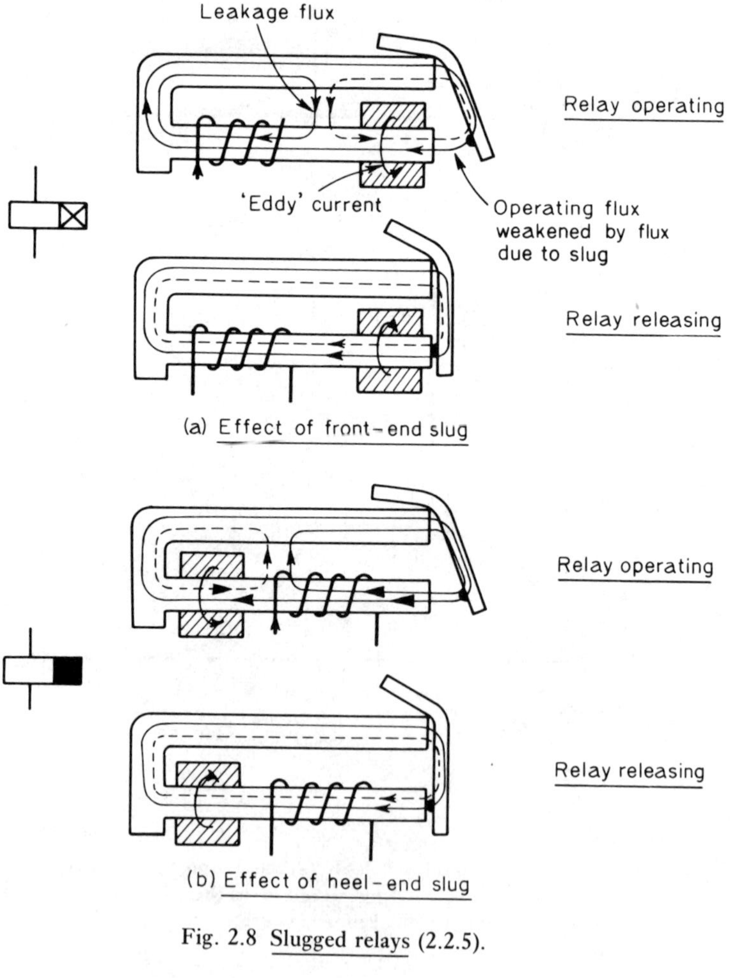

Fig. 2.8 Slugged relays (2.2.5).

2.2.5 Slugged relays. In many circuits it is necessary for some of the relays to have particularly long operate lags or release lags. One commonly used example of this is the *B* relay used in automatic selector circuits which has a release lag of about 200 ms so that it will not release in the short period ($66\frac{2}{3}$ ms) that the subscriber's dial disconnects the line current on each pulse and yet will release when the current is disconnected for good at the end of the call.

Relays can be made slow-to-release or slow-to-operate by providing a circuit in which the magnetic flux of the coil can induce currents, the magnetic effect of which is to oppose the change (either increase or decrease) taking place in the main magnetic flux. For the greatest effect, this circuit should have as little resistance as possible, a condition which can be met by making it in the form of a thick 'collar' of copper around one end of the core. This is known as a copper *slug* and relays fitted with it are called *slugged relays*.

A slug may be fitted to the end of the core nearest to the armature, in which case it is a *front-end* slug; or it may be at the end farthest from the armature when it is known as a *heel-end slug*. These are also sometimes known as fore-end or armature-end slugs and rear-end slugs respectively. The difference between the two is that the front-end slug makes the relay both slow to operate and slow to release, while the heel-end slug makes it slow to release, but has practically no effect on the operate lag. Both are illustrated in Fig. 2.8.

The effect of a front-end slug is as follows:

Operation of relay. When the coil current is switched on, it sets up a steadily increasing flux as shown by the unbroken lines in Fig. 2.8a. This flux is linked with the slug and therefore induces a current in the slug. The induced (eddy) current in the slug sets up an opposing flux about the slug as shown by the broken lines in Fig. 2.8a. The combined effect of the two fluxes is that the operating flux in the air-gap (**2.2.1**) is less than it would be without the use of the slug but this state of affairs lasts only while the induced eddy current is flowing in the slug. As the coil current reaches its full value, the rate at which it and the coil flux are growing falls off. The reduced rate of change of flux results in a reduced e.m.f. being induced in the slug and thus a smaller induced eddy current. The flux produced by the slug is therefore losing strength while the coil flux is increasing. Eventually, the flux in the air-gap which consists of the difference between the coil flux and the slug or eddy current flux, becomes sufficient to operate the armature. The effect of the slug has thus been to delay but not to prevent the operation of the relay.

Release of relay. When the coil current is switched off, the flux starts to decrease. As it changes, however, it again induces current in the slug. The change taking place in the flux is now the opposite of what it was when the relay was operating so the induced current is in the opposite direction to what it was before. The eddy current flux therefore adds its strength to that of the coil flux in the air-gap and continues to hold the armature after the strength of the coil flux has fallen below the value required to maintain the armature operated on its own. In due course the eddy current dies away as before and the armature finally releases.

The effect of the heel-end slug is as follows:

Operation of relay. An eddy current flux is set up in exactly the same way as already described for the fore-end slug. The difference in the position of the slug, however, allows the flux produced by the slug to take a path which does not pass through the armature and its air-gap as shown in Fig. 2.8b. A very small part of it does in fact go via the air-gap at the armature but its effect is negligible in practice so that the heel-end slug has no effect on the operate lag of the relay.

Release of relay. Once again current is induced in the slug to set up a magnetic flux. The armature is already operated and the armature air-gap is short so that the reluctance (see **2.2.4**) of the magnetic circuit of the relay is lower than it was before the relay was operated. The eddy current flux now finds it easier to take the path right round the magnetic circuit and through the armature air-gap. By passing through the air-gap in this way, it delays the release of the relay in the same way as a front-end slug does.

2.2.6 Circuit methods of controlling relay timings. The use of slugs is one way of providing a path for eddy currents to delay the operation or release of a relay. Another way is to connect the two ends of a coil winding together to provide a complete circuit in which induced currents can circulate. If the ends of the winding are directly connected to one another, the winding is said to be short-circuited; if connected together through resistance it is said to be shunted. The lower the resistance of the shunt the longer the lag produced. Relay windings are normally spread along the whole length of the core and so the effect of a short circuit or shunted winding is the same on both the operate and the release lag. The effect can be restricted to either the operate or the release condition by using contacts of the same relay or of other relays to switch the shunt or short circuit in or out at the appropriate times. Examples of this technique are shown in Fig. 2.9.

One particularly useful way of obtaining a slow release feature without making the relay slow to operate is to connect a permanent shunt across the operating coil of the relay in the form of either a resistor or a rectifier.

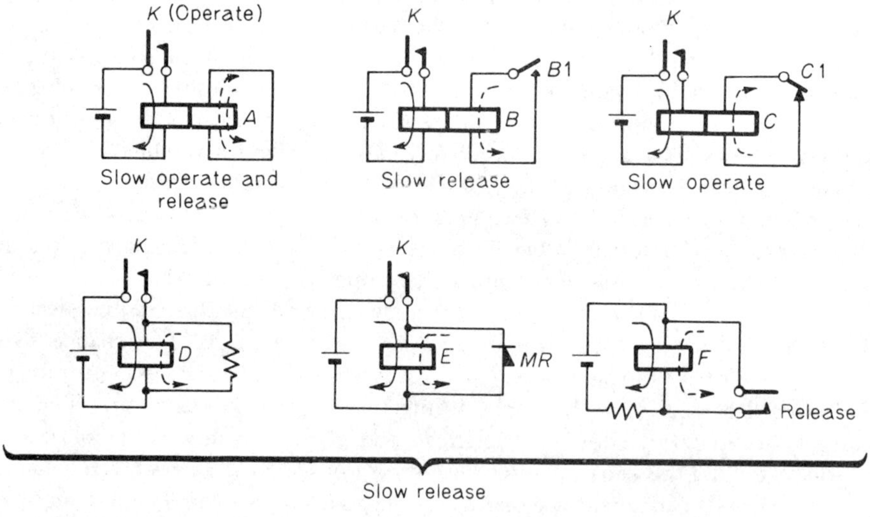

Fig. 2.9 Control of relay lags. The operating current paths are shown by the full arrows, the induced currents by the dotted arrows (2.2.6).

2.2.7 Isthmus armatures. An important application of the non-polarized 'telephone' type of relay is its use for converting the 'break' pulses received from the subscriber's dial (**1.4.2**) to 'make' pulses of current to the electromagnet of a selector. The relay must not 'distort' the pulses too much by altering either the length of the pulses or the spaces between the pulses.

This distortion can occur if, for any reason, the operate lag of the relay differs from its release lag.

Although it is not difficult to make the lags equal for a given relay under constant operating conditions, in practice the relay would have to receive pulses over any one of several lines, each of different resistance. It is desirable therefore that the operate and release lags of the relay should vary as little as possible under different line conditions.

To make the operate lag substantially independent of circuit current and therefore independent of line length, the operate current should be made as low as possible (see Fig. 2.10). This can be done by using minimum spring load, maximum number of turns etc. Unfortunately these measures also result in a low value of release current which both increases the release lag and its dependence on variations in relay adjustments and tolerances. The use of an *isthmus armature* (Fig. 2.11) overcomes this by increasing the release current without significantly increasing the operate current. The cut away portions reduce the cross-sectional area of the armature, giving it a higher reluctance. The armature reluctance is still small compared with the air-gap reluctance so that increasing its value has little effect on the total reluctance of the magnetic circuit with the relay unoperated. The isthmus armature thus has little effect on the operate current value.

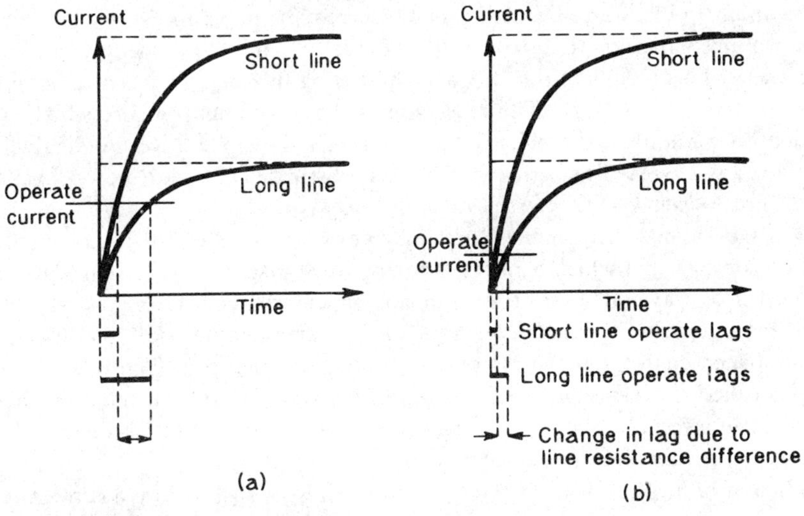

Fig. 2.10 Variation in operate lag with line resistance. The difference in operate lag due to differences between the resistance of long and short lines is greater for the relay with a high value of operate current (a) than for the relay with the lower value (b) (2.2.7).

Once the relay is operated, however, the air-gap reluctance is small. The increase in the armature reluctance resulting from the use of an isthmus armature is therefore a greater proportion of the total reluctance of the magnetic circuit under this condition. The result is that the isthmus armature produces a greater increase in the release current than it does in the operate current.

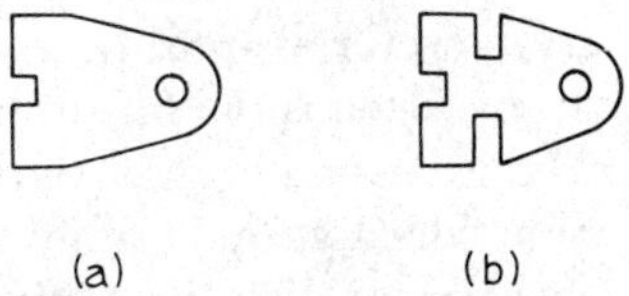

Fig. 2.11 Isthmus armature (b) compared with an ordinary armature at (a) (2.2.7).

The isthmus armature also reduces the effect of the line conditions on the release lag by limiting the flux in the relay on short lines to little more than is produced by the smaller current on the long lines.

2.2.8 Nickel-iron sleeves. Relays connected to the speech wires in circuits providing transmission bridge facilities (**1.3.6**) are required to offer a high impedance to currents at speech frequencies but a relatively low resistance to d.c. To achieve this the relay must have a high inductance.

If the relay has an ordinary soft-iron core, the eddy current and hysteresis losses reduce the effective impedance at speech frequencies. The eddy current also produces a magnetic 'skin effect' which reduces the impedance at these frequencies by reducing the inductance.

Alternating current in the relay coil sets up an alternating magnetic flux in the iron core which induces an alternating eddy current in the iron. The eddy current sets up an opposing magnetic flux at the centre of the core which neutralizes the original flux there. This results in the effective alternating flux being concentrated in the outer layer or 'skin' of the core. The cross-sectional area of the iron core which is effective under this condition is therefore reduced and the inductance correspondingly decreased.

The eddy current flowing in the core also dissipates electrical power in the form of heat. This power loss is referred to as the *eddy current loss*.

A magnetic material, such as iron, may be regarded as consisting of small particles or 'domains' each of which behaves as a permanent magnet. Each cycle of an alternating current in a relay coil causes these domains in the iron core to point first one way and then the other way, power being expended in overcoming the internal forces which oppose such changes in the molecular structure of the iron. The power used in this way is called the *hysteresis loss*. The term hysteresis on its own refers to the effect of these same internal forces in causing the magnetic changes to lag behind the current changes, but this does not concern us here.

The power for the eddy current loss and the hysteresis loss has to come from the a.c. in the coil. The greater the losses, the larger must be the current. Now, impedence is by definition the ratio of voltage to current in an a.c. circuit and therefore any increase in the current for a given voltage means that the impedance has decreased. To obtain *a high impedance, eddy currents and hysteresis losses should therefore be kept to a minimum.*

The nickel-iron alloy (**2.2.4**) used for the cores of fast operating relays has a high resistivity which reduces eddy currents and it also has a low hysteresis loss. In practical circuits, however, a relay coil carries d.c. as well as a.c. At the values of d.c. energization normally employed in relays, nickel-iron has a lower permeability than soft-iron. This reduces the inductance of the coil and also reduces the tractive effort on the armature. For this reason, the nickel-iron is used in the form of sleeves round the outside of a soft-iron core as shown in Fig. 2.12. The sleeves are split down their length so that they can be fitted over the core easily and to reduce eddy currents. The skin effect, due mainly to the iron core, tends to confine alternating magnetic fluxes at speech frequencies to the nickel-iron sleeves to give the required high impedence. The d.c. flux, however, is spread evenly through the whole cross section of the core, the greater part of which is the soft-iron.

Transmission bridge relays often require two windings of equal impedance. These are provided in the form of a *sandwich winding* as shown in Fig 2.13 which gives the same average diameter for both coils so that they both have the same number of turns and the same resistance.

2.2.9 Polarization of telephone relays. As will be described in more detail in Chapter 3, it is sometimes necessary to transmit signals over junctions between

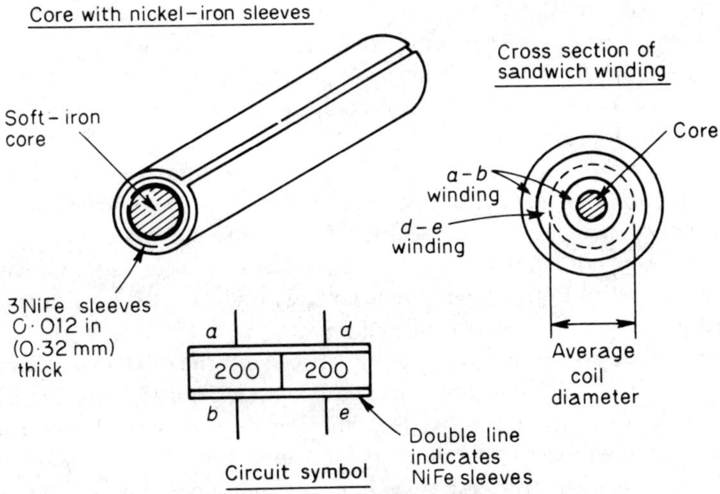

Fig. 2.12 High impedance relay coils for use in transmission bridge circuits (2.2.8).

telephone exchanges in the form of reversals of the normal polarity of the battery and earth connexions at one end of the line. To receive these signals at the other end, a relay is required which will operate when the line polarity is reversed in this way but will release when the normal polarity is connected. This means that the relay must operate on current in one direction and release on current in the other direction.

One way of polarizing a telephone relay for this type of signalling is shown in Fig. 2.13. An additional magnetic circuit is provided in parallel with the main relay operating coil magnetic circuit. In use the relay has this additional *shunt-field* energized

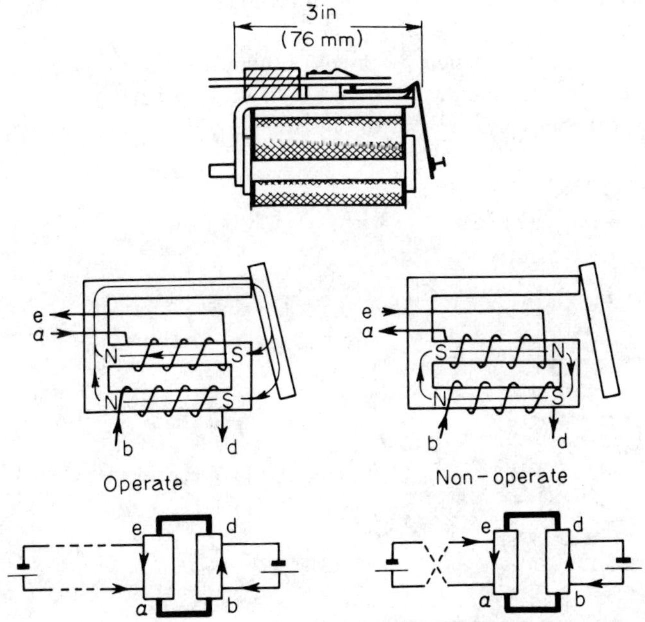

Fig. 2.13 Shunt-field relay. Non-operate current direction is shown by arrows on circuit symbol (2.2.9).

locally. The relay will only operate when the operating coil (connected to the line) is energized by a current in such a direction as to produce a flux in the same direction as that of the shunt field. Current in the opposite direction in the operating winding, or no current, produces a magnetic flux wholly within the iron of the magnetic circuit so that the relay does not operate.

The non-operation, or release, of the relay depends on there being no magnetic flux in the armature air-gap. A faulty joint between the cores of the two coil cores at the armature end can give rise to a leakage flux outside the iron circuit in the air-gap. In practical designs of shunt-field relay it is difficult to ensure a good magnetic joint at this point, so relays of this type have been largely superseded by ordinary telephone relays used in conjunction with rectifiers or diodes.

The rectifier may either be connected in series with the relay or in parallel with it. In a practical circuit, the parallel rectifier is used because it maintains a path for current in the direction in which the relay does not operate (i.e. normal junction polarity). This current is required to operate relays in the system. Copper oxide rectifiers do not provide a sufficiently low resistance to ensure non-operation of the relay and so a second rectifier is usually connected in series with the relay (see **3.6.1** and Fig. 3.17). In later circuits using semiconductor diodes, only the parallel diode is used.

2.3 Telegraph relays

2.3.1 General features. Telegraph relays are used for repeating telegraph signal pulses with minimum distortion at pulsing speeds greater than those used in telephony. They must therefore operate fast, and function reliably over a large number of operations. To receive pulses over long (i.e. high resistance) lines, the relays must be sensitive.

The measures adopted to obtain the required sensitivity include the use of just a single contact, this being directly attached to the armature, a large cross-sectional area in the magnetic circuit to reduce the reluctance and the use of permanent magnet polarization.

Reliable operation is obtained by precision manufacture, by using large contacts to reduce wear and by minimizing the number of moving parts.

To ensure that the relays are fast to operate, eddy currents and hysteresis losses are

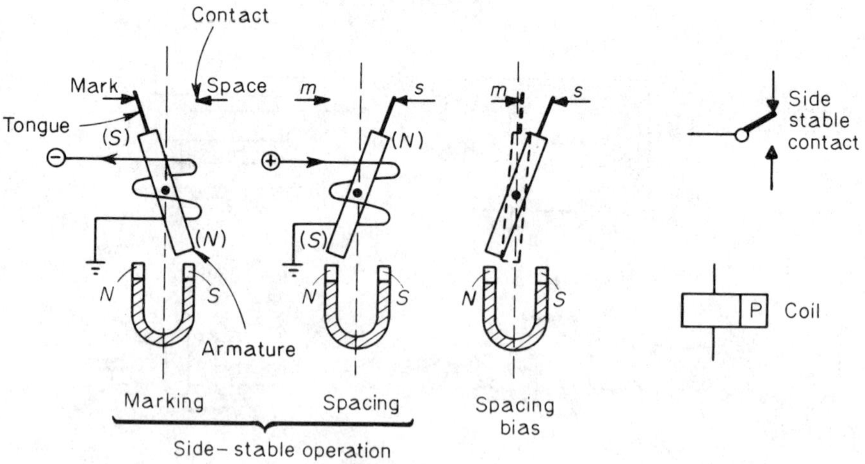

Fig. 2.14 Principle of polarized telegraph relay (2.3.2).

kept to a minimum by the use of special magnetic materials such as nickel-iron and by constructing the magnetic circuit in laminated form (i.e. in layers) to restrict the flow of eddy currents. Increased sensitivity (low operate current) also helps make the relay fast.

Distortion is minimized by the fast operating features and by the use of adjustable contact screws and pole pieces to permit accurate adjustment of the relay's operating characteristics. The relays are polarized in such a way that the armature moves to one side on current in one direction and moves in an identical manner in the opposite direction when the current is reversed. Because the two operations occur in the same manner, they take place at the same speed. Provided, therefore, that signals are transmitted in the form of positive and negative pulses, the relay lags cancel each other out to give virtually no distortion (2.2.7). This type of operation is called *double-current signalling* (3.3.6).

2.3.2 Polarization of telegraph relays. Telegraph relays are polarized by means of permanent magnets, the effect of which is to increase the sensitivity by adding to the tractive effort of the coil (see Appendix 3) and to reduce distortion by providing a symmetrical operating characteristic. The essential principle of such relays is illustrated in Fig. 2.14.

The armature is mounted within the coil so that it is magnetized in a direction which depends on the direction of the current in the coil. If the current is such as to induce a north pole, for example, at the end of the armature nearest to the permanent magnet, the armature will be attracted to the south pole of the magnet. If the current is then switched off, the armature will remain held in this position by the magnet. A current in the opposite direction will magnetize the armature so that it moves to the north pole of the permanent magnet and stays there. A relay operating in this way is said to be *side stable*.

If the armature is mounted on a spring or similar flexible device, instead of pivoted freely, it can be arranged to return, under the action of the spring, to the centre position when the current is switched off. It is then said to be *centre stable*.

In some types of relay, the contacts can be positioned so that, in one direction of operation, the armature is prevented from passing the centre position. When the current is switched off, therefore, instead of remaining operated at this contact, the armature is attracted back to the other contact. The relay is then said to be *biassed*. If the contact to which it returns when the current is disconnected is that corresponding to a 'marking' signal, the relay is said to have a *marking bias*. If the contacts are positioned to restrain the armature to the other side of the centre position, the relay has a *spacing bias*.

Practical relays must, of course, have yokes and residual gaps as we have seen for telephone relays. The yoke usually takes the form of iron extensions attached to the magnet poles. These may also serve to provide an operating force at both ends of the armature. The residual gap can be provided by suitably positioning the contacts or by incorporating non-magnetic spacers, e.g., brass at the armature mounting or pivot. Typical magnetic circuits of this sort are shown in Figs 2.22 and 2.24.

The arrangement of Fig. 2.24 which shows a relay used for signalling in both telegraphy and telephony is of particular importance. In this the armature moves towards whichever side has the stronger total magnetic flux. This will be the side on which the coil flux is in the same direction as the permanent magnet's flux. As there are two permanent magnets, one providing a magnetic field on each side of the armature, the armature can be operated to either side depending on the current direction, and the relay is normally side stable. By using a spring-mounted armature, the relay can be made centre stable.

2.4 Relay circuits and diagrams

In practical applications, relays are not usually employed singly. Each circuit consists of a group of relays connected together in such a way as to provide the required facilities. Circuits of this sort are often quite complex, and it is essential to have diagrams which show the wiring connexions in such a form that the sequence of operations by which the facilities are provided is easily followed for the purpose of locating faults for maintenance work.

A typical use of relays which illustrates this type of diagram is the provision of a transmission bridge (**1.3.6**) at an automatic exchange for calls switched via a junction to another automatic exchange. The facilities provided would be, microphone current to the calling telephone, repeating gravity switch (loop) and dialling signals to the next exchange, and maintaining an earth on the *P*-wire for holding the control relays in the selectors.

The basic circuit symbols used for relay coils and contacts are illustrated in Figs 2.2, 2.3, 2.8, 2.12, 2.13, and 2.14. From the point of view of someone wiring the circuit in the first place, probably the most convenient way is to draw the diagram as much like the physical layout of the actual apparatus as possible. This means, among other things, showing the relay contacts next to the coil which operates them. The transmission bridge referred to above is shown in this form in Fig. 2.15. Even though components such as the capacitors *C*1 and *C*2 are shown in the most convenient place rather than attempting to indicate their mounting position relative to the relays, some of the wires are very difficult to follow. The coil *I*, for example, is connected directly between the two capacitors and has 'make' contacts of relays *B* and *CD* across it. To establish these relatively simple facts it is necessary to trace wires backwards and forwards along the diagram, turning through right angles at no fewer than nine points.

This method of representing a circuit has been replaced by a system known as a *detached contact* system in which each relay is designated by suitable letters. All contact units on the relay are then designated by the same letters and can be drawn anywhere on the diagram. To help in establishing all the functions of a given relay, the total number of contacts on it is indicated under the designation (letter) beside the coil and each contact unit on the diagram is numbered in sequence.

The use of these labels, or designations, can be seen by referring to Fig. 2.16 which is the same circuit as Fig. 2.15. Consider what happens when relay *A* operates, for

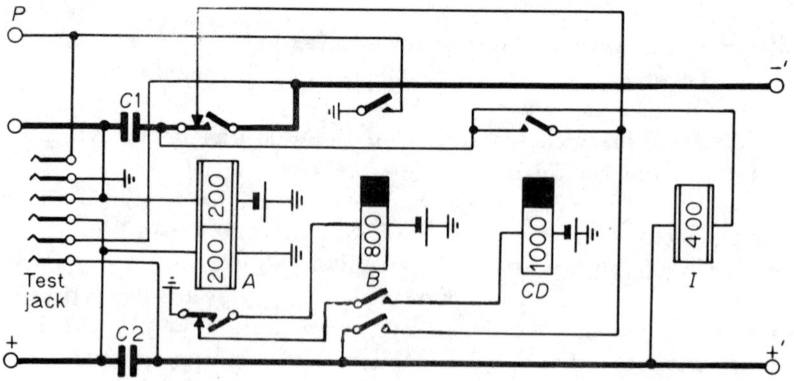

Fig. 2.15 Typical attached contact diagram. Compare this with Fig. 2.16 (2.4).

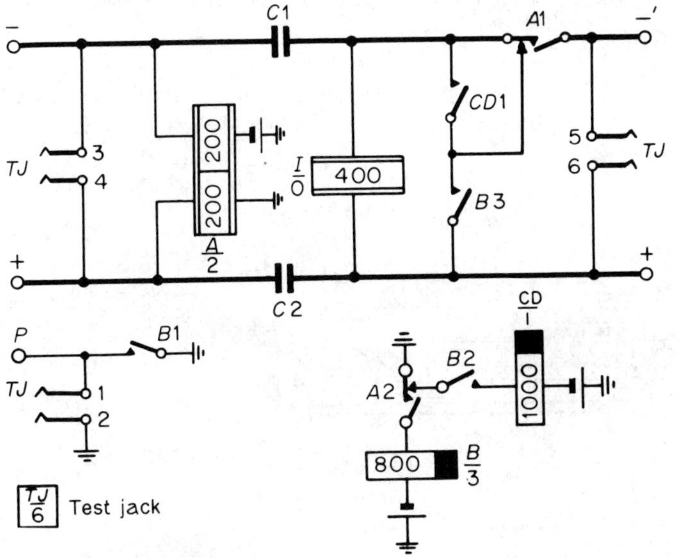

Fig. 2.16 Typical detached contact diagram. This represents exactly the same circuit connexions as Fig. 2.15 (2.4).

example; it is labelled $\frac{A}{2}$ so there are two contact units. Look first for contact unit *A*1; it is in the top right-hand corner, and when it operates it connects *I* across the –′ and +′ wires. Next look for *A*2; it is lower down, between the coils of relays *B* and *CD*, and when it operates, it completes a circuit to operate relay *B*.

Similar labelling can be used for other components like the test jack on this circuit, for example. Notice, too, that the coils and contacts can be drawn whichever way round gives the simplest layout. A well drawn diagram has few, if any, wires crossing over one another and few very long wires. All the items concerned with a particular facility are grouped near one another, and as a result the diagram presents a neat appearance which does not confuse the eye.

Compare Figs 2.15 and 2.16 with these points in mind and it will be clear why the detached contact system is superior even for this relatively simple circuit. For large circuits consisting of perhaps twenty or more relays, the attached contact type of diagram becomes quite unreadable unless used in conjunction with a detailed description or timing chart.

Now turn to page 56 and test yourself on questions 1–20 there before reading the next part.

PART 3 TYPICAL PRACTICAL RELAYS

2.5 Typical telephone relays

2.5.1 British Post Office 3000 type. This is the basic standard relay used for most switching purposes in telephone and telegraph automatic exchanges. Its main features are illustrated in Fig. 2.17.

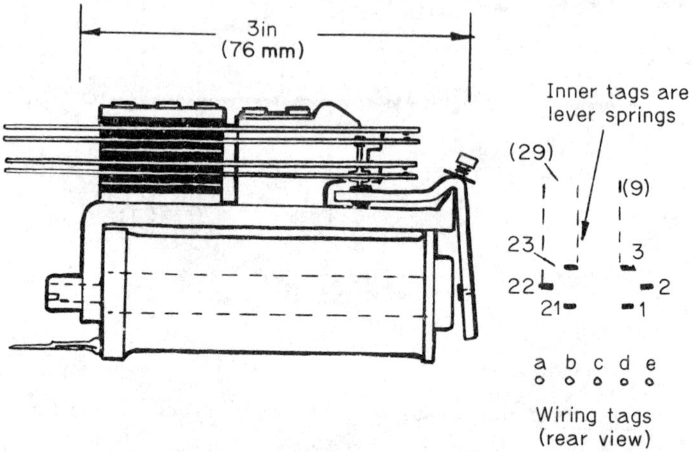

Fig. 2.17 B.P.O. 3000-type relay (2.5.1).

The magnetic circuit consists firstly of a soft-iron core which is fitted to the soft-iron yoke by means of the core nut which screws on to the end of the core. Tightening this nut ensures a good magnetic joint between core and yoke. The core normally has an enlarged pole face at the armature end to provide a large cross-sectional area at the air-gap and thus a low reluctance. A nickel-iron core may be fitted to relays which are required to be fast to operate, in which case the pole face has the same diameter as the core. Nickel-iron sleeves may be fitted if a high impedance is required (**2.2.8**). The armature end of the yoke has a triangular knife-edge about which the armature pivots. The armature, also of soft-iron, is shaped so that it is effectively wrapped round the knife-edge to provide a large cross-sectional area at the air space between the armature and the yoke, both when the relay is operated and when it is released. This provides a uniformly small reluctance at this point in both conditions with little leakage flux. The armature may be either plain or isthmus (**2.2.7**).

The coil is normally wound of copper wire. The resistance depends on the number of turns and on the thickness of the wire used, but the standard value for general purposes is 2 000 Ω. Other values of resistance are used when required to meet special circuit conditions, additional turns of special alloy resistance wire being used, where necessary, to give the exact resistance needed. Up to five coil tags can be provided and any number of separate windings may be wound on to the spool up to a maximum of four. In addition to the windings which contribute to the magnetic effect, other windings may be provided for the purpose of producing circuit resistance. These are equivalent to resistors and are called non-inductive windings. They consist of resistance wire and have too few turns to produce any appreciable magnetic field. In special cases, they may consist of two portions wound in opposite directions so that the magnetism of each part cancels the other.

The spool on which the coil is wound consists of the iron core having cheeks fitted at each end. At the heel end, the cheek is made of a laminated plastic. At the armature end it is normally made of copper. Originally this was done so that the eddy currents

Fig. 2.18 Relay mounting (3000-type). The particular relay shown here has a $1\frac{1}{2}$ in (4 cm) heel-end slug and an adjustable residual screw (2.5.1).

in the copper cheek would slightly slow the speed of operation and release to reduce the tendency of the contacts to 'bounce' and momentarily open immediately after closing. Where fast operation is needed, however, this is dispensed with and a plastic front cheek is used.

Either heel-end or front-end slugs may be fitted if slow operation and/or release is wanted. Slugs can be $\frac{1}{2}$ in, 1 in, or $1\frac{1}{2}$ inches in length depending on the lag required. Without slugs the relay will operate in about 15 ms and release in about the same time, depending of course on the spring load etc. With a $1\frac{1}{2}$ in slug these times would be increased to about 60 ms operate and about 400 ms release.

The armature residual pin may either be a fixed length stud of 4 mils, 12 mils or 20 mils (thousandths of an inch), or it may be an adjustable screw fitted with a locking nut. The adjustable residual screw is provided when it is necessary accurately to adjust either the release lag or the release current of the relay.

Two springsets are provided and each may have up to nine springs arranged to give any number of make, break, change-over or K-action contact units within this maximum number of springs. The standard relay uses pin-operated springs but where long life is necessary, comb-operated springs may be used and these provide for a maximum of ten springs per springset. The normal spring thickness on pin-operated relays is 14 mils, thinner springs of 12 mils being used when a lighter spring load is required.

If required, one contact unit (make or break) can be arranged to function before any others on the relay on operation, and function later than them on release. Such a contact is indicated on circuit diagrams by the letter *x* beside it. Alternatively one contact can be arranged to function last on operation and first on release. This is a *y* action.

The normal contact material is silver and is used for currents up to 300 mA. Platinum (Pt) or palladium (Pd) is used for currents up to 1 A and either tungsten (W) or nickel-silver (AgNi) for currents up to 5 A.

Retards (coils retardation) used as inductors, in transmission bridge circuits for

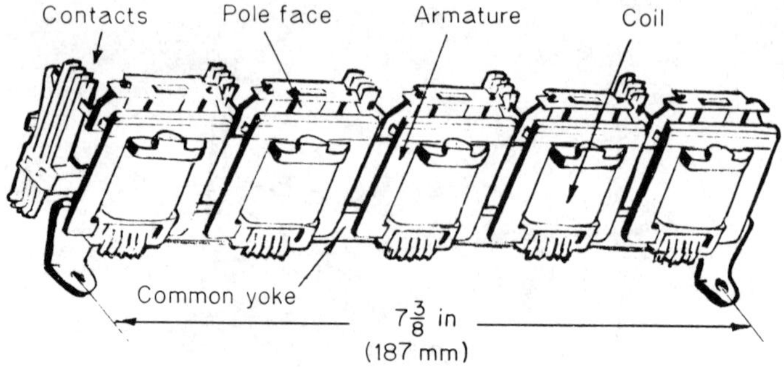

Fig. 2.19 B.P.O. relay unit, type 12 (2.5.3).

example, consist of a 3000 type relay with no springset and with the armature bent to fit close to the pole face. They usually have nickel-iron sleeves.

The relays are normally mounted on plates having a standard set of holes pierced in them as shown in Fig. 2.18. To simplify the mounting of other apparatus, most other components used in exchange switching circuits are designed to fit these same piercings.

2.5.2 British Post Office relays type 12 and 600. In automatic telephone exchanges, two relays must be provided in the 'line circuit' connected to every subscriber's line. Because each line is used relatively infrequently, compared with many other circuits in an exchange, a cheaper and less robust relay can be used for this purpose.

A smaller and simpler version of the basic 3000 type relay known as the 600 type relay is one such relay. Another type of relay used for some designs of line circuit in the Strowger system and in TXE 2 is shown in Fig. 2.19. These relays are assembled with up to five relays on a common yoke, only one screw being used per relay in the construction, a method which considerably reduces assembly costs. The residual air-gap is provided by a strip of adhesive plastic tape stuck to the armature.

2.5.3 High speed relays. The relays described so far (B.P.O. 3000 type, etc.) have operate lags of about 10 ms to 20 ms when not fitted with slugs and can be termed 'fast operating'. Shorter operate times can be obtained from *high speed* relays (Fig. 2.21).

Operate and release times as low as 0·5 ms to 1·0 ms can be provided by the relay shown in Fig. 2.20 under favourable circuit conditions. The high speed is obtained by restricting the spring load to one contact and providing screw adjustment of the contacts to enable a very short armature travel to be used. The armature is extremely light to reduce the mechanical inertia, and is directly attached to the spring to reduce friction. These measures permit the use of few turns on the coil, giving a small bobbin and therefore a short magnetic circuit. The result is a very low inductance.

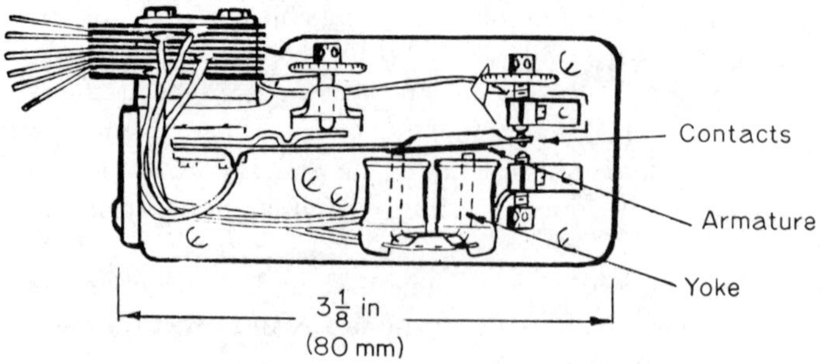

Fig. 2.20 Single-contact high speed relay (2.5.3).

Another relay based on the same principles has two armatures, each attached to a separate contact spring but resting on the same yoke and energized by the same coil (or coils). The additional load of the second contact necessitates larger coils etc. and slows the relay down. In spite of this, operate and release times of the order of 2 ms can be obtained.

2.6 Typical telegraph relays

2.6.1 Polarized telegraph relay, type 299 etc. The relay 299AN as illustrated in Fig. 2.21 is a typical relay of the type having a flexible armature fixed at one end.

The permanent magnet used in the earlier versions is a horseshoe magnet of tungsten steel as in Fig. 2.21, but later versions have a smaller but more powerful bar magnet of Alnico, an alloy of aluminium, nickel and cobalt as shown in Fig. 2.22.

The yoke, armature and pole pieces are made of Permalloy, an alloy of nickel and iron which gives a low reluctance magnetic circuit and low hysteresis loss. The lower end of the armature is clamped within a non-magnetic spacer which also holds the two halves of the yoke apart to avoid short circuiting the flux of the permanent magnet.

The narrow section or waist near the lower end of the armature is provided to make it sufficiently flexible for operation as a side stable relay. The armature is free to move within the stationary spool on which the coils are wound.

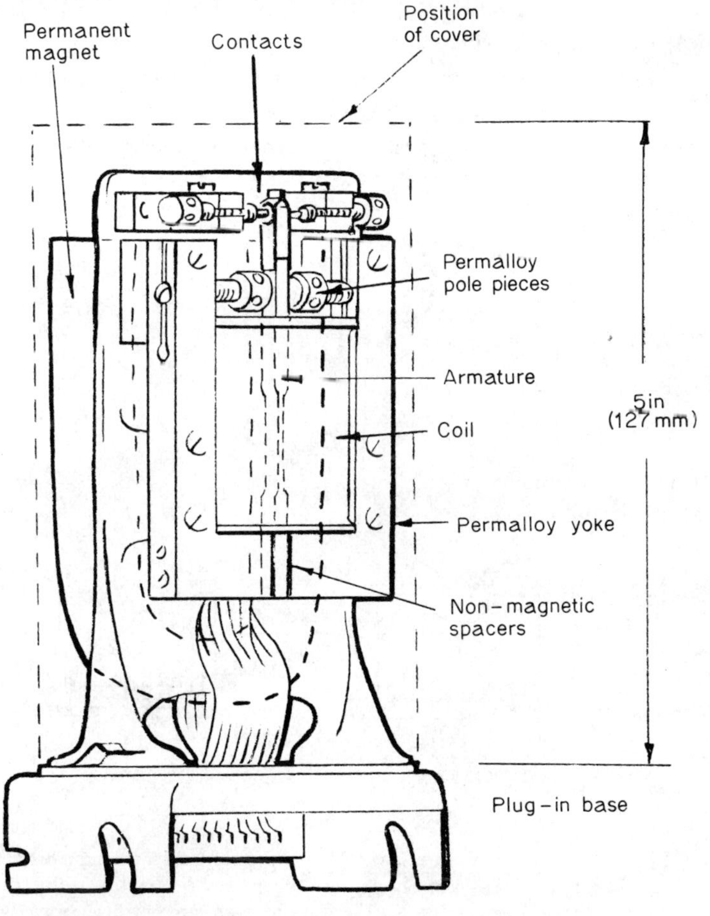

Fig. 2.21 Polarized telegraph relay (B.P.O. type 299 AN). Compare this with Fig. 2.22 (2.6.1).

The contact tongue is attached to the armature and consists of a pair of springs tensioned against each other. This forms an *anti-chatter* arrangement, the springs bending as the tongue strikes the contacts causing the ends of the springs to slide on one another. The friction between them dissipates most of the kinetic energy of the moving armature and spring assembly, preventing the tongue from bouncing off the contact screw. The contact bounce which occurs if rigid tongues are used causes sparking which damages the contacts, and also delays the proper closure of the contact, so distorting the signal.

The material used for the contacts is an alloy of 60% palladium, 40% copper.

The principle of operation of the relay is shown in Fig. 2.22. From this diagram, it can be seen that if the current is in the direction designated there as 'marking', the fluxes of the coil and of the permanent magnet are aiding one another on the left-hand side of the armature, but are opposing each other on the right-hand side of the armature. The magnetic attraction is therefore greatest on the left so that the armature moves in that direction.

Another way of looking at the same thing is that the current in the coil induces a 'south' pole at the upper end of the armature which is attracted towards the left-hand pole piece on the 'north' pole of the permanent magnet.

If the current is in the 'spacing' direction, the armature moves to the right.

With windings connected to provide 3 800 turns (370 Ω), the relay will operate on about 5 mA. Its speed is suitable for repeating pulses as short as 12·5 ms, (i.e. 0·0125 s) which is a telegraph speed of $\frac{1}{0{\cdot}0125} = 80$ baud (see also 3.2.3). Mechanical resonance of the armature prevents its use at higher speeds.

A cheaper version of the same basic relay (known as relay 328) has an armature which is not waisted and is thus less flexible making the relay less sensitive. The magnet

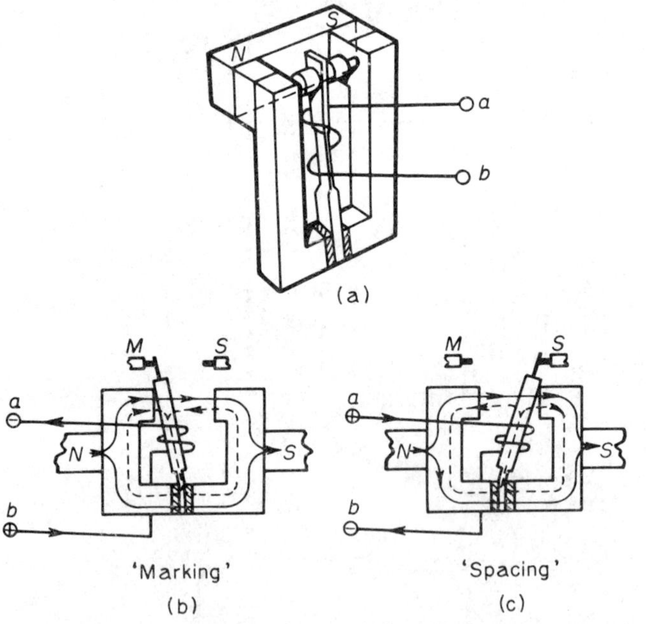

Fig. 2.22 Magnetic circuit of relay 299AN. At (a) is shown the bar magnet used on the later design in place of the horseshoe magnet of Fig. 2.21. At (b) and (c) are shown the flux for each current direction. The full lines indicate the flux due to the permanent magnet and the broken lines the coil (electromagnet) flux (2.6.1).

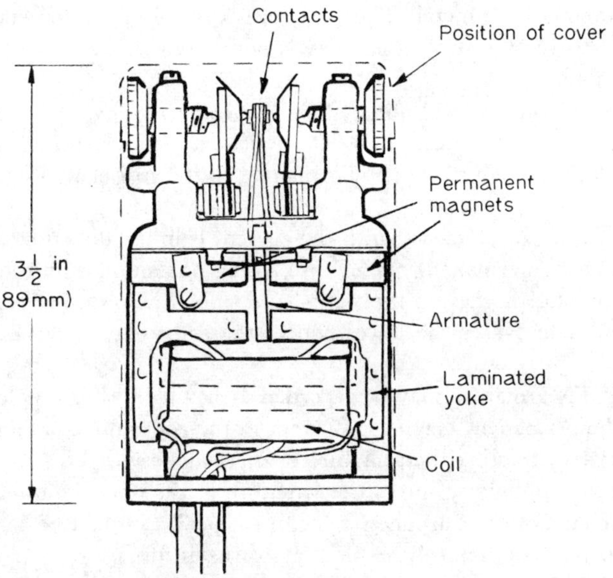

Fig. 2.23 Carpenter relay type 4 (B.P.O. polarized relay 2B). Compare this with Fig. 2.24 (2.6.2).

is also smaller and less powerful which together with the stiffer armature makes the relay centre stable, and small enough to mount on a 3000 type telephone relay mounting plate. There is virtually no mechanical resonance and it will repeat pulses at a speed of up to 150 baud (66 ms pulses).

2.6.2 The Carpenter relay (B.P.O. polarized relay 2B). This is available in a number of versions, including one used in both telephony and telegraphy which is suitable for mounting in the standard 3000 type telephone relay mounting as either a wired or a plug-in unit.

The permanent magnets are of Alnico, the pole pieces and yoke are built up in

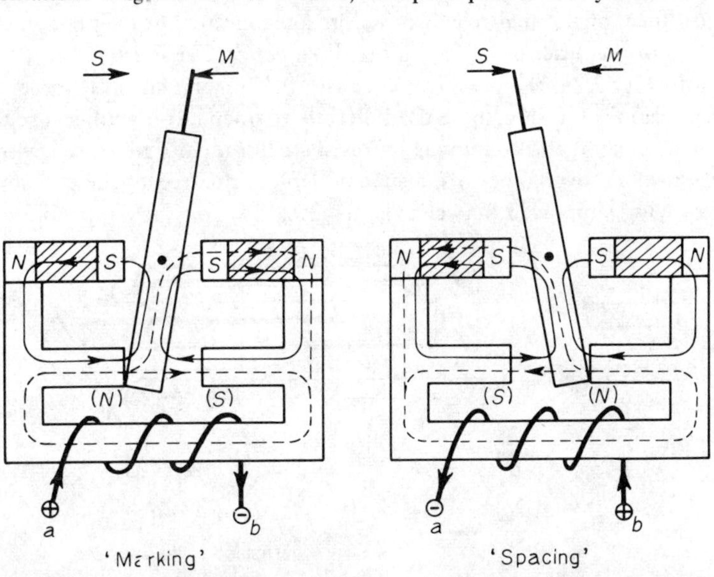

Fig. 2.24 Magnetic circuit of Carpenter relay showing flux for each current direction (2.6.3).

layers (laminations) of Mumetal. The armature is made of Radiometal and is spring mounted at its centre of gravity. The relay is normally side stable, but a centre stable version is available.

Contact chatter (bounce) is eliminated by mounting the 'fixed' contacts on springs tensioned against buffer plates, the friction between these springs and the plates dissipating the kinetic energy in a similar manner to the anti-chatter springs described in **2.6.1**.

From Fig. 2.24, it can be seen that, if the current is in the direction shown as mark- ing', the coil and the permanent magnet are aiding one another on the left-hand side and are opposing one another on the right-hand side. The resulting greater attraction on the left-hand side causes the lower end of the armature to move to the left (see also **2.3.2**).

Alternatively, the armature may be regarded as having a 'south' pole induced at its lower end by the permanent magnets. Current in the coil induces a 'north' pole at the left-hand pole face attracting the armature to that side.

Spacing current, of course, moves the armature in the opposite direction.

The larger version of the Carpenter relay, the type 3, can be used at speeds of some 700 baud on currents of about 4 mA in 3 600 turns at 96 Ω.

The type 4, used for signalling in conjunction with telephone and telegraph exchange switching, requires about 9 ampere-turns to operate with adequate contact pressure. This is typically achieved by a current of 2·25 mA in a 100 Ω winding of 4 000 turns.

2.7 Special purpose relays

2.7.1 Thermal relays. Not all relays employ the magnetic effect of an electric current for their operation. Thermal relays, for example, use the heating effect.

A current in a heater winding produces thermal expansion of the component layers of a bimetal strip. This bimetal strip consists of two layers of different metals having a different coefficient of expansion. The heat therefore causes one layer to expand more than the other so that the strip bends when it is heated. This bending is used to move the contact.

A practical form of thermal relay as used in some circuits in telephone exchanges to provide delays of the order of half a minute, is shown in Fig. 2.25.

Referring to Fig. 2.25, the heating coil causes the bimetal strip (3) to bend upwards slowly. This action is too slow to be used directly to open a contact because the current would dissipate heat at the contact as its resistance gradually increased with reducing contact pressure. To overcome this, a snap action is imparted to the contact operation by a bowed spring supported between the fixed tag (4) and the adjustable support (7).

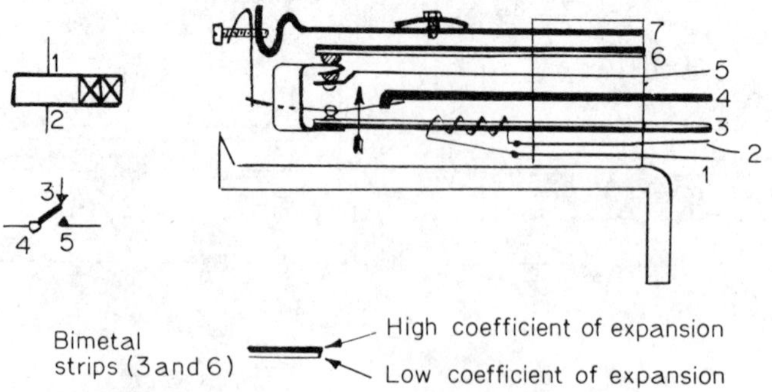

Fig. 2.25 Thermal springset. The springset shown here is fitted to the yoke of a 3000-type relay (2.7.1).

A second bimetal strip (6) is provided which is unheated to compensate for changes in the ambient temperature (the temperature of the surrounding air).

Changes in the ambient temperature tend to bend the two bimetal strips in opposite directions so that they pull against one another and there is no resultant movement.

2.7.2 Dry reed relay. This is of increasing importance in exchange switching applications (**4.2.2**). It is called 'dry' because some other types of reed relay employ mercury to wet the contact surface to obtain a good contact. The dry reed relay does not.

The reed contacts are usually made of 50% nickel/50% iron in the form of wire flattened at one end to provide a contact surface which is gold plated to a thickness of 0·1 mil to 0·2 mil. After heat treatment to diffuse the gold into the reed, the two reeds are sealed into a glass tube. The tube usually contains nitrogen at ordinary atmospheric pressure.

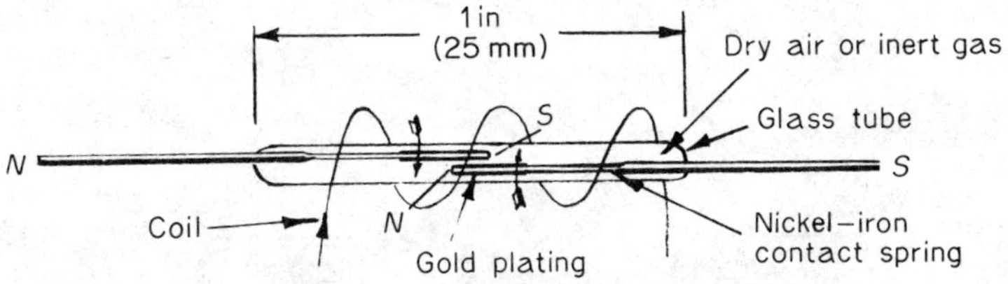

Fig. 2.26 Dry reed relay (2.7.2).

In use, this sealed contact unit is placed inside the relay coil spool in a similar manner to the armature of the telegraph relay (**2.6.1**). Current in the coil magnetizes the two reeds so that they attract one another bringing their surfaces firmly together to close the contact. When the current is disconnected, they spring apart to open the contact.

In practical applications, four such contact units are commonly mounted in one coil to form a four-contact relay.

Now turn to page 57 and read the first paragraph of 2.9.4.

PART 4 SUMMARY

2.8.1 Non-polarized relays.

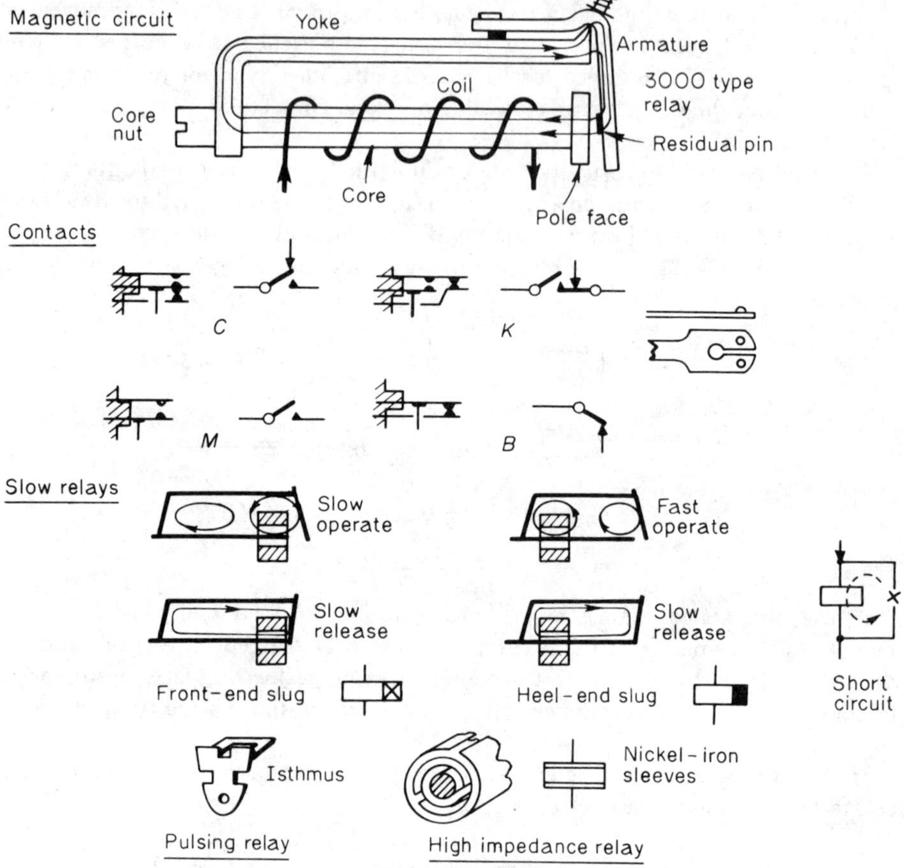

Fig. 2.27 Non-polarized relays—summary.

For high operate current value:	many, thick springs. Few turns, long air-gap (travel), and high reluctance.
For high release current value:	many, thick springs. Few turns, long air-gap (residual), and high reluctance.
For fast operate:	small inertia, low L/R, large circuit current (low resistance, low operate current), reduced eddy currents (Ni-Fe core, plastic front cheek).

2.8.2 Polarized relays

Polarization by permanent magnet or separate winding increases sensitivity, operation depends on current direction.

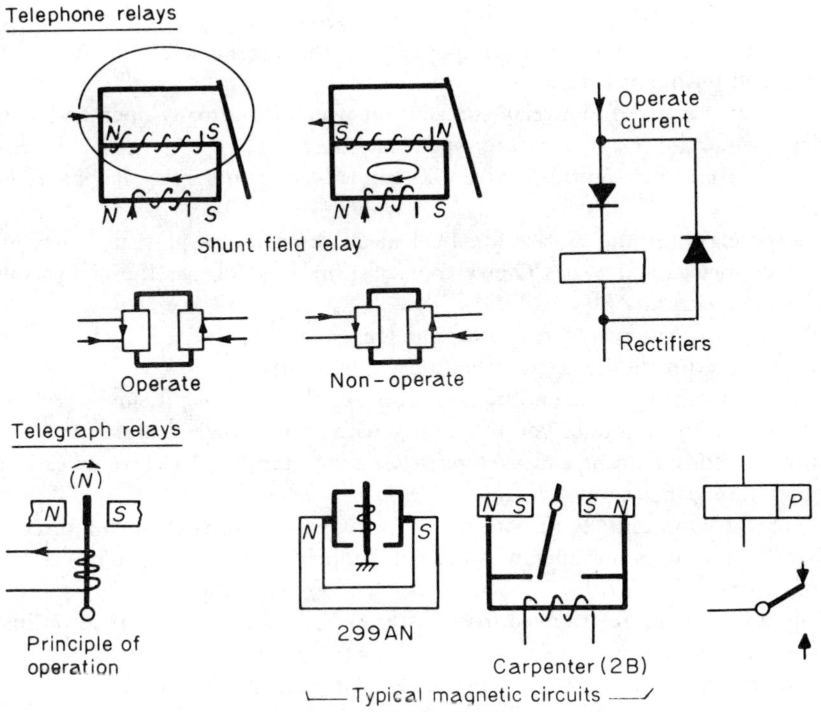

Fig. 2.28 Polarized relays—summary.

2.8.3 Special purpose relays

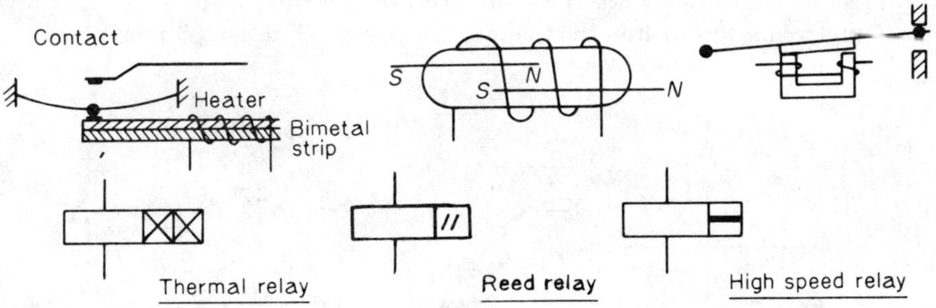

Fig. 2.29 Special purpose relays—summary.

Try the rest of the questions on page 57 before leaving this chapter.

PART 5 SELF-TEACHING NOTES

2.9.1 First read Parts 1 and 2 of Chapter 2, that is **2.1.1** to **2.4**.

2.9.2 Having read **2.1.1** to **2.4**, try the following twenty questions. Write all the answers down before turning to the answer page.

Q.1 Name the three main iron components of the magnetic circuit of a non-polarized telephone relay.

Q.2 What name is given to a relay contact unit which is normally open and which closes when the relay (armature) operates?

Q.3 What current is required to give 100 ampere-turns in a relay having 10 000 turns?

Q.4 If two relays, *A* and *B*, are identical in all respects except that *A* has four contact springs and *B* has twelve contact springs, which has the higher value of operate current?

Q.5 Which of the two relays in Q.4 has the longer operate lag?

Q.6 What is the function of a front-end (armature-end) slug?

Q.7 A relay has one winding and no slug. How would you make it slow to release?

Q.8 For what purpose would you use a relay with an isthmus armature?

Q.9 How can eddy current and hysteresis losses be minimized to give a relay coil a high impedance at speech frequencies?

Q.10 If current flows initially in both coils of a shunt-field relay in such a direction that the relay does not operate, what will happen if we reverse the direction of both currents?

Q.11 Why do telegraph relays need to be faster to operate and release than ordinary telephone relays?

Q.12 Name two advantages of using permanent magnets in telegraph relays.

Q.13 A given current in the coil of a side-stable telegraph relay moves the tongue to one of the two fixed contacts. What happens if we reverse the direction of this current?

Q.14 In the same relay as Q.13, what happens if we next disconnect the current?

Q.15 Nickel-iron has a lower permeability than soft-iron. Why then, is it used in preference to soft-iron for the magnetic circuit of telegraph relays?

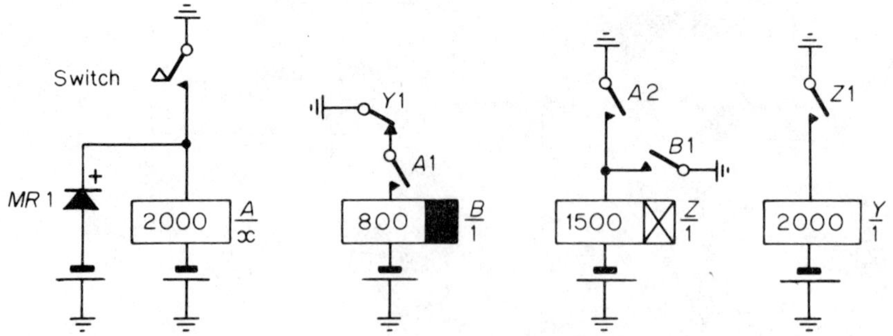

Fig. 2.30 Relay circuit for questions 16 to 20 (2.9.2).

The next five questions all refer to Fig. 2.30.

Q.16 What number should be written in place of *x* in the designation of relay *A*?

Q.17 Which of the four relays are slow to release?

Q.18 When the switch is closed, it operates relay *A*. What then operates relay *B*?

Q.19 What later releases relay *B*?

Q.20 Which relays, if any, remain operated after the switch has been opened for, say, 10 min?

Now check your answers on page 258.

2.9.3 You should have answered most of questions 1 to 15 correctly. If you got fewer than ten of the fifteen right, see if the notes on the answer page are any help and then study **2.1.1** to **2.4** again and try these fifteen questions once more.

For questions 16 to 20, see the note on the answer page.

Once you have at least ten of questions 1 to 15 right and have attempted questions 16 to 20, read the rest of the chapter (i.e. **2.5.1** to **2.8.3**).

2.9.4 The whole of Chapter 2 should have been studied before answering the following questions. A few days should preferably have passed since reading Parts 1, 2 and 3 (**2.1.1** to **2.7.2**).

Q.21 Draw the magnetic circuits of
(*a*) a receiving delay for double-current pulses from a teleprinter and (*b*) a receiving relay for loop-disconnect pulses from a subscriber's dial.
Indicate the component parts of (*b*). (C & G 1963).

Check your answer on page 259.

Q.22 Describe, with the aid of a sketch, the construction of a telephone relay with a heel-end slug.
Explain why the slug makes the relay slow to release. Indicate another way in which a relay can be made slow to release. (C & G 1962)

Check your answer on page 260.

Q.23 Sketch the magnetic circuit of a telephone relay and explain the reason for
(*a*) residual gap
and (*b*) an enlarged pole-face.
What factors influence the pull on the armature?

Check your answer on page 260.

3 · SIGNALLING

For self-teaching notes, see page 78.

PART 1 INTRODUCTION

3.1.1 Why is signalling necessary? Suppose you want to make a telephone call. You know whom you want to call and from where. The connexion will be set up through one or more telephone exchanges and disconnected when you have finished your call. To have this done, you must be able to let the exchange know what you want by signalling over the line wires which connect your telephone to its exchange. If the call goes through more than one exchange, the signals must be passed from one exchange to the next over the junction between the exchanges.

If you had been using a teleprinter instead of a telephone, the message itself would be sent in the form of another set of signals for each of the letters in the words of the message.

3.1.2 In this chapter, there are thus three sorts of signalling you will learn about: control signals on subscribers' lines, control signals on junctions and signals for sending telegraph messages.

Before looking at particular practical systems, we shall first see what information the systems have to signal and then what forms the signals themselves can have and the sort of signalling paths available. It will then be easy to understand and remember how each of the practical systems works.

PART 2 SIGNALLING PRINCIPLES

3.2 Signals required

3.2.1 Subscribers' lines. There are at least three signals we must have on any system. First of all the subscriber must be able to send a *calling* signal to the exchange when he wants to make a call. Similarly the exchange must be able to let the subscriber know that a connexion has been made to his line when someone else wants to call him. This signal usually has to ring a bell at the called subscriber's telephone so it is called a *ringing signal* or just *ringing*. Finally, the subscriber must be able to let the exchange know when he has finished the call so that the exchange can remove the connexion and free his line to make or receive other calls. This requires a *clearing* signal.

As well as these signals, on automatic systems, more signals are needed to do some things which can be arranged on manual systems by talking to the switchboard operator. The subscriber has to signal the number of the telephone he wants to call by *dialling* and the exchange signals back such information on the progress of the call as 'number unobtainable', 'line busy' etc. by means of tones which can be heard by the subscriber. It is also necessary to send an *answer* signal to the exchange when a called subscriber answers. This normally takes the same form as the calling signal and it causes the ringing signal to be disconnected. On most public telephone systems the answer signal also causes the call to be recorded automatically so that the calling subscriber can be charged for it. This is called *metering*.

Some of the signals referred to here as being necessary on automatic systems can also be used on manual systems to improve the speed of service. The use of an answer signal, for example, enables an operator to deal with another call in the time which might otherwise be spent listening for a called subscriber to answer after being rung.

3.2.2 Junctions between exchanges. Signals are needed in both directions. The directions are referred to as *forward* or *backward* as shown in Fig. 3.1.

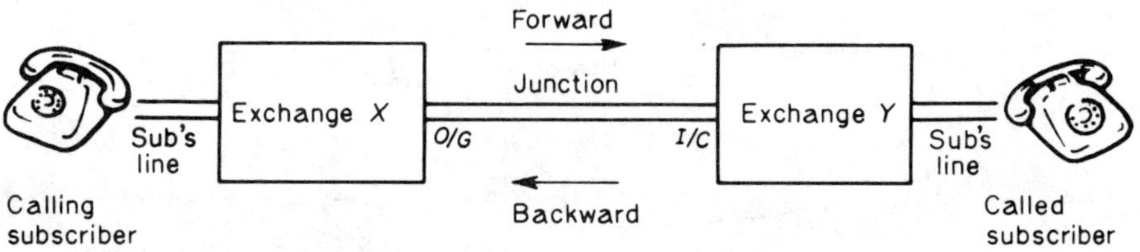

Fig. 3.1 Junction signals from X to Y are said to be 'forward' signals, those from Y to X are 'backward'. The outgoing end (O/G) is at X, the incoming end (I/C) at Y (3.2.2).

A junction may be arranged to carry calls in one direction only in which case we call it a *unidirectional* junction. In other cases it may be possible to switch the connexions to a junction so that it can carry calls from X to Y or from Y to X at different times. Junctions of this sort are called *bothway* junctions.

Forward signals include calling, dialling, and clearing (sometimes called *clear forward* or *forward clear*). An *answer* signal is needed in the backward direction to control metering at exchange X. The called subscriber's clearing signal is also usually sent as a *clear back* signal from Y to X where it may be needed if the connexion has been set up through a manual switchboard.

The answer and clear back signals on a manual exchange assist the operator in supervising the setting up and control of a connexion and are thus often referred to as *supervisory* signals.

3.2.3 Telegraph signals. Telegraph messages consist of letters, figures and punctuation marks such as full stops, etc. Each of these is called a *character* and each is represented by its own signal consisting of a combination of shorter *signal elements.* The complete set of character signals used in a particular telegraph system is called the *signal code* for that system. An example of this is the well-known Morse code where the signal elements are dots, dashes and spaces (see **3.4.1**).

The control signals required for switching calls on a Telex (telegraph exchange) network are similar to those required on a telephone network. An important difference is that a teleprinter is not designed for speech operation and therefore has no listening facilities, so tone signals cannot be used. Information on the progress of a call must therefore be signalled to the calling station as groups of character signals such as '*NC*' for 'trunks busy'.

Telegraph signalling speeds are expressed in a unit called the *baud*, which is the inverse of the duration of the shortest signal element. The speed in bauds is thus numerically equal to the number of such elements transmitted per second (see also **2.6.1**).

3.3 Types of signal

To illustrate and compare the different types of signal which could be used for some of the purposes described so far, let us consider two places, call them X and Y, some distance apart. They could be two exchanges, for example, connected by a junction as in Fig. 3.1. At exchange X there would be a sending or transmitting contact $A1$

and at exchange Y there would be some suitable receiving device LS which could be a relay, for example (see Fig. 3.2). The signal to be sent could be any of those already mentioned such as a calling signal on a telephone exchange junction. A signal can consist of $A1$ operating or releasing and causing the operation or release of LS as required.

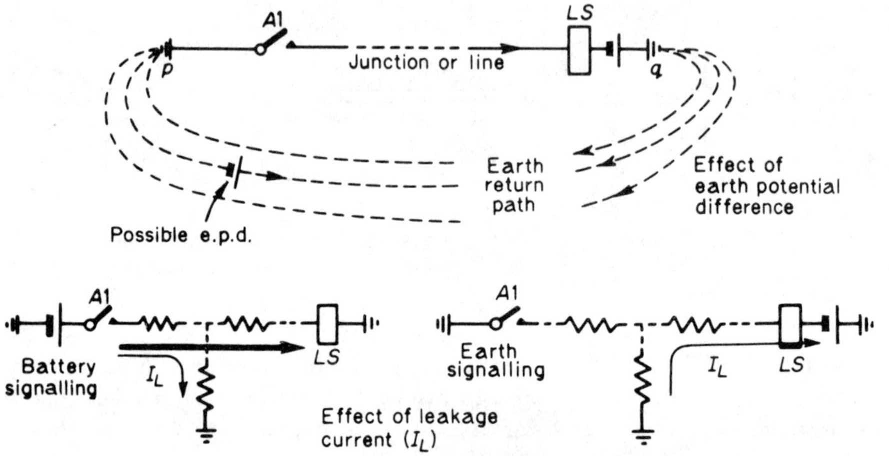

Fig. 3.2 Single-wire d.c. signalling. When $A1$ closes, current flows from the positive plate of the battery to earth at q, then through the very low resistance of the earth to p, through contact $A1$, along the line wire and through the coil of LS. This current operates relay LS (3.3.1).

3.3.1 Single-wire signals. In Fig. 3.2, the operation of $A1$ completes a circuit for LS using conducting paths in the earth as part of the circuit, that is *earth return* is used for the signalling currents. This earth return path has such a large cross-sectional area that its resistance is practically zero. If the connexions at p and q are at the same potential such a system has the advantage that there is the resistance of only one wire of the line in series with the signalling currents (**3.3.3**). In most practical cases, however, the earth connexions at p and q are not at the same potential, that is, there is an *earth potential difference* or 'e.p.d.'. If the e.p.d. is in the opposite direction to the battery e.m.f. (as in Fig. 3.2) it can interfere with signalling by reducing the current flowing so that LS either operates too slowly or does not operate at all. This e.p.d. is shown in Fig. 3.2 in the form of a battery to illustrate the idea. It is really a voltage drop (current × resistance) caused by the earth return currents of other circuits or the effects of electrical storms, etc. In some parts of the world, e.p.d. makes single-wire signalling unsuitable even for the shortest lines. In the UK, however, there are many cases where single-wire signalling can be used.

The line wire is, of course, insulated from earth but practical insulating materials are not perfect and they do pass some current. In a d.c. system, if the battery potential is connected at the receiving relay LS, this 'leakage' current may prevent, or at least delay, the release of the relay when the signalling contact $A1$ opens. If the battery is connected at the signalling contact, however, the effect of the leakage current is to reduce the operating current available at the relay when $A1$ is closed, but only by a small amount because the insulation resistance will be high compared with the line resistance (typically 50 000 Ω compared with 1 000 Ω).

Practical examples of single-wire d.c. signalling are C.B. junctions and telegraph circuits, and in both cases the battery potential is connected at the signalling contact, the receiving relay or electromagnet being connected to earth.

3.3.2 A.C. and d.c. signalling. In the example of single-wire signalling shown in Fig. 3.2, the signal current is obtained from a battery and is therefore direct current. If we had used alternating current, the e.p.d. would not have affected the signalling so long as we had made *LS* a receiving device which would respond to a.c. but not to d.c. This can be achieved by connecting a capacitor in series with it as shown in Fig. 3.3. Another advantage of using a.c. for signalling is that a.c. receiving devices can often be made more sensitive (i.e. operate on smaller currents) than d.c. devices. A magneto bell, for example, is more sensitive than a d.c. trembler bell. This means that a.c. signalling can be used over longer distances. The transmitting arrangements are usually more complicated than d.c. systems and if a.c. signals are sent on telephone circuits at the same time as speech, they may interfere with conversation.

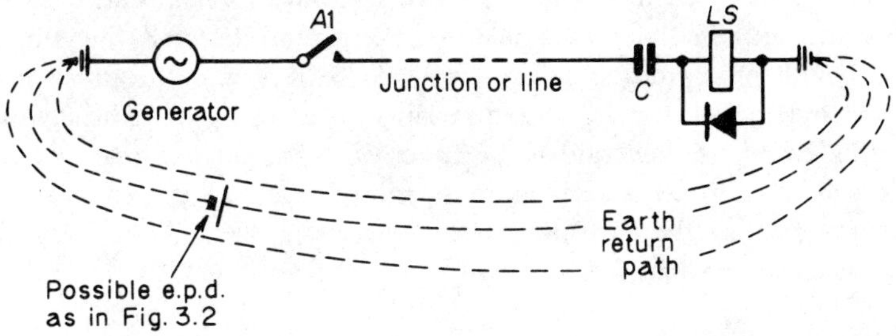

Fig. 3.3 Single-wire a.c. signalling. The capacitor *C* in series with *LS*, prevents any d.c. from any earth potential difference from interfering with its operation (3.3.2).

In a circuit carrying a.c. or a rapidly changing d.c., the varying current produces a changing magnetic field and any conductor cut or linked by the lines of flux of this field will have an e.m.f. induced in it. Another property of a.c. is that it can pass through a capacitor. Any two conductors or wires placed near to one another act as though they are the two plates of a capacitor, i.e. there is 'capacitance' between them.

If a line using single-wire signalling runs near another telecommunication circuit using single-wire signalling or near an a.c. power line unbalanced by an earth fault, it receives interference currents from the line insulation resistance, from the capacitance between the lines and from magnetically induced e.m.f.'s as shown in Fig. 3.4. This

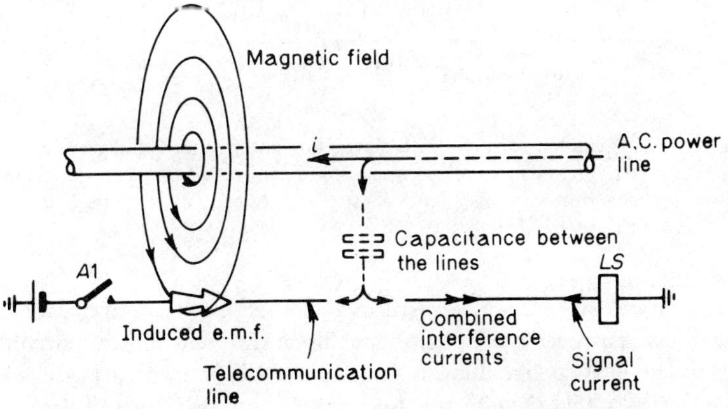

Fig. 3.4 Interference on single-wire circuits. The interference currents in the direction shown here (at an instant when *i* is increasing) are opposing the signalling current and could make *LS* either operate too slowly or fail to operate at all when *A*1 closes. When *A*1 is open, they could cause false operation of *LS* (3.3.2).

can interfere with signalling either by falsely operating the signal receiving device or falsely releasing it, and it can interfere with speech by causing noise in the receiver at the subscriber's telephone.

3.3.3 Loop signalling. For this, a second wire is used in place of the earth return, the two wires usually being referred to as the *A* and *B* wires or as the 'positive' and 'negative' wires.

The earth connexions *p* and *q* are at the same end of the line so that there is no e.p.d. in the path of the signalling currents.

Any interfering e.m.f. induced in one wire of the pair is cancelled by a similar e.m.f. induced in the other wire because the two e.m.f.'s are acting in opposite directions around the loop.

The receiving relay *LS* in Fig. 3.5 has two windings each wound so as to have the same inductance and the same resistance as the other and therefore the same impedance (**2.2.8**). Although one winding is connected to battery and the other to earth, the battery resistance is very low so the impedance to earth is the same for each wire of the line. The circuit is then said to be *balanced*. This ensures that any interference currents received from other telecommunication circuits or power lines are equal in the two wires and, as they are in opposite directions around the loop, they cancel each other as shown in Fig. 3.5.

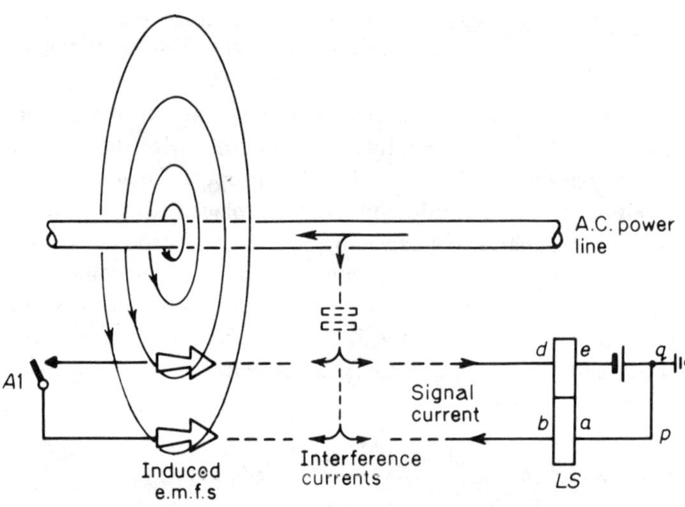

Fig. 3.5 Balanced loop circuit. This shows how interference effects are eliminated. The relay windings oppose one another when the current flows in the direction *d*—*e* and *b*—*a* (see Chapter 2). Compare this with Fig. 3.4 (3.3.3).

Balanced loop circuits are widely used in all forms of telecommunication to prevent interference between speech currents or signals in different lines or circuits.

A disadvantage of loop signalling is that the signalling current passes over the two line wires in series so that the effective line resistance is twice that of the corresponding single-wire system. Also, because the battery potential is connected at the signal receiving relay, the effect of leakage currents via the line insulation between the wires or from the 'negative' wire and earth is greater than with battery signalling on a single-wire circuit (**3.3.1**).

3.3.4 Loop-disconnect pulsing. For controlling the operation of the selectors at an automatic exchange, the contact *A*1 of a loop signalling system (Fig. 3.5) is made to pulse giving a series of 'breaks' in the loop. This contact could be the pulsing contact springs of a dial, for example, (**1.4.2** and **1.6**) in which case the number of breaks would be equal to the digit dialled. These breaks are usually $66\frac{2}{3}$ ms in length and are separated by 'make' periods of $33\frac{1}{3}$ ms (see Fig. 3.6). Each 100 ms period so formed is termed a *dial pulse* so that there are thus 10 dial pulses per second. Such pulses would be described as having a *speed* of 10 p.p.s. and a *ratio* of $66\frac{2}{3}$% break (or $33\frac{1}{3}$% make).

The inter-digital pause (I.D.P.) produced, for example, by the 'wind-up' time of a dial plus its 'lost-motion period' (**1.6**), enables the exchange selectors to complete each stage of selection before receiving the next digit of a number. Without such a pause, a number having more than one digit (e.g. digit 2 followed by digit 3) would be indistinguishable from a single digit number having the same total number of pulses (e.g. a single digit 5).

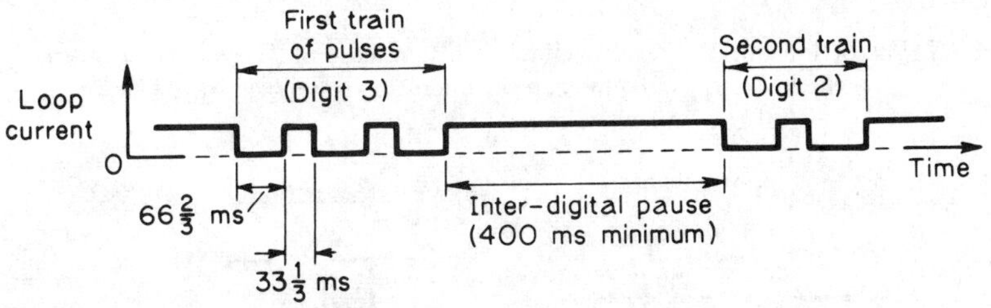

Fig. 3.6 Loop-disconnect pulsing. This illustrates the pulsing of the number 32. The circuit for this is shown in Fig. 3.5 (**3.3.4**).

3.3.5 Single-current signals. Both the single-wire d.c. signals and the loop d.c. signals so far described use signalling currents which flow in one direction only. Signals of this sort are known as *single-current* signals.

Where signals are to be sent in the form of short pulses as in telegraphy or when dialling, the time duration of each pulse must be kept within certain limits. This is necessary to ensure satisfactory operation of the apparatus such as teleprinters or selectors which the pulses are used to control. The electromagnets of selectors require too much current for them to be operated directly by the line signals, so these signals are always repeated to them by the contacts of signal receiving relays. It is also necessary to use relays to repeat teleprinter signals. If a signal receiving relay has unequal operate and release lags (**2.2.4** and **2.2.7**), it will shorten either the pulses or the intervals between the pulses. This effect is known as *pulse distortion* and may be due either to the type of relay used or to the effect of the line on the signalling current used to operate the relay.

One property of a line which can cause pulse distortion is capacitance between the line and earth, the line being regarded as one plate of a capacitor with the earth as the other plate. This is illustrated for a single-current, single-wire system in Fig. 3.7.

When *A*1 closes, the line capacitance charges quickly via *A*1 (low resistance, large current) and only slightly delays the operation of *LS*. When *A*1 opens, the line discharges slowly via *LS* (higher resistance, smaller current) delaying the release of *LS*. The release lag of *LS* is thus increased by a greater amount than its operate lag.

This effect is also shown in the current graph in Fig. 3.8a where it can be seen that the length of the output pulse from the relay also depends on the value of the relay operate and release currents (**2.2.3**) in relation to the line current i_1. The distortion produced by any given relay will therefore depend on the resistance of the line to which it is connected.

Similar effects occur on d.c. loop signalling except that it is the charging rather

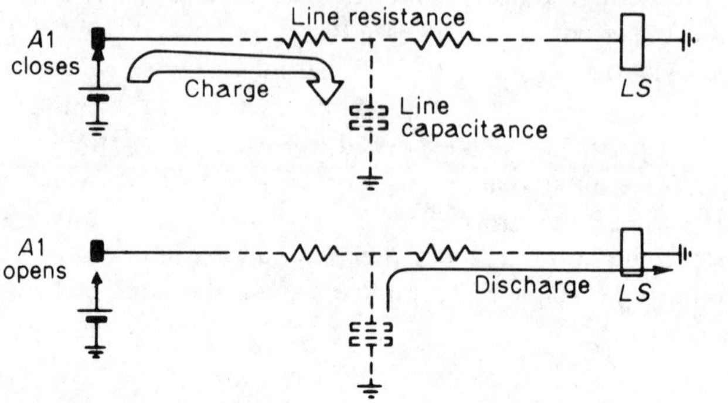

Fig. 3.7 Distortion due to line capacitance. The release of *LS* is delayed more by the discharge current than the operation is delayed by the charging current of the line capacitance (3.3.5).

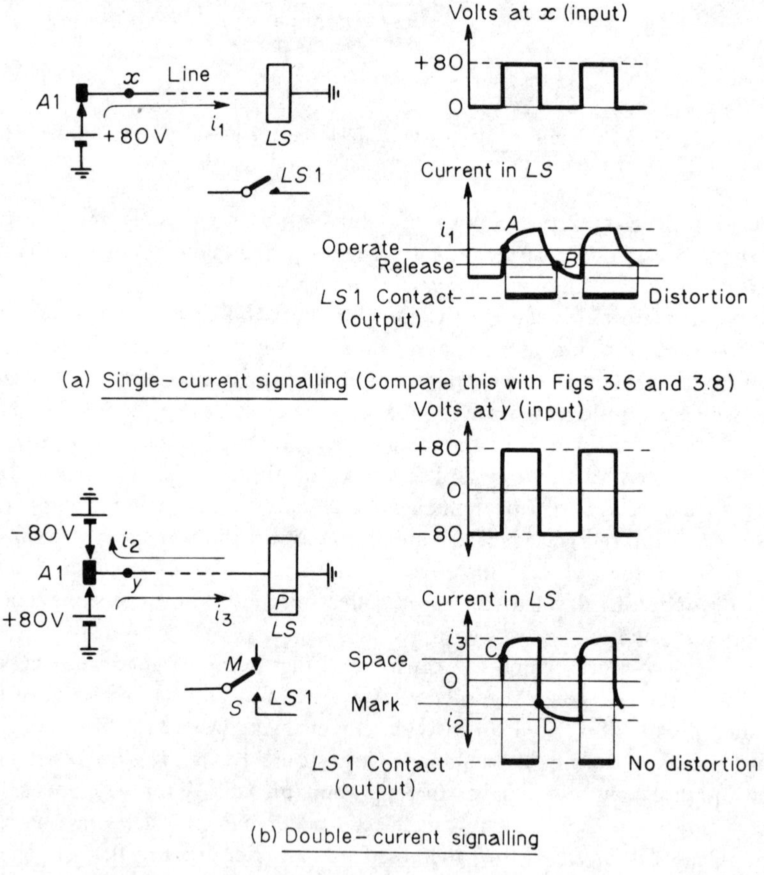

Fig. 3.8 Single- and double-current signals. The effects of line capacitance can be seen at *A* and *B* for single-current signals and at *C* and *D* for double-current signals (3.3.6).

than the discharging of the line capacitance which delays the release of the relay.

3.3.6 Double-current signals. Two directions of current are used, the signal current being switched from one direction to the other instead of being switched on and off as in single-current signalling (see Fig. 3.8).

The advantages of double-current over single-current signalling are as follows:

(*a*) A polarized receiving device such as a telegraph relay is used which operates to one side on current in one direction and to the other side on current in the other direction. The operate lags in each direction are practically equal when used in this way so that there is very little pulse distortion caused by the relay itself.

(*b*) Distortion due to line capacitance is also reduced because, although it charges at the start of a pulse via one side of contact *A*1 as in single-current signalling, the line capacitance is charged to the opposite potential through an equally low resistance at the other side of *A*1 at the end of the pulse. The delay which the line capacitance causes in the operation of the relay is the same in each case.

(*c*) Interference is reduced because *A*1 provides a low impedance to earth on both positive and negative pulses (through the low internal resistance of the batteries). Interference currents such as those shown in Fig. 3.4 tend to flow through this low impedance path rather than via relay *LS*.

(*d*) The polarized relays used are more sensitive and faster in operation than the non-polarized relays normally used in single-current signalling. Polarized relays can be biassed electrically or by adjustment (**2.3.2**) and used for single-current signalling, but twice the line voltage must then be used to obtain the same change of ampere-turns in the relay between the two signalling conditions (mark and space, **1.4.3** and **3.4**).

Double-current signalling thus gives reduced distortion, increased sensitivity and increased speed of operation. These improvements can either be used to increase the signalling speed so that lines can carry more telegraph messages in a given time or they can be used to tolerate more distortion and delay to be caused by the line which means that longer lines can be used.

3.3.7 Tones. As we have already seen (**3.2.1**), an automatic exchange provides information to a calling subscriber in the form of distinctive sounds which are called 'tones'. In order to produce these tones, a low voltage a.c. supply is connected to the subscriber's line through a transformer. This a.c. supply is produced by a tone generator serving the whole exchange. The alternating current itself is usually called a 'tone' because of the sound which it produces in the receiver.

Two typical circuit arrangements for connecting tone to line are shown in Fig. 3.9. These are both 'balanced' tone feeds, the impedance to earth being the same for each wire of the line. This avoids the possibility of the tone causing interference on other lines or of its being interfered with by tones or speech currents in other lines.

The first of the two arrangements shown is used in circuits which have a transformer type transmission bridge. An additional winding *WX* is provided on the transformer and the tone is connected to this winding when required by relay contact *TS*1. This current in winding *WX* induces equal e.m.f.'s in each of the two parts (*Y* and *Z*) of the line winding to produce a line current as indicated by the arrows on the diagram.

The second arrangement shown in Fig. 3.9 is used in circuits without a transformer type transmission bridge. In these circuits, an additional winding of 570 Ω is provided on the signal receiving relay to which the tone is connected when required by contact *TS*1. The relay coil acts as a transformer, as current in the 570 Ω winding induces e.m.f.'s in the two 200 Ω line windings. This produces a current in the line as indicated by the arrows in the diagram (Fig. 3.9).

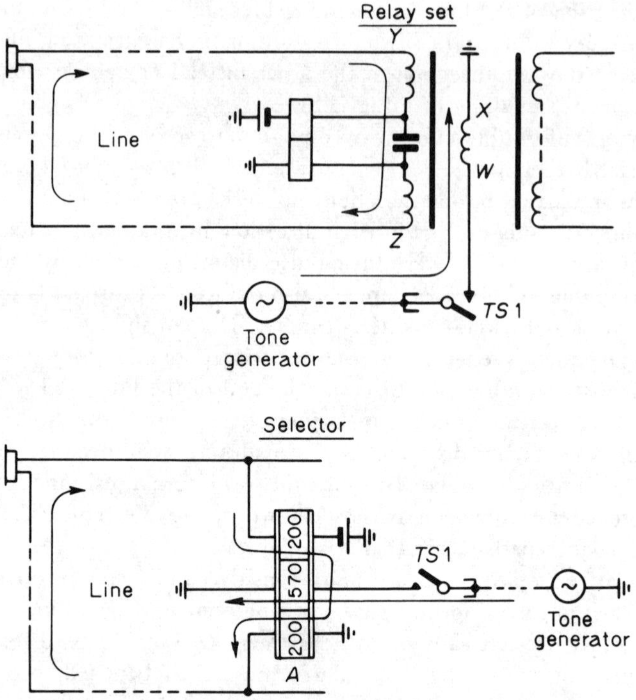

Fig. 3.9 Balanced tone circuits. The 200 Ω windings of relay *A* are connected so that loop signalling currents produce the same polarity of magnetism in the core. The a.c. tone induced in these windings by the magnetic field of the 570 Ω windings is thus in the same direction *round the loop* for both windings (3.3.7).

In both cases, the tone is induced in a pair of equal line windings so that the circuit is balanced (**3.3.3**).

3.3.8 Differential signalling. Balanced and unbalanced signalling can be combined to enable independent signals to be transmitted on the same line. One such circuit arrangement is shown in Fig. 3.10. A typical application of this is for sending forward a special signal to indicate that the call originates from a manual switchboard on a circuit used primarily for loop signalling.

The loop signalling relay corresponding to relay *LS* in Fig. 3.5 is in this case relay *A*. In series with it is a second relay *YO* which has differentially connected windings, i.e. one winding is reversed so that the relay is not operated by the normal loop current

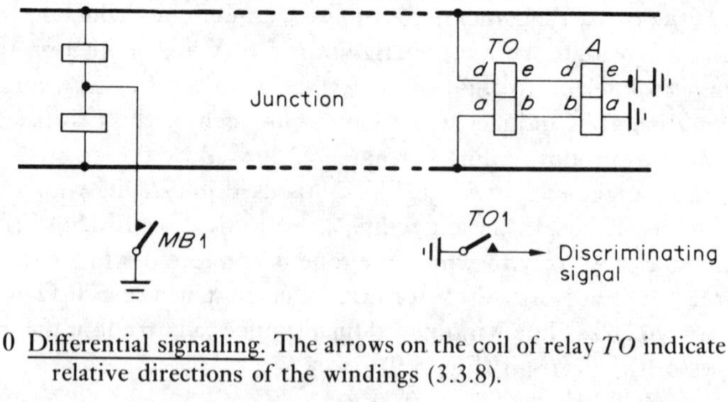

Fig. 3.10 Differential signalling. The arrows on the coil of relay *TO* indicate the relative directions of the windings (3.3.8).

because the effects of the two windings cancel each other.

To send the special discriminating signal, the circuit is unbalanced at the sending end by connecting a suitable potential to the loop. In Fig. 3.10 an earth connected for this purpose at contact *MB*1 which completely short circuits one winding of *TO* (and also *A*). Relay *A* remains operated under this condition. Relay *TO* operates and its contact *TO*1 is used to establish the appropriate circuit conditions for a manual board call at the receiving exchange.

3.3.9 Voice frequency (v.f.) signalling. On very long junctions or trunk lines, the speech currents need amplifying at intervals along the line by equipment known as *repeaters.* Special apparatus called *carrier equipment* is often used for sending several conversations at the same time over one pair of wires or over a radio link, and one of the repeaters may even be on a space satellite. In such cases, there is no possibility of physically providing a wire or pair of wires as a signalling path for each call or conversation. We must therefore use a.c. signals at a frequency which the amplifiers, carrier equipment, and radio can handle. This means using signals within the same frequency range as speech and is called voice frequency (v.f.) signalling.

Systems operating in the telephone speech band, 300–3400 Hz (**1.2.1**), are also called *in band* systems and can be used on any circuit capable of carrying speech, including physical lines where the line resistance is too high for other forms of signalling. The repeaters can then amplify the signal currents along with the speech currents.

On modern v.f. systems, electronic oscillators are used to generate the signalling currents and tuned electronic receivers distinguish between speech and signal currents, typically by checking that unwanted frequencies are not present at the same time as the signal frequency. This ability to prevent speech imitation of signals is known as *voice immunity.*

The v.f. signalling current produces an easily heard note if connected to a telephone receiver and for this reason is often referred to as v.f. tone.

Single frequency v.f. can be used to transmit dial pulses by switching the v.f. tone on for the duration of each $66\frac{2}{3}$ ms pulse.

Multi-frequency v.f. can transmit numbers more rapidly by representing each digit by a different combination of more than one frequency at a time. This is used on some types of push-button telephone and for signalling between certain types of common-control exchanges (**4.3.5**).

3.4 Telegraph signalling codes

3.4.1 Morse code uses two distinctive signal elements, the dot and the dash, which differ from each other in length (i.e. time duration). The dash is three times the length of the dot and each character consists of a number of dots and dashes. The dot and dash elements making up a character signal are separated from one another by an interval equal in duration to one dot. The characters in a word are separated by an interval equal to three dots. The words of a message are separated by an interval equal to seven dots (see Fig. 3.11 and Appendix 1).

The Morse code requires two basic signalling conditions, one for the dots and dashes and the other for the intervals between them. The 'dot/dash' condition is called *marking* and the 'interval' condition is called *spacing*. These terms were first used because telegraph signals were recorded by 'marking' the dots and dashes in ink on paper tape and leaving 'spaces' between the marks but the descriptions are still used for any telegraph system using two basic signalling conditions.

Morse code can be used on land line systems for a.c. or single-current d.c. signalling by using current 'on' for marking and current 'off' for spacing. For double-current

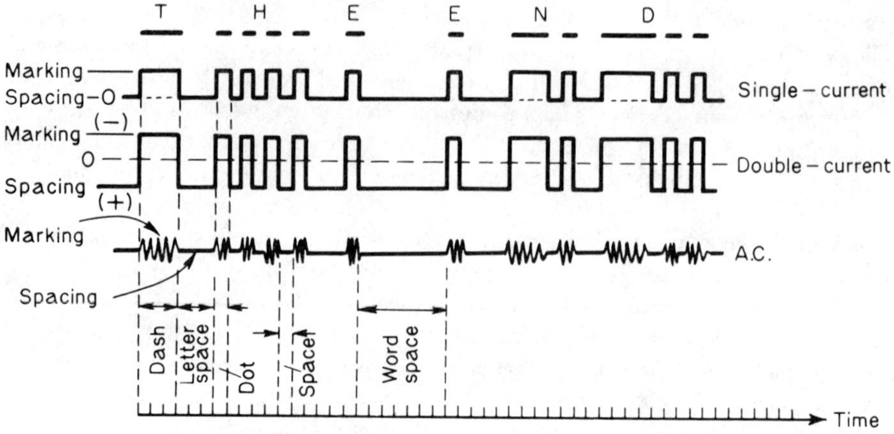

Fig. 3.11 Morse code. Examples of use with different signalling systems (see Appendix 1 for complete code). Notice that for double-current signalling, '*S*pacing' condition is a positive battery connected to line (3.4.1).

d.c. signalling, positive potential can be used for spacing and negative for marking. One system used for radio telegraphy is to transmit a radio wave for marking and to switch it off for spacing. In any of these the mark and space signal conditions could be reversed, using, for example, current 'on' for spacing and 'off' for marking.

The total duration of the signal combination is not the same for each character. The longest is the numeral 0 which has a total duration nineteen times that of the shortest character E, which is also the most frequently occurring character in messages.

3.4.2 Five-unit code. This is the code used for teleprinter operation. Each character consists of five signal elements of equal duration (20 ms) and each element consists of either of two signalling conditions (marking and spacing). This means that, for each of the two possible conditions for the first element, there can be two possible conditions for the next element and so on. The total number of two signalling conditions in the five elements is thus $2 \times 2 \times 2 \times 2 \times 2 = 2^5 = 32$. The code can thus give only thirty-two characters. This is not enough for the twenty-six letters of the alphabet and ten numerals, let alone punctuation marks and teleprinter control signals.

To increase the number of characters available, two of the thirty-two combinations are set aside for use as *shift signals*, also called *case-shift* signals. One of these is *letter-shift* and the other is *figure-shift*. When a letter-shift signal is transmitted, it causes the receiving instrument to set itself into a condition such that any signals afterwards received will be printed out as letters (letter-case). When a figure-shift is transmitted, any signals are afterwards printed out as numerals or punctuation marks (figure-case). Teleprinter control signals are unaffected by shift signals. The letters are often referred to as *lower-case* or *primary* characters and the figures etc. as *upper-case* or *secondary* characters.

As already explained in Chapter 1, the receiving and transmitting teleprinter instruments must run at the same speed and this synchronization is most easily achieved by starting and stopping the receiving instrument for each character. For this purpose, each character is preceded by a 'start' signal (20 ms spacing) and followed by a 'stop' signal (either 20 ms or 30 ms marking) so that the transmission of each character requires *seven* signal elements.

The principle of the five-unit code is illustrated in Fig. 3.12 and the complete code is shown in Appendix 2.

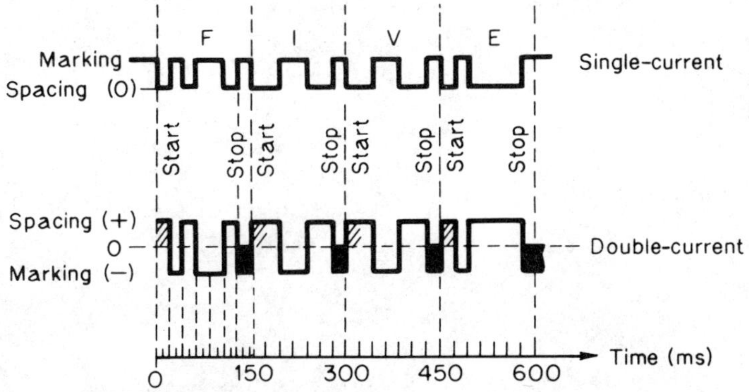

Fig. 3.12 Five-unit code. Typical word in five-unit code on single-current and double-current signalling systems (all signal elements 20 ms, start element 20 ms and stop element 30 ms) (3.4.2).

Now turn to page 78 and test yourself on questions 1–15 there before reading further.

PART 3 TYPICAL SIGNALLING SYSTEMS

3.5 Subscriber's line signalling

3.5.1 Dial telephone system. This employs d.c. loop signalling similar to that shown in Fig. 3.5, the battery at the exchange providing current for signalling and for the carbon microphone so that answer and clearing signals are provided automatically at the gravity switch contacts on the subscriber's instrument.

Dial pulsing contacts *DL* in Fig. 3.13 are provided at the subscriber's instrument to interrupt the calling loop sending loop-disconnect pulses to provide remote control of the exchange selectors. A slow-releasing relay *B* is provided at the exchange to distinguish between the short disconnexions which are dial pulses and the long disconnexion which is a clearing signal.

Another relay *F* is provided to disconnect the ringing from the called line when the called subscriber answers.

The ringing supply consists of alternate 'ringing' and 'silent' periods to give the characteristic interrupted ringing of the subscriber's bell. During the ringing period, a machine-driven alternator (ringing machine) having one terminal earthed delivers a.c. ringing current (1.3.3). During the 'silent' period, the supply lead is connected to earth by cam operated interrupter contacts at the ringing machine (see also Fig. 1.11).

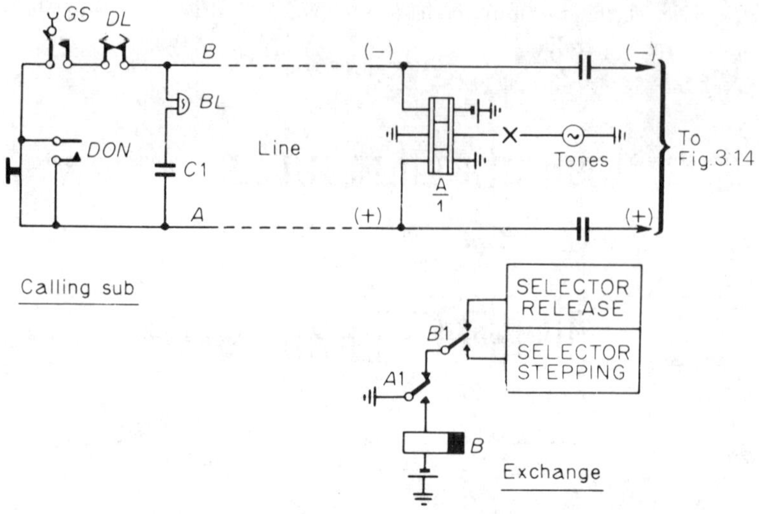

Fig. 3.13 Automatic system. Signalling between calling subscriber and exchange (3.5.1).

The basic circuit elements are shown in Figs 3.13 and 3.14 and the operation is as follows:

Subscriber calls exchange by lifting his handset. This operates *GS* (Fig. 3.13), completing the loop so that current flows from earth via one coil of relay *A*, over the *A* wire of the line, via *GS*, over the *B* wire to the battery via the other coil of relay *A*. This current operates relay *A*.

*A*1 operates relay *B*.

*B*1 prepares selector stepping circuit.

Subscriber dials and relay *A* releases for each 'break' pulse and re-operates on each 'make' pulse. It remains operated at the end of dialling.

*A*1 pulsing pulses the selector via *B*1, causing a connexion to be made to the required called line.

The slow-releasing relay *B* is not disconnected for long enough on any *break* pulse (*A* released) for it to release during dialling.

Tones are returned as required via relay *A* (see **3.3.7**). Busy tone (400 Hz, 750 ms on, 750 ms off) is heard if no free selector is available or the called line is engaged; N.U. (number unobtainable) tone (400 Hz continuous) is heard if an unused number is dialled.

Exchange calls subscriber. A.C. ringing is connected to the called line. Current flows via relay *F*, *F*1, line, bell, *F*2, *R*1, and the exchange battery to earth and thus to the other terminal of the generator. The capacitor in series with the bell prevents d.c. from flowing in the line at this time (*GS* is open).

No a.c. or d.c. flows during the 'silent' period.

Relay *F* cannot operate on the a.c. ringing alone because its slug and short-circuit winding make it slow to operate, so that it does not have time to operate on one half-cycle before the current changes direction for the next half-cycle.

Called subscriber answers by lifting the handset, looping the line. If the subscriber answers during the ringing period, there is now d.c. from earth, through the ringing generator, relay *F*, *F*1, line, *F*2, to battery via *R*1 in addition to the a.c. ringing. If the subscriber answers during the 'silent' period, there is d.c. only. In either case *F*

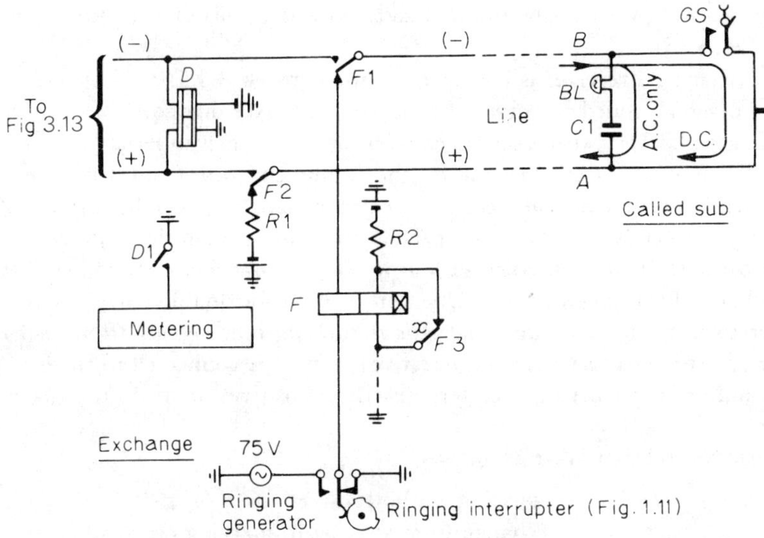

Fig. 3.14 Automatic system. Signalling between called subscriber and exchange (3.5.1).

now starts to operate on this line current and the armature moves far enough to open contact *F*3(*x*).

*F*3 removes the short circuit from the other coil of *F*. Relay *F* now completes its operation on a local circuit on this coil to operate its other contacts:

*F*1 and *F*2 disconnect ringing and connect relay *D*.

The use of an *x* (early break) contact at *F*3 ensures that the second winding of *F* is operative *before* *F*1 and *F*2 disconnect the initial operating path. Also the line coil has only to move the armature far enough to operate the *x* action, a heavier current being available then to complete the operation on the other coil.

Relay *D* now operates and *D*1 controls metering.

During conversation, relays *A* and *D* remain operated and form part of the transmission bridge.

3.5.2 Multi-frequency (m.f.) signalling telephone. Line signalling for calling, clearing, and ringing is the same as for dial telephones (3.5.1). The dial, however, is

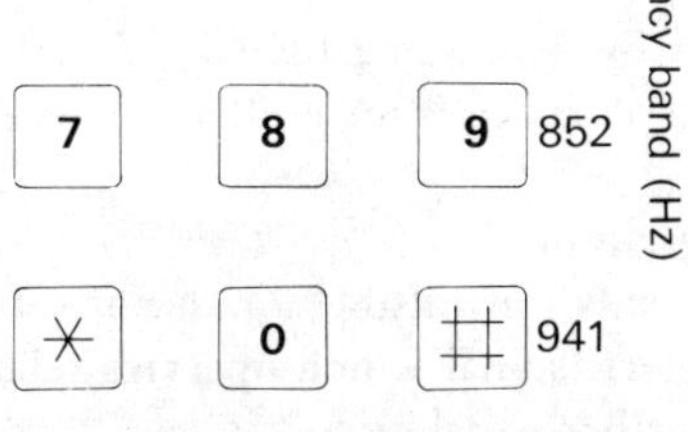

Fig. 3.15 Push-button (mf.) dialling. Pressing any button causes a pair of frequencies to be generated as shown, eg: button 5 generates 1336 + 770 Hz (3.5.2).

replaced by a set of twelve push buttons associated with oscillators having frequencies as shown in Fig. 3.15.

At the exchange, a receiver is connected at the register (4.3.5) and converts the m.f. signals to d.c. signals coded as required by the particular exchange equipment. In step-by-step exchanges (4.3.2) it is necessary to convert the m.f. signals to serial pulses (Fig. 3.6) and the caller has to wait after keying for the equipment to respond. Common-control (4.3.5) exchanges (crossbar and electronic) employ signals on parallel groups of wires and can respond at virtually the same speed as the caller keys in the number.

3.5.3 Automatic Telex subscriber's line. The subscriber's station is connected to the exchange by a two-wire line. One wire is used for double-current signals from subscriber to exchange and the other for double-current signals from exchange to subscriber. Positive battery (earthed negative) at 80 V is connected to the line for the spacing condition and negative battery (earthed positive) at 80 V for the marking condition.

The basic control signals are as follows:

Line disengaged —spacing (in both directions).
Call —change from spacing to marking on 'send' wire.
Dialling —interrupt the marking with 60 ms pulses of spacing at 10 p.p.s.
Teleprint (*five-unit code*)—start, 20 ms spacing.
Code elements, 20 ms spacing or marking.
Stop, 30 ms marking.
Clearing —spacing for at least 325 ms.
Call connected (*answer*) —change from spacing to marking followed by station identity code.

In addition to these, certain other signals are sent by the exchange equipment which can be considered as equivalent to the tones used in telephony as follows:

Proceed to select —100 ms marking (used as dial tone).
Immediate release IR —250 ms marking, 1 250 ms spacing, repeated continuously to release the connexion.

Some of these *service signals* consist of a 200 ms marking signal followed by groups of characters in five-unit code as follows:

OCC —Subscriber busy.
NC —Trunk busy.
NP —Spare number (used as N.U. tone).
DER —Called line faulty.
ABS —Station closed.

(Each service signal is followed by the *IR* signal.)

The circuit elements in Fig. 3.16 are used to send and receive these signals as follows:

Subscriber calls exchange by depressing the *CALL* key operating relay *DC*.

*DC*1 changes the *S* wire potential from +80 V (*S*) to −80 V (*M*), operating relay *A* in the exchange.

*DC*2 prepares *PD* relay circuit.

When the selectors are ready for pulsing, a pulse of −80 V (*M*) is returned on the *R* wire as a 'proceed to select' signal which operates relay *PD*.

*PD*1 connects the subscriber's dial circuit.

Subscriber dials. The dial is fitted with 'change-over' pulsing contacts to give double-current dial pulses which pulse relay *A*.

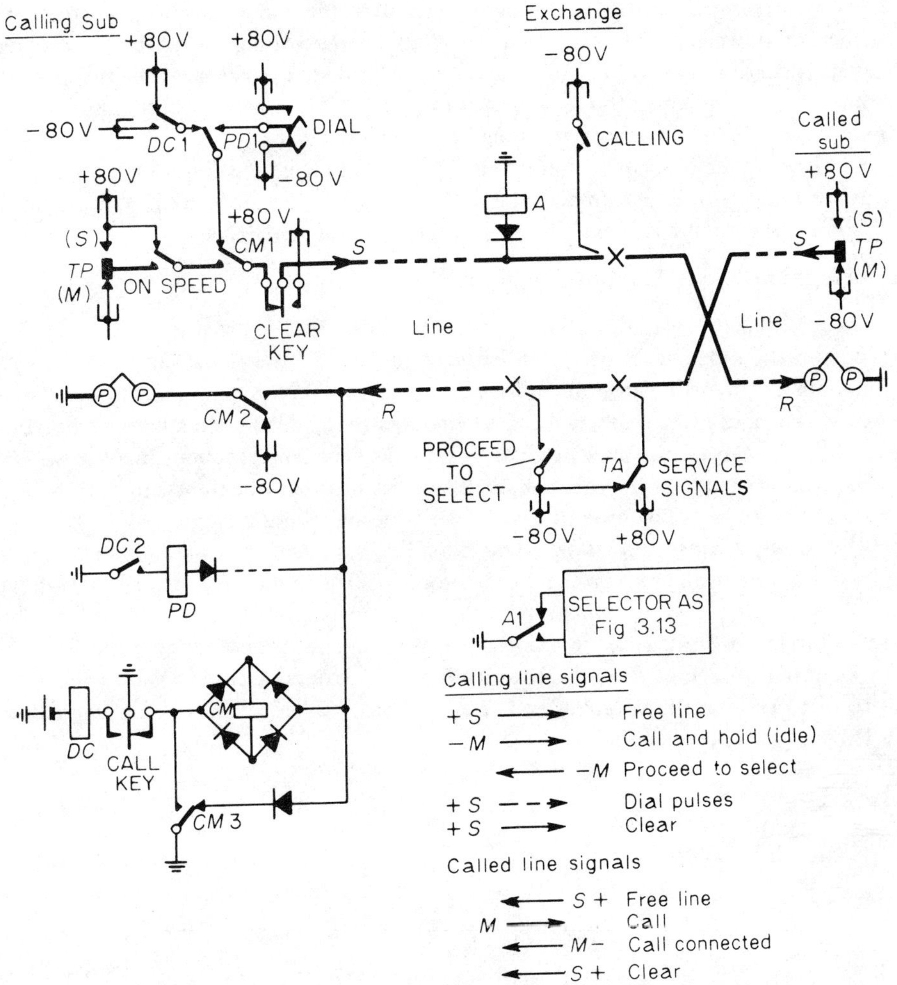

Fig. 3.16 Automatic telex system. Subscriber's station signalling elements (3.5.2).

*A*1 controls the exchange selectors which make connexion to the required line.

Service signals are returned as required by the exchange equipment at the contact *TA*.

Exchange calls subscriber by connecting −80 V (*M*) on the called subscriber's *R* wire operating relay *CM*.

*CM*1 and *CM*2 connect the called teleprinter to line.

*CM*3 disconnects the rectifier shunting *CM* so that *CM* now holds on both spacing (−) and marking (+) signals.

Another contact of *CM* starts the teleprinter motor. When the called subscriber's teleprinter motor reaches operating speed, its *ON SPEED* contact changes the *S* wire from +80 V (*S*) to − 80 V (*M*) as a 'call connected' signal. This operates the *CM* relay at the calling station to connect the calling teleprinter to line.

The called subscriber's teleprinter then automatically transmits an identifying code consisting of a pre-determined set of up to twenty characters in five-unit code so that the calling subscriber knows that the correct connexion has been established and sending can commence.

Character signals can then be sent directly over the connexion between the subscribers. The selector *A* relay is also pulsed by the calling subscriber's 'spacing' signals (negative pulses) but control relays are provided which prevent this pulsing from having any effect on the selectors. The *CM* relays at both subscribers' instruments are held on both marking and spacing signals.

Clearing. The subscriber operates the *CLEAR* key, connecting +80 V to the *S* wire for long enough for contact *A*1 to release the exchange selectors. The exchange disconnects both subscribers' *R* wires to release their *CM* relays.

3.6 Telephone exchange junction signalling.

As in the case of subscribers' lines, the signalling system provided on a junction depends on the types of exchange which it connects. An additional factor in the case of junctions is the wide variation in length between different junctions. The distance between exchanges in a densely populated city area may be less than a kilometre; trunk lines, which we can for the present regard simply as very long junctions, may be several hundred km long; submarine cable or space satellite links may be thousands of km long, and each of these has to be provided with a suitable signalling system.

3.6.1 Loop signalling junctions. Loop signalling similar to that used on subscribers' lines is employed on all junctions between automatic exchanges up to about 1 500 Ω loop resistance (750 Ω per wire) which is about 27 km in length (using 20 lb cable). This includes practically all local junctions. Answer signals are sent by reversing the connexions of the selector *A* relay at the incoming end to operate either a shunt-field relay or a rectified relay at the outgoing end (2.2.9). This is called *reverse battery supervision* (3.5.1).

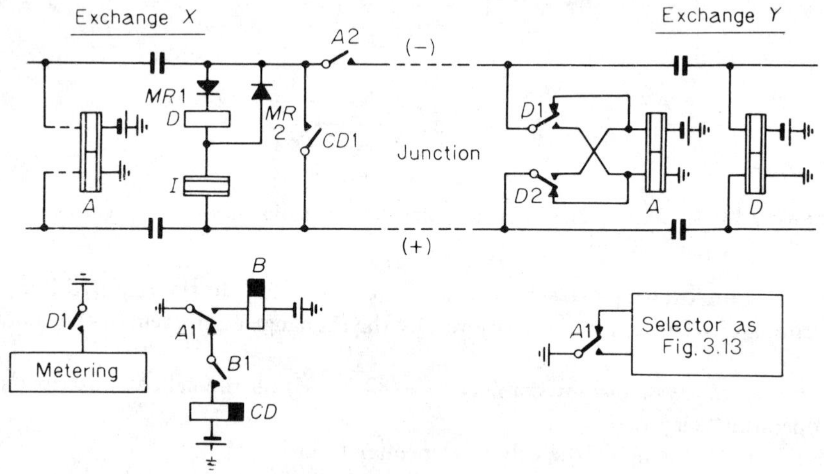

Fig. 3.17 Loop-disconnect (auto) junction (3.6.1).

The main circuit elements are shown in Fig. 3.19 and the operation is as follows:

X calling Y: Calling subscriber's loop operates *A* at exchange *X*.

*A*1 operates *B*.

*A*2 loops junction to operate relay *A* at exchange *Y* via *D*2, the retard *I*, rectifier *MR*2, *D*1.

(Rectifier *MR*1 has a high resistance to current in this direction so *D* does not operate).

At exchange Y, $A1$ prepares the selector circuit (3.5.1).

Dialling: At exchange X, A releases once on each 'break' pulse from the calling subscriber.

$A2$ opens the loop once for each pulse so that relay A at exchange Y also releases once for each pulse.

$A1$ releasing, operates CD on the first pulse of each train (digit), (3.3.3).

The slow releasing relay B is not disconnected at $A1$ for long enough on any 'break' pulse for it to release during dialling. CD is not disconnected for long enough on any 'make' pulse for it to release during dialling either. (See also 6.3).

At exchange Y, $A1$ steps the selector (3.5.1).

At exchange X, $CD1$ provides a signalling path during dialling which does not include the retard I. If this were not done, the inductance of I would slow the operation of A at the end of each pulse.

End of dial pulse train. Relay A remains operated on the subscriber's loop at exchange X.

$A1$ holds B, releases CD.

$A2$ holds A at exchange Y.

$CD1$ removes short circuit from junction wires.

Called subscriber answers. D operates to the called subscriber's loop at exchange Y (3.5.1).

$D1$ and $D2$ reverse the direction of current flow in the junction wires.

At exchange X, D operates to current in this direction via $MR1$ (rectifier $MR2$ has a high resistance to current in this direction).

$D1$ controls metering.

Release (clearing): Opening of the calling loop releases A at exchange X.

$A1$ releases B, operates CD.

$A2$ releases relay A at exchange Y.

$B1$ releases CD.

At exchange Y, $A1$ releases the selectors.

3.6.2 Signalling system AC9 is a single frequency in-band v.f. system (3.3.9) used on trunk lines in the U.K. Signals consist of different length pulses at 2280 Hz as follows;

Forward		
Seizure (calling)	65 ms	(50–80 ms)
Dial pulses	55 ms	(40–70 ms)
Clear forward (release)	900 ms	(700 ms minimum)
Backward		
Answer	250 ms	(200–300 ms)
Clear back (called sub clear)	250 ms	(200 – 300 ms)
Release guard	1000 ms	(750 ms minimum)
Backward busy (blocking)		continuous

A block diagram of the system is shown in Fig. 3.18. It is designed for use between exchanges employing ordinary two-wire loop signalling within the exchange. The four-wire trunk line provides separate 'go' and 'return' channels, each connected as a pair of wires at the exchange but usually consisting of radio or carrier channels (3.3.9). Conversion from two to four wire is by means of *terminating sets* or hybrids operating in basically the same way as the induction coil in a telephone instrument (1.3.5).

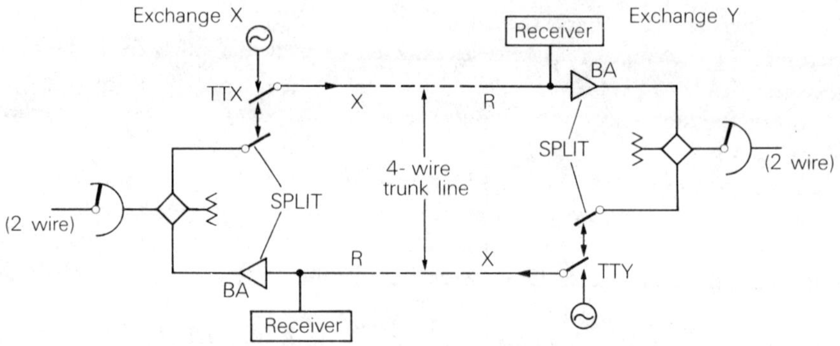

Fig. 3.18 Block diagram of AC9 system. Splitting is achieved by opening the relay contacts in the X (transmit) pair and biassing off the buffer amplifier (BA) in the R pair. (3.6.2).

D.C. signals at the exchange are received at the transmission bridge associated with the hybrid and the corresponding v.f. pulses sent on the go channel. Pulses are timed by relays in earlier versions of the system and by integrated circuit logic in the more modern equipment.

The duration of signals detected by the voice-immune receiver is checked by the logic (or relays) and converted to the appropriate d.c. signals for connection to the two-wire line at the hybrid.

Each receiver is protected from near-end interference by a buffer amplifier which permits transmission in one direction only. Speech (or noise) from the caller, for example, could otherwise mask the reception of the answer signal and suppress metering.

The speech path is split, by contacts and by switching off the buffer amplifier, when signals are being sent or received. This prevents speech or noise from the sending end interfering with signals, and stops signals spilling over into the two-wire connection to the exchange. Such spillover could interfere with signalling on other junctions in the complete connection or be mistaken for tones if heard by the subscribers.

The operation on a call is as follows:

X calling Y. Calling loop at exchange X causes TTX to operate for 50–80 ms to send the seizure signal which is converted to a loop again at exchange Y.

Dialling. Each break pulse detected at exchange X is converted to a fixed length pulse operating TTX for 40–70 ms. This is received at exchange Y and converted to a fixed length break pulse of 57 ± 4 ms.

Called subscriber answers. The reversal (**3.6.1**) from the exchange equipment at exchange Y causes TTY to operate for 200–300 ms. The v.f. signal is received at exchange X and the reversal passed back in the same way as the operation of relay D in Fig. 3.17.

Called subscriber clears. Restoration of normal polarity at exchange Y again causes a 200–300 ms pulse of v.f. tone to be sent and the polarity restored at exchange X.

Caller clears. Removal of the calling loop causes a 700 ms signal to be sent. If this is effective, a release guard signal of 400–600 ms is returned from exchange Y and the circuits are restored for use on further calls. In the event of a temporary channel failure or noise interference preventing either the clear forward or release guard signal from being effective, the outgoing circuit at X is arranged to repeat a sequence of seizure followed by clear forward at approximately 50-second intervals.

Now turn to page 79 (**3.8.4**).

PART 4 SUMMARY

3.7.1 Signalling principles

Signals required: Calling, ringing, clearing.
Dialling, answer, tones.
Telegraph characters.

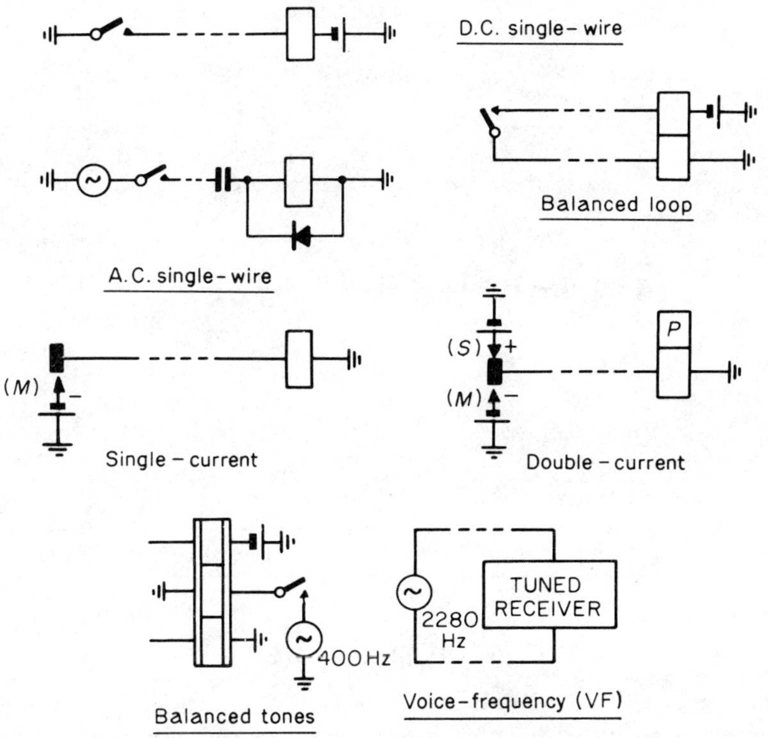

Fig. 3.19 Types of signal summary.

3.7.2 Signalling systems

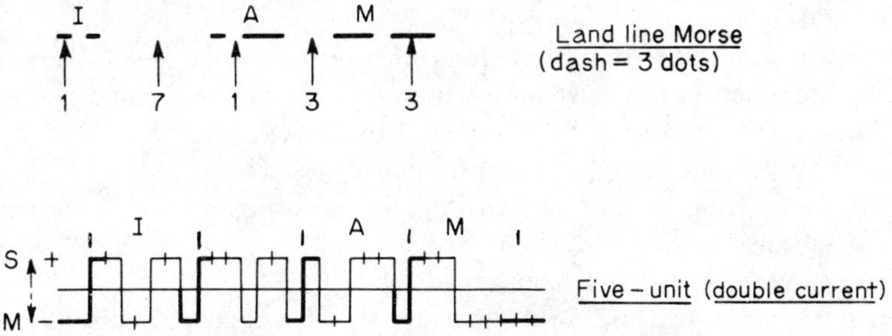

Fig. 3.20 Telegraph codes—summary.

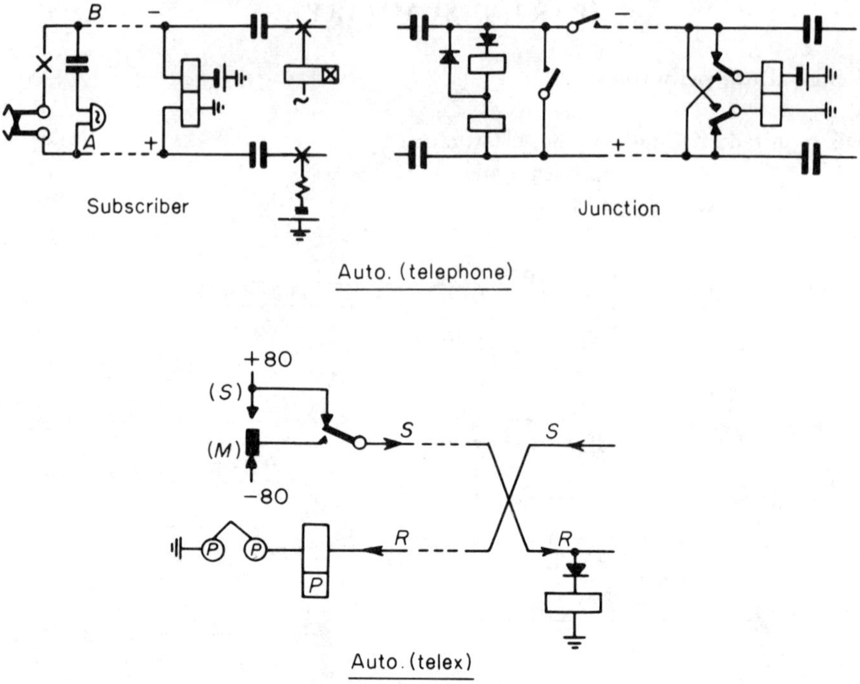

Fig. 3.21 Typical systems—summary.

Now turn to page 79 and try the remaining questions there before leaving this chapter.

PART 5 SELF-TEACHING NOTES

3.8.1 First read **3.1.1** to **3.4.2** (inclusive) carefully. This chapter concerns signalling circuits and these are better described by diagrams than by words. Study the diagrams carefully therefore.

3.8.2 When you have read **3.1.1** to **3.4.2**, try the following fifteen questions. Remember now, *write* all the answers down.

Q.1 What three signals are essential on both automatic and manual exchange subscribers' lines?

Q.2 What other signals are needed from the subscriber's telephone on an automatic exchange?

Q.3 What forward signals must be used on a junction between two manual exchanges?

Q.4 Why are tones used to give information on the progress of an automatic call?

Q.5 What is used in place of these tones on the Telex system?

Q.6 For what purpose is loop-disconnect pulsing used?

Q.7 What is the name given to a signalling system using an earth return circuit?

Q.8 What do the initials e.p.d. stand for?

Q.9 State three advantages of loop signalling.

Q.10 State two ways in which double-current signalling reduces pulse distortion.

Q.11 In what type of signalling does the transmitter contact change over between a positive and negative potential?

Q.12 What property must a component have if it is to present a high impedance to speech currents but a low resistance to d.c.?

Q.13 What device would you use as a transformer to provide a balanced tone feed on a circuit not equipped with a transformer type transmission bridge?

Q.14 What is the length of the interval between elements, letters and words in the International Morse Code?

Q.15 How many signal elements are used for each character in the five-unit code?

Now check your answers on page 260.

3.8.3 You should have answered most of these fifteen questions correctly.

If you scored less than ten out of fifteen, see if the notes on the answer page help, then read **3.1.1** to **3.5.3** once more and try again.

If you scored ten or more, you should be able to deal with a longer question now so try question 16.

Write a complete answer, but spend no longer than 20 minutes to half an hour on it.

Q.16 Describe, with the aid of simple circuit elements, the d.c. signalling arrangements which would be used over a junction between two automatic exchanges which are spaced 3 mile (5 km) apart.
Why would this method of signalling not be suitable if the exchanges were 300 mile apart (500 km)? What other method of signalling would be used in such circumstances? (C & G 1963)

Now check your answers against the points listed on page 261.

Then read the rest of Chapter 3 (**3.6** to **3.7.2**).

3.8.4 A few days after reading **3.6** to **3.7.2**, try the following two questions.

Q.17 A teleprinter station connected by a physical circuit to a teleprinter automatic exchange sends distinctive signals for each of the following purposes:
(*a*) calling the exchange,
(*b*) selecting the desired connexion by dialling,
(*c*) transmitting a message to the selected teleprinter station and
(*d*) clearing down the connexion.
What are the line signals used for each of these purposes? Illustrate your answer by means of approximate timing charts.
What signals might be received by the calling station whilst setting up a connexion? (C & G 1961)

Now check your answer on page 261.

Q.18 Describe the five-unit telegraph code used in teleprinter working, making clear the need for (*a*) start and stop elements, and (*b*) letter and figure shift characters. Give details of the time durations of each code element and state the maximum number of characters that may be transmitted in one minute. (C & G 1963)

Now check your answer on page 261.

4 SWITCHING SYSTEMS

For self-teaching notes, see page 114.

PART 1 INTRODUCTION

4.1.1 Manual exchanges. The useful thing about telephones connected to the public network is that they can be used to communicate with any other subscriber on the system. The same thing applies to teleprinters on the Telex network.

This is made possible, as we saw in **1.2.2**, by manual or automatic exchanges which provide centralized switching facilities.

The early exchanges were relatively simple affairs serving small numbers of subscribers. They were, of course, manual exchanges, the connexions being set up by hand at the exchanges as requested verbally by the subscribers (or telegraphically by telegraph stations). Later on, the growth of the system required larger exchanges than could be provided by the original simple switching techniques and this resulted in the introduction of a variety of different types of manual exchange to suit different numbers of subscribers and different calling rates.

Even in automatic exchanges, some services are still provided by manual switching facilities at what is called an *auto-manual* switchboard. This is used to provide assistance to subscribers having difficulty in obtaining a connexion by dialling, and for providing essentially personal services such as reverse charge calls, enquiries and complaints.

4.1.2 Automatic exchanges. The first public automatic exchange opened at La Porte, Indiana (U.S.A.) in 1892 using rotary switches of a type originally invented in 1891 by an undertaker in Kansas City (U.S.A.) called Almon B. Strowger, after whom the system is named.

Since then, a variety of automatic switching systems has been developed. The most successful of these use either arrays of relays or mechanical devices called crossbar switches, which achieve the same result using fewer operating coils. More recently electronic exchanges have been introduced, using these types of switching in conjunction with control circuits based on electronic devices such as transistors. We may expect these in their turn to be superseded by fully electronic exchanges.

For the present there are still more exchanges of the Strowger type than of any other system.

4.1.3 In this chapter you will learn about the general switching principles of manual exchanges and how automatic switching systems have developed from them. We shall then consider the principal features of crossbar and similar common-control systems.

In the last part of the chapter some typical practical systems are described.

The Strowger (step-by-step) system will be dealt with in greater detail in the subsequent chapters.

PART 2 SWITCHING SYSTEM PRINCIPLES

4.2 Manual exchanges

4.2.1 General facilities. The equipment provided at a manual exchange for setting up the required connexions between subscribers' lines is known as a *switchboard*, and the staff employed to make and disconnect these connexions are known as switchboard *operators* or *telephonists*.

Certain basic features are common to all types of manual switchboard. Associated with each subscriber's line there must be some form of calling indicator, e.g. relay and lamp. On receipt of a calling indication, the operator connects to the calling line using whatever type of connecting equipment is provided on the particular switchboard and also connects the position telephone (or teleprinter) to the same connecting circuit. The means of connecting the telephone could in principle be any switching device but in practice keys are universally employed for this purpose. This key is designated the 'speak' key.

A key is a manually operated switch in which contacts rather like relay contacts are operated by a lever attached to a small handle instead of by a magnetically operated armature.

The calling subscriber tells the operator what number he wishes to call and the operator connects to the called subscriber using the same connecting circuit as for the calling subscriber. Ringing current is then connected to the called line by means of the 'ringing' key, which in practice may be part of the same key as the 'speak' key.

In addition to these facilities, supervisory indication of subscriber clearing conditions are provided as described in **3.5.1**, so that the operator knows when to disconnect the call.

4.2.2 Co-ordinate switching. The term *co-ordinate* is used to describe systems of the type shown in Fig. 4.1 because of the similarity of the connexions to the co-ordinates used to locate the points on a graph. The switching devices are provided at the intersections of the 'vertical' and 'horizontal' circuits. These intersections and their switches are sometimes called *crosspoints*. The complete array is called a *matrix*.

The insets (a), (b) and (c) in Fig. 4.1 illustrate three typical forms of crosspoint.

Figure 4.1a shows a simple crossbar peg switch originally used as a 'concentrator' for interconnecting relatively small numbers of telegraph circuits. It consists of two sets of brass bars, insulated from one another, fixed at right angles and with holes drilled through both sets so that any of the upper bars can be connected to any of the lower bars by inserting brass pegs in the holes. Its application to telephone switching is limited by the fact that it provides for switching only one wire through at each crosspoint.

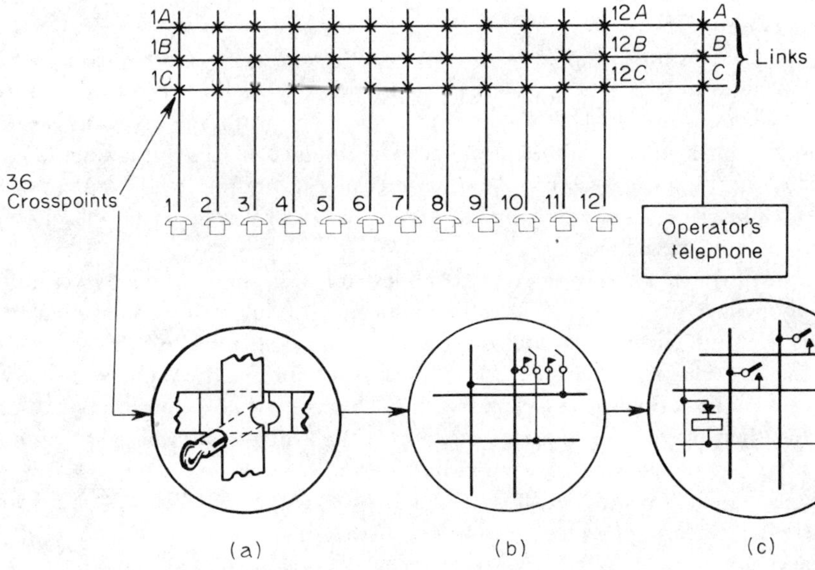

Fig. 4.1 Co-ordinate switching (4.2.2).

A more suitable method for telephone switching is shown in Fig. 4.1b. This employs keys to switch the two speech wires at each crosspoint. This system is used on the 'cordless' switchboards provided for interconnecting a number of telephones (extensions) and lines (exchange lines) in the type of subscriber's installation known as a *Private Manual Branch Exchange* or *P.M.B.X.*

Co-ordinate switching, which was one of the earliest exchange systems, now forms the basis of some of the latest types of electronic automatic exchange. The way in which it is applied is illustrated by Fig. 4.1c, where the relay contacts perform exactly the same function as the key contacts of the previous example. One relay is provided for each crosspoint, the coils being connected at the intersections of vertical and horizontal 'marking leads'. The reed relay as shown in Fig. 2.26 is used for this purpose.

Referring now to the main diagram in Fig. 4.1, it can be seen that, to connect any subscriber to any other subscriber, or to the operator, two crosspoints on the same horizontal 'link' must be operated. For example, operating 1*C* and 12*C* would connect subscriber 1 to subscriber 12 using link *C*. Any two subscribers can thus be connected together by connecting them both to the same free link. It is important that the link should be free, of course, otherwise the two conversations would interfere with each other.

The maximum number of calls which we can have in progress at one time on this system is equal to the number of links. Increasing the number of subscribers, therefore, will increase the number of links required because the larger number of subscribers will make more calls between them. The number of crosspoints is equal to the number of subscribers multiplied by the number of links and so increases roughly as the square of the number of subscribers. Doubling the number of subscribers, for example, requires about four times as many crosspoints.

The size of the crosspoints together with the large number required, make co-ordinate switching too cumbersome for manual exchanges much larger than that shown in Fig. 4.1. The cost is also high because each crosspoint is provided with a pair of contacts for each wire to be switched and yet only a few can ever be in use at one time (six out of thirty-six in Fig. 4.1 for example).

The practical solution adopted is to provide only one contact per wire in a fixed position in the form of a 'jack' or socket and provide the other contact in the form of a plug on the end of a flexible 'cord' or wire. This is then still a link system in which the number of links equals the total number of simultaneous calls but it is no longer a co-ordinate system and has no component increasing in number as the square of the number of subscribers. The principles of plug and cord systems of this sort are described in the next section.

4.2.3 Plug-and-cord switching. In its simplest form, a plug-and-cord switchboard consists of one jack per subscriber, a calling indicator per subscriber and as many cord circuits as the number of calls requires. These cord circuits are the equivalent of the links in the co-ordinate system described in the preceding section and each consists of two inter-connected cords associated with a 'speak' key. To connect any two subscribers together, one cord of a pair is inserted in the calling subscriber's jack and the other cord of the pair is inserted in the called subscriber's jack.

The basic switching plan is shown in Fig. 4.2.

The essential features of a plug-and-cord switchboard are shown in Fig. 4.3.

The jacks and calling indicators are arranged in numerical order on the vertical portion face of the switchboard forming what is known as the *jackfield*. The cord circuits are equipped on the horizontal *keyshelf*.

Each cord consists of three conductors connected to the three contact points of the plug: the *tip, ring and sleeve*.

The number of lines which can be accommodated on a switchboard is limited both by the size of the jacks and indicators and by the size of the jackfield. The jacks and indicators must not be too small or the operator will be unable to use them; the jackfield must not be too large or the operator will be unable to reach all the jacks. (You could say that it is the thickness of the operator's fingers and the length of an operator's arms that limits the number of lines.) The maximum number of lines on a practical switchboard is 10 000.

The number of subscribers to which one operator can attend, however, depends upon the traffic, i.e. the number of calls originated per subscriber and the length of time taken to deal with each call. An operator can normally handle about 200 simple calls in an hour. If the subscribers originate more calls than this, more than one operator's position must be provided.

Two positions can be provided relatively easily. They can simply be placed side by side with half the total number of lines equipped on each board. Each operator can then reach the jacks on her own position and on the adjacent position.

It is not a practical proposition, however, to provide a third position in this way. It might just be possible for the two operators on the two end positions to reach across to each others' jackfields, but this would need excessively long cords and would interfere with the work of the operator on the centre position.

The practical solution to the problem of providing for traffic in excess of the load of two operators is to repeat every subscriber's jack at intervals along the face of the switchboard so that every operator has within reach a jack connected to each subscriber on the exchange. This is called a *multiple*. A three-position switchboard needs only a partial multiple because the jacks on the centre position can be reached by all three operators and do not need to be multipled.

The object of connecting more than one jack to each subscriber's line to form a

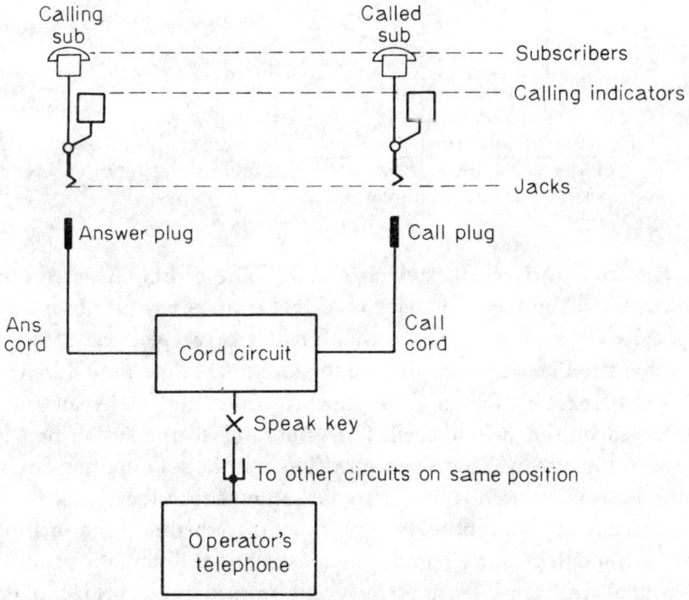

Fig. 4.2 Plug and cord switching (4.2.3).

multiple is to enable any operator on the switchboard to complete a call *to* any line on the exchange. It is not necessary for every operator to be able to *receive* calls from every subscriber. It is sufficient, therefore, to multiple the jacks without also multipling the calling indicators or lamps.

Each operator's position thus has relatively few answering jacks and associated calling indicators or lamps. The number of these on each position is such as to provide a reasonable working load for one operator (about 150 lines) and these need not be arranged in numerical order.

The usual arrangement is for each subscriber's line to be connected to only one answering jack and calling indicator (or lamp and relay). All calls outgoing from the subscriber are therefore answered at this 'home' position, either by the operator on that position or by the operator on one of the adjacent positions.

In addition to the answering jack, two or more multiple jacks are connected to each line. These are placed at intervals of either two or three operators' positions along the switchboard so that there is at least one within reach of every operator. The number of multiple jacks per line thus depends on the number of positions on the switchboard.

The multiple jacks are arranged in numerical order at each appearance so that all operators can readily locate the number asked for by a caller.

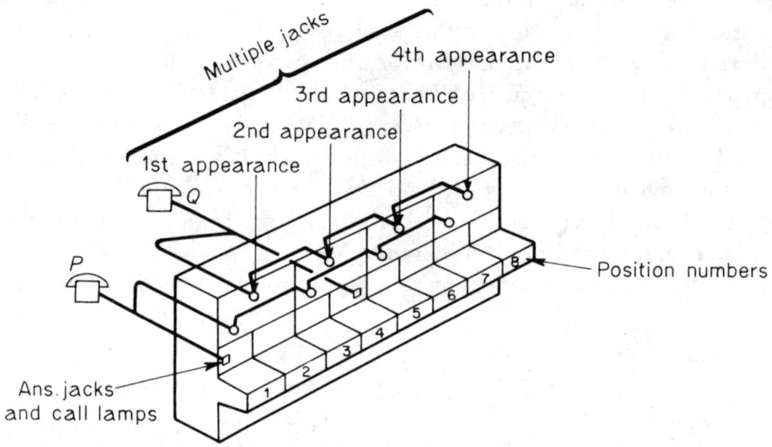

Fig. 4.3 Principle of subscriber's multiple. The multiple is only shown for two of the subscribers, *P* and *Q*. The multiple is repeated every two positions in this example (4.2.3).

4.2.4 Auto-manual and trunk switchboards. These differ from ordinary manual exchange switchboards in that they have no direct connexions to subscribers' lines.

Incoming calls at an auto-manual switchboard are received via the selectors of an automatic exchange. The number of incoming circuits is determined by the traffic (the number of simultaneous calls) and is thus considerably less than the number of subscribers served by the switchboard. Calls outgoing to the subscribers are made by dialling via the automatic exchange selectors. The number of outgoing circuits provided for its purpose is also determined by the traffic and no subscribers' multiple is provided. Cord-type auto-manual switchboards are therefore smaller than manual exchange switchboards. Although generally similar in construction, different circuit principles are employed so that the cord circuits have no transmission bridge apparatus. The supervisory lamps are controlled by relays in the incoming and outgoing jack circuits by means of signals over the sleeve connexions. This is known as *sleeve control* and enables a

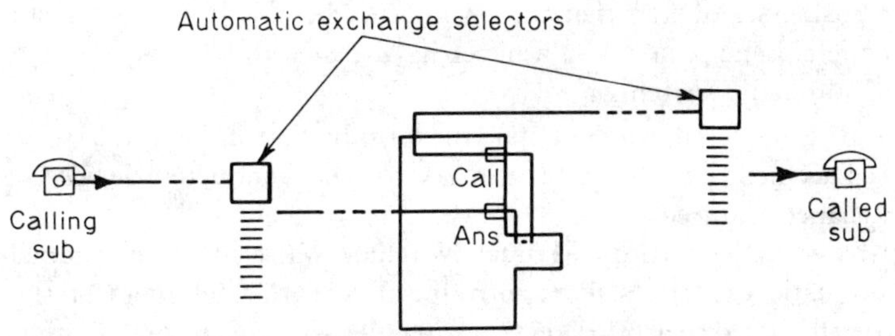

Fig. 4.4 Auto-manual switchboard principle (4.2.4).

variety of different signalling systems to be used as required on different junction circuits connected to the switchboard.

To distinguish them from this type of system, manual boards, in which the supervisory lamps are controlled by the transmission bridge relays in the cord circuit, are called *bridge control* systems.

In the case of trunk manual switchboards, incoming calls are received over junctions or trunk lines from other exchanges. The number of such circuits depends both on the number of exchanges from which the traffic originates and on the traffic from each. The same is true of the outgoing junctions and trunk lines.

Apart from the larger number of trunk and junction groups, which makes the switchboard generally bigger, cord type trunk switchboards are similar in principle to auto-manual switchboards.

4.3 Automatic exchanges

4.3.1 Advantages of automatic switching. Although automatic exchanges employ switching apparatus which is more complicated than the plugs and jacks used in a manual exchange, they have a number of advantages which make them worth the extra initial cost and the higher cost of maintenance. These advantages are as follows:

(*a*) The cost of setting up each call is less on an automatic exchange. This is because each call on a manual exchange occupies some of an operator's time and this must be paid for.

(*b*) Calls are set up more quickly on an automatic exchange because only one person, the caller, is involved in setting up the call. On a manually switched call there is one operator involved at each exchange through which the call is routed. Each operator in turn has to be told what number is required and this takes time.

(*c*) An automatic exchange connexion is released as soon as the caller clears, resulting in more efficient use of the available plant and greater convenience to the subscriber. In the case of a manual exchange, the cord circuit and any junctions used are not available for further use until the operator sees the clearing indication and withdraws the plugs.

(*d*) An automatic exchange provides a uniform service at all times. An unexpected rush of calls at a normally quiet time of the day would almost certainly find a manual exchange with too few operators on duty. An automatic exchange never sleeps or goes off duty

(*e*) Greater privacy is provided by an automatic system. It was for this reason, in fact, that Strowger invented his switch—the wife of a rival undertaker was employed as an operator at the local telephone exchange.

(*f*) One last point; if all the telephone calls made nowadays had to be switched manually, it would not be possible to find enough operators without so depriving other businesses of staff that they would not be able to make the calls anyway.

The telephone system could never have grown to its present size without the use of automatic switching.

Future growth is dependent upon the introduction of electronic systems in which the replacement of the moving parts by semiconductor devices reduces the maintenance required.

4.3.2 Step-by-step operation. Manual switching was a highly developed art before the first automatic exchanges were introduced, and the designers of the automatic systems naturally based their work on the principles which were already well established for manual switching. They were, after all, still dealing with the same sort of subscribers and the new automatic exchanges had to fit into the existing manually switched network.

Strowger's automatic exchange used ratchet-driven rotary switches remotely controlled by pulses from the calling subscriber. In effect this switch was a mechanical version of the plug-and cord switchboard with the subscriber doing the operator's job. The wipers were connected by short flexible cords and performed a similar function to the switchboard plugs. The place of the jack field was taken by a bank of fixed contacts and the wipers moved one step from each contact to the next for each pulse.

At first, each subscriber's line had its own switch and the exchange was limited to 100 lines by the size of the switch. Later it was found possible to build larger exchange units with the same size of switch by connecting several stages of switches in cascade The expense of providing one large switch or selector for each subscriber was avoided by using a stage of smaller cheaper lineswitches to connect the subscribers to the selectors as and when they required them.

The setting up of a call progressed through the exchange stage by stage. Each switch connected the call to a free selector in the required group in the next stage in time for the next train of pulses to set that selector. This is known as *step-by-step* operation and, when used with Strowger selectors, it needs only simple control circuits which can be provided economically at each selector.

4.3.3 Strowger selectors. Two main forms of selector mechanisms are used as illustrated in Fig. 4.5.

The simpler of these, the *uniselector*, has a set of moving contact arms called *wipers* which rotate over a bank of fixed contacts arranged in the form of an arc.

The *two-motion selector*, or bi-motional selector as it is sometimes called, also has a set of wipers which move over a bank of contacts. The difference is, however, that the bank contacts are arranged as ten separate *levels*, one above the other. To make a particular selection, the wipers are first lifted vertically to the required level, and then rotated to the required position in the level.

While there are a number of variations on the basic designs used for different purposes, the most commonly used uniselector has twenty-five contacts available to each of its four or five wipers. This is described as being an *availability* of 25 if all outlets can be used. For reasons we shall discuss in more detail in Chapter 6 (circuits), the availability is likely to be only 23 in practice (**6.6.1**).

The two-motion selector has ten levels of eleven contacts available to each wiper; 110 contacts in all. Only ten of the eleven contacts are normally used as working outlets, the eleventh being used for busy (tone) conditions. The outlet availability per level is therefore 10, and the total number of working outlets is 100.

By arranging the contacts in a double row on each level, an outlet availability of 20 can be provided, the wipers also being provided in pairs. Relays are used to switch the input to whichever of the two wipers in each pair is required and this then gives

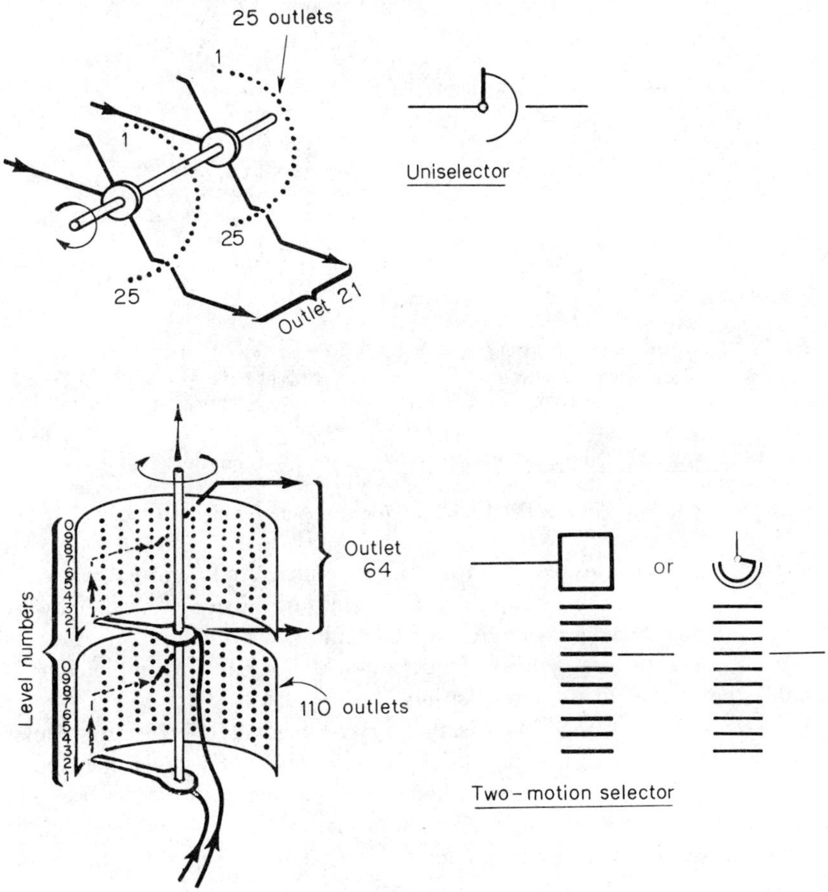

Fig. 4.5 Selector types used in the Strowger system (4.3.3).

us a 200 outlet selector.

The movement of the wipers on both types of switch proceeds as a series of steps produced by pulses of current in an electromagnet driving a ratchet and pawl mechanism. If these pulses of current are produced under the control of a device such as a dial, the selector is said to be 'stepped'. If the electromagnet circuit is connected in such a way that the wipers step continuously until the circuit is disconnected, the selector is said to be 'self-driving'. Both types of switch may use either of these modes of operation.

The driving force for stepping the wipers of a selector is provided by an electromagnet similar in principle to that of a relay. The armature of this magnet is linked to a *pawl* which engages with the *ratchet* teeth to shift the ratchet and, therefore, the wiper assembly to which the ratchet is attached, one step for each operation of the magnet.

There are two basic types of ratchet and pawl mechanism, the main features of which are shown in Fig. 4.6. A *forward-acting* mechanism steps the wipers as the armature is operated by the magnet; a *reverse-acting* mechanism steps the wipers as the armature releases under the action of its restoring spring after operating.

Reverse-acting mechanism. With the armature at normal, the pawl rests in one tooth of the ratchet. Energizing the coil causes the armature to move the pawl into the next tooth. The detent prevents the ratchet from moving in this direction. When

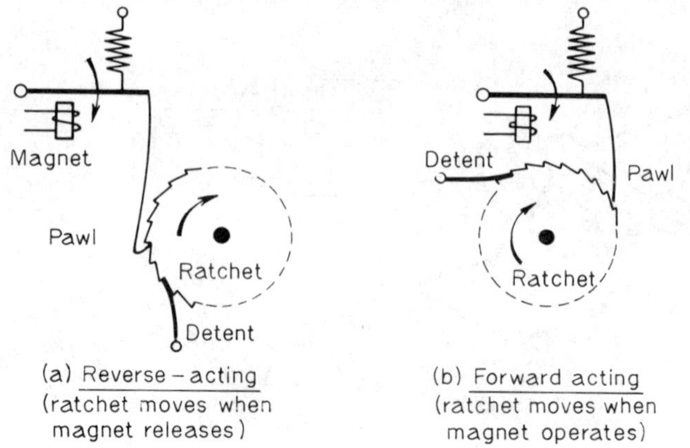

Fig. 4.6 Basic ratchet and pawl mechanism. Both the pawl and the detent in these examples are in the form of flat springs. The stops and guides employed in practice have been omitted here for clarity (4.3.3).

the coil current is disconnected, the armature is returned to its normal rest position by its restoring spring, taking the pawl with it. On this return motion the pawl pulls the ratchet, moving the wipers one step. The detent, which is held against the ratchet under spring pressure, drops into the next tooth as the ratchet completes its step. This holds the ratchet in its new position.

Forward-acting mechanism. Energization of the magnet operates the armature which pushes the pawl against the ratchet tooth to shift the ratchet one step. The detent holds the ratchet in its new position. Disconnexion of the coil current allows the armature restoring spring to return the armature to its normal rest position, taking the pawl with it in readiness for the next step. The detent prevents the ratchet from moving back with the pawl.

The shape of the ratchet teeth is important. Each has one short, steep side against which the driving force of the pawl and the retaining action of the detent operate. The other side of each tooth is longer and shallower, allowing the ratchet to move freely under the detent in the foward direction and the pawl to ride over the tooth in the other direction.

Uniselectors usually have reverse-acting mechanisms and two-motion selectors have forward-acting mechanisms (**5.2.1**, **5.3.1**).

4.3.4 Strowger trunking principles. The general arrangement of switches in a typical four-digit Strowger type exchange is shown in Fig. 4.7. The operation of this when a subscriber makes a call is briefly as follows:

The subscriber's *line uniselector* wipers are first rotated to seize a free *first selector*.

The first selector wipers are stepped vertically by the first digit dialled by the subscriber and they are then rotated to seize a free *second selector* in the required *thousands* group.

The second selector wipers are stepped vertically by the second digit dialled and they are then rotated to seize a free *final selector* in the required *hundreds* group.

The final selector wipers are stepped vertically by the third dialled digit and rotated by the fourth digit to select the required subscriber within the hundreds group.

It can be seen from this that the selection of the required connexion under the control

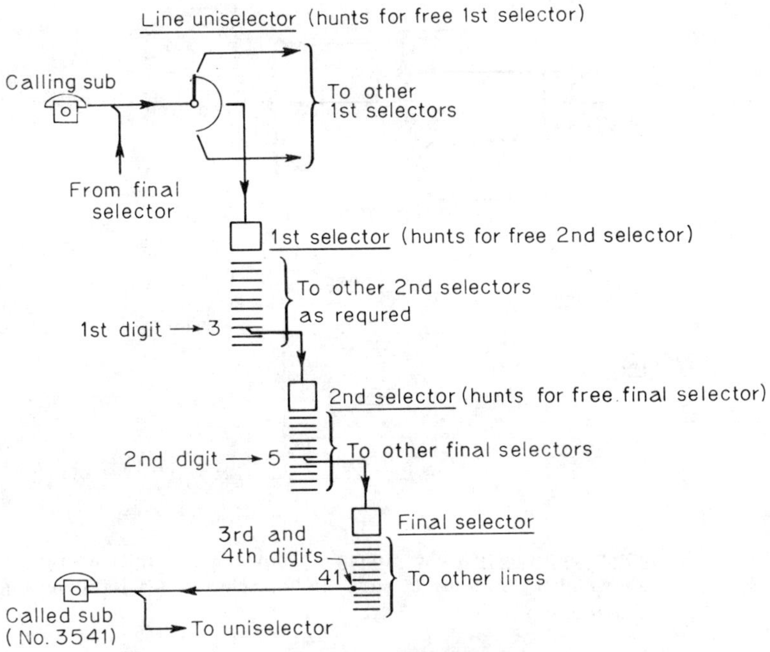

Fig. 4.7 Strowger trunking principle (four digit exchange). Notice that it is not necessary to draw every selector in the exchange in a trunking diagram. There would be more than 1 000 uniselectors, for example, in this exchange; the connexions to them all are adequately represented here by one uniselector symbol (4.3.4).

of the subscriber's dial is carried out by selectors of the two-motion type. It would, however, be uneconomic to provide every subscriber with his own complete set of such selectors capable of establishing all the calls he might ever want to make. This is why each subscriber is only connected to a first selector as and when he wishes to make a call. It is then only necessary to provide enough selectors to carry the normal maximum number of calls expected at one time on the exchange as a whole, in much the same way as cord circuits in a manual exchange.

In general, this is achieved by switches which self-drive to the appropriate outlet on receipt of the subscriber's calling signal. Such switches may function either as *hunters* or as *finders*. The subscribers' line uniselectors in Fig. 4.7 are being used as hunters. The difference between hunters and finders is illustrated in Fig. 4.8.

When used as a *hunter*, a switch has its wipers connected to the calling circuit. When self-driving, it searches for a free outlet. We call this hunting because, just as with a human hunter, there will be some occasions when it searches unsuccessfully. This is when all the selectors to which it has access are already in use by other subscribers.

When used as a *finder*, a switch has its wipers connected to the required (called) circuit and the calling circuits are connected to the bank contacts. When self-driving, it searches for the circuit on which the calling signal originated. Since the switch self-drives in this way only when there is such a calling signal, it is always possible for it to *find* the calling circuit.

When the *hunter* principle is used for concentrating the traffic from a large number of lines on to a smaller number of first selectors, one switch is required per line. Two-

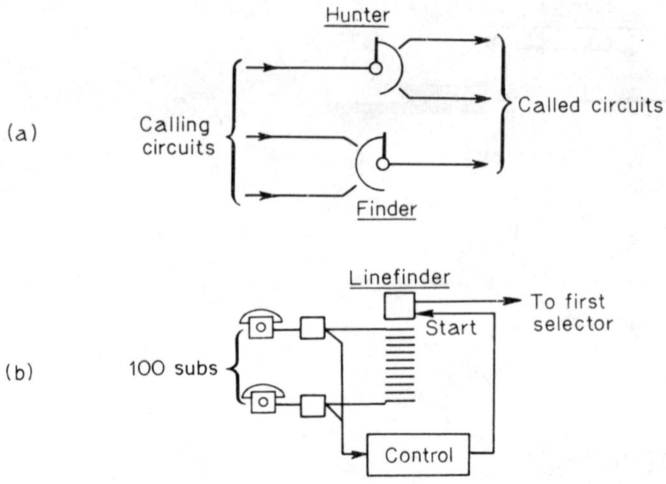

Fig. 4.8 Hunters and finders. The principle of finder and hunter operation is shown at (a). The use of a two-motion selector as a 100 point linefinder is shown at (b) (4.3.4).

motion selectors are too expensive for this purpose and in any case would provide more outlets than are necessary. For this reason, uniselectors are used as we have already seen. Two-motion selectors do, however, operate as hunters when they are rotated to seize a free selector in the next switching rank.

When the *finder* principle is used for concentrating traffic from the lines on to the smaller number of first (group) selectors, only one finder switch is required per first selector. When used for this purpose, the switches are called *linefinders* and to use them economically, each should have access to a relatively large number of lines. It is therefore both convenient and economic to use two-motion selectors as linefinders on at least the larger linefinder exchanges.

The choice between providing linefinders or subscribers' uniselectors (hunters) depends on the calling rate. If each subscriber makes a lot of calls (i.e. a high calling rate) uniselectors are generally used. Linefinders are more economical for low calling rate lines. The traffic from a group of low calling rate lines is often concentrated on to a smaller group of line uniselectors by means of 50-point uniselector linefinders.

On a manual switchboard, we found that no more than two operators could normally obtain access to one set of jacks and so a multiple was necessary. Automatic selectors find it even more difficult—only one switch can obtain access to one set of contacts. A bank contact multiple is thus essential at each switching stage.

Consider first the subscribers' uniselectors. Each one has access to a number of first selectors but it does not have exclusive use of them. They are shared with other uniselectors. The two wires from the line are connected to the wipers of the uniselector and two wires are connected from the bank contacts at each outlet to a first selector. Each first selector is also connected in the same way to the bank contacts of a number of uniselectors as shown in Fig. 4.9. In practice, at least one more wire (6.4.1) must be switched to the first selector and is multipled in the same way as the two that are shown.

Now suppose the exchange is required to serve only 100 numbers. The selectors connected to the outlets of the uniselectors can each have their outlets wired to the

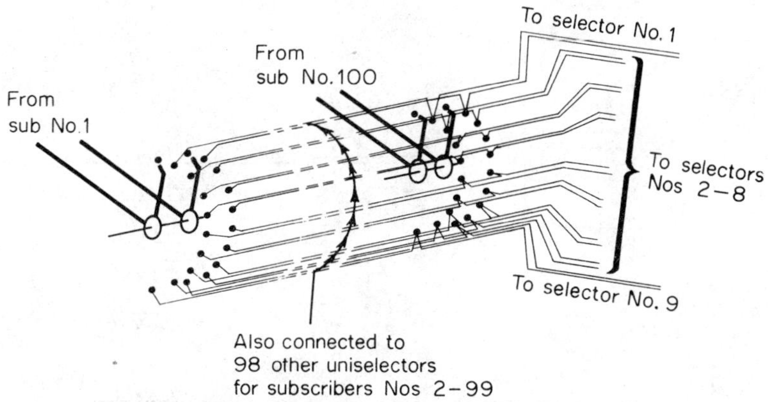

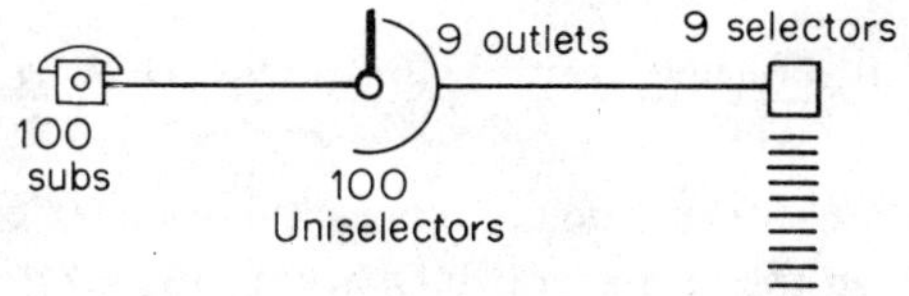

Fig. 4.9 Principle of uniselector multiple. In this example, any of the 100 subscribers can be connected by his uniselector to any one of the nine selectors. Control circuits are provided in practice to ensure that only one subscriber is connected to any given selector at one time, as described in Chapter 6 (4.3.4).

subscribers' lines at the rate of one outlet per line.

Since the selectors are giving direct access to the lines in this case, they are final selectors. They must all have access to all the lines in exactly the same way and so their contacts must be connected in multiple.

As it stands, this arrangement only provides capacity for 100 lines. To extend it to say 1 000 lines requires ten groups of final selectors and an extra digit on each subscriber's number. This means another rank of selectors to select the appropriate group of final selectors. These we therefore call *group selectors*. In particular, because they are controlled by the first dialled digit, as we have seen, they are first (group) selectors.

We do not provide each of these first group selectors with its own exclusive set of ten final selectors. This would be wasteful because it could never use them all at one time. Instead, it shares the final selectors in each group with other first selectors. This is why the group selectors are arranged to hunt by rotation of the wipers after they have been stepped vertically by the dial pulses. As in the case of the uniselectors, shared access means providing a multiple.

If more than 1 000 numbers are required, another rank of group selectors can be used as shown in Fig. 4.7.

A hunting switch connects a call via its later contacts only if the first ones it searches over are already engaged. This means that the selectors connected on the contacts farthest round the bank receive the fewest calls from it. It is therefore possible to use fewer switches by sharing these later choice selectors between a greater number of switches than the earlier choice selectors. This involves a method of inter-connecting the multiples of a number of groups of either uniselectors or group selectors which is called *grading*. We shall consider this in more detail in Chapter 7.

In each of the trunking diagrams given in Figs. 4.7 to 4.10, the symbols for

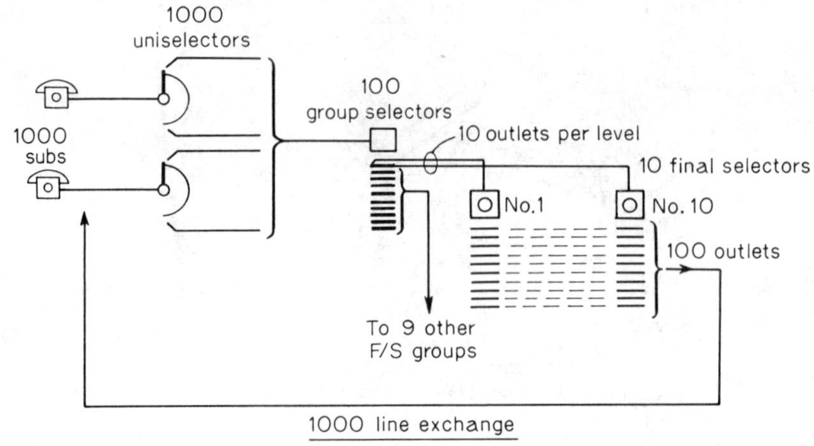

Fig. 4.10 Principle of group selector multiple (4.3.4).

uni-selectors and two-motion selectors do not represent just one selector each. They each represent all the selectors in the exchange which are connected in that particular way. Where it is important to know how many selectors are represented by a given symbol on a diagram, this information can be written beside it.

4.3.5 Common control. Variations on the step-by-step principle have been used with several different types of switch including motor-driven rotary switches and crossbar switches (**4.3.6**). Such mechanisms are not directly pulse-operated like Strowger switches so they need more complex control circuits, which cannot really be provided economically at every switch.

The control circuits are shared over more than one switch in most modern systems of this type, a method of operation known as *common-control*. It may be compared with a manual switchboard where the plugs and cords obviously cannot be directly controlled by the subscribers and each operator provides common control for all the cord circuits on a position. The control (or operator) need only handle one call at a time and is employed on each call for only a small part of the total duration.

The principal functions of the common control circuits are to detect requests for calls, receive and count the pulses from the subscriber, examine the network to decide which path is to be used, and finally to provide the appropriate instruction signals to the switches. The division of these functions between the control circuits varies from system to system, but the circuits which receive and count the pulses are usually known as *registers* and those which instruct the switches are known as either *markers* or *controls*.

To understand how the common control circuits are arranged we need to look more closely at the switches and how they are interconnected. The principles of crossbar switch operation and trunking are described in the following sections.

4.3.6 The crossbar switch is a mechanical device based on the same principle as the co-ordinate switchboard (**4.2.2**). Each crosspoint consists of a set of relay type contacts but, instead of a separate coil and armature at each crosspoint as in a relay matrix (Fig. 4.1c), there is only one coil and armature for each vertical and horizontal row of crosspoints. To operate a particular contact, the two coils on the relevant horizontal and vertical need both to be used.

Fig. 4.11 shows a general view of a complete switch and a much simplified diagram of a single crosspoint on a crossbar switch. The *flexible selecting finger* should be

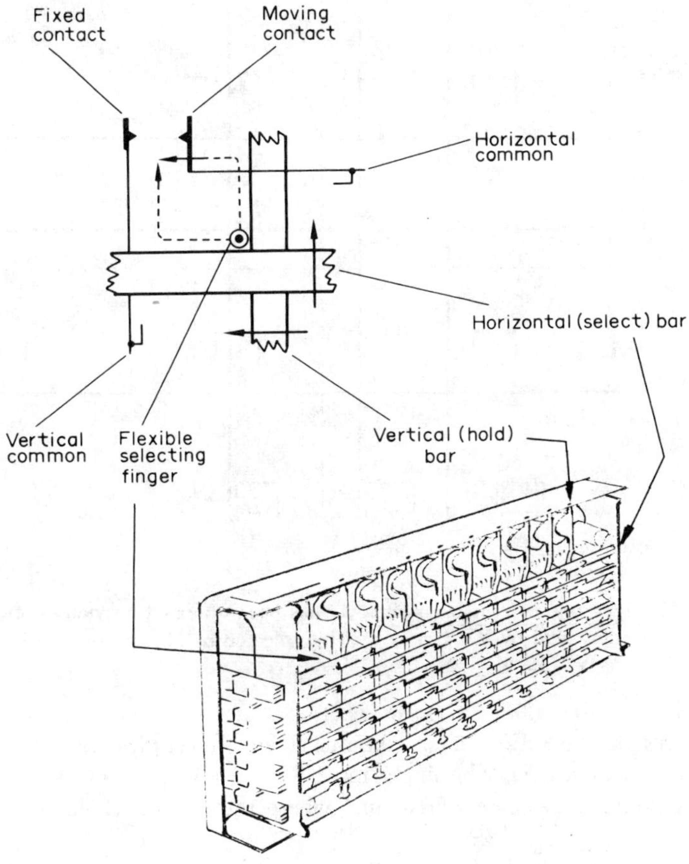

Fig. 4.11 Principle of crossbar switch. This shows a single crosspoint. The finger is perpendicular to the paper. Movement of the bars is shown by the arrows (4.3.6).

thought of as a stiff piece of wire sticking up out of the paper. The horizontal (*select*) and vertical (*hold*) bars from which the switch gets its name move in the direction shown by the arrows when their associated coils are energized.

In a practical switch, the finger is usually physically attached to the horizontal bar, which operates by rotating a few degrees about its axis to provide the vertical displacement of the finger. This arrangement permits the use of two coils for each horizontal bar so that the fingers can swing up or down to control two rows of crosspoints.

The dotted arrows in Fig. 4.11 indicate the movement of the finger when the bars are operated. It can be seen that the contact will be operated by the finger only when the horizontal bar 'selects' before the vertical bar operates. Once operated, the contact remains held by the vertical bar which traps the finger and holds it even when the horizontal bar has been restored.

The operation of the switch is illustrated in Fig. 4.12 which shows nine crosspoints controlled by three vertical bars and three horizontal bars. Practical switches have typically ten or more verticals and five or more horizontal bars (ten horizontal rows of crosspoints). They also have a complete springset at each crosspoint providing two or more 'make' contacts. The sequence of operations is as follows:

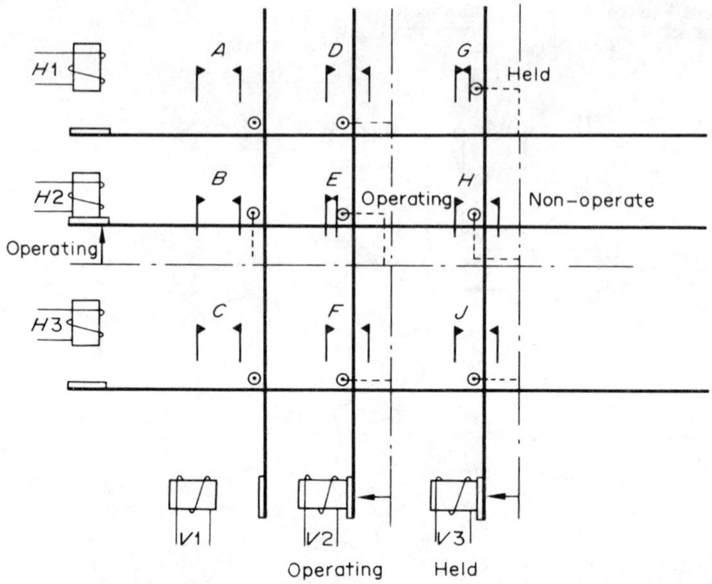

Fig. 4.12 Operation of crossbar switch. G has been operated previously and is now held by V3. E is now being selected by H2 and then operated by V2 (4.3.6).

Operate required horizontal (select) bar (H2).
This prepares the operation of crosspoints where the vertical (hold) bar has not already been operated (*B*). The finger moves into a position from which it cannot operate the contacts at those crosspoints where the vertical (hold) bar is already operated.
Operate required vertical (hold) bar (V2).
This operates the crosspoint where the horizontal (select) bar has already been operated (*E*). The fingers do not engage the contacts at those crosspoints where the horizontal select bar has not previously been operated. (*D*,*F*).
Release the horizontal (select) bar (H1).
This leaves the finger trapped by the vertical (hold) bar (*V*3) and the contact spring tension (*G*). Any horizontal bar may now be operated to select another crosspoint and will have no effect on the crosspoint already operated and held. The finger is sufficiently flexible to permit movement of its associated horizontal bar in either direction without itself being dislodged from the operated crosspoint.
Release the vertical (hold) bar (V1).
The vertical (hold) bar remains operated throughout the call. When the crosspoint is to be released it is only necessary to release this bar and the spring finger restores (*A*,*C*).

The most commonly used types of crossbar switch operate in this way. There are however other types. Some are arranged so that the crosspoint is mechanically latched when operated so that it remains held without the need for current in the coil. Release is effected by operating the select bar for a short period.

4.3.7 Use of crossbar switches. We saw in **4.2.2** that the simple co-ordinate switch is uneconomic for large networks. This limitation can be overcome in automatic exchanges by using more than one stage of switching.

To illustrate this, Fig. 4.13a shows how a single switching stage can be used to provide for nine paths between nine inlets and nine outlets using 81 crosspoints. In Fig. 4.13b

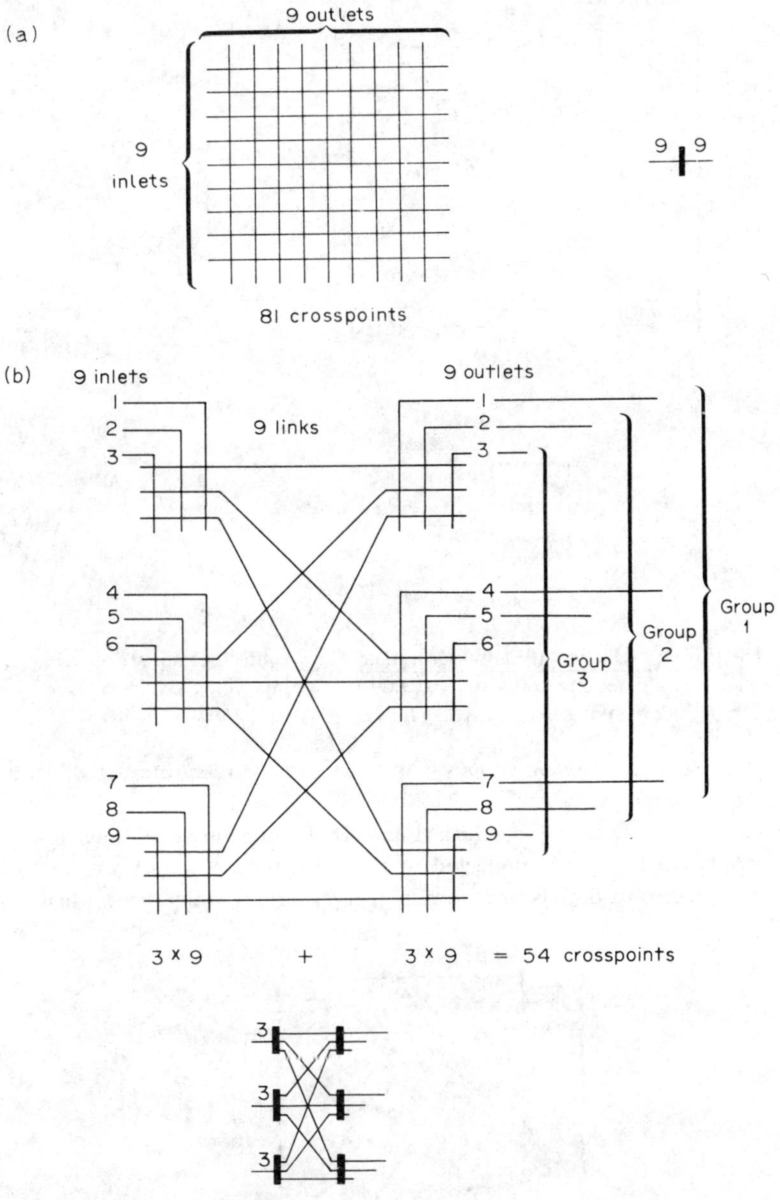

Fig. 4.13 Two-stage link trunking. Compare this with the single stage trunking of Fig. 4.1 (4.3.7).

however nine paths are also provided for the same purpose but using only 54 crosspoints by connecting six separate switches as two switching stages with transposed links. The advantage of using this method is even greater for larger networks as shown in Fig. 4.14.

These simple examples provide only for the *distribution* of calls from the inlets to an equal number of outlets via the same number of links. In practical exchanges we need provide only as many paths through the switching network as the number of calls expected to be in progress at one time, e.g. three links for 12 subscribers on the small switchboard in Fig. 4.1.

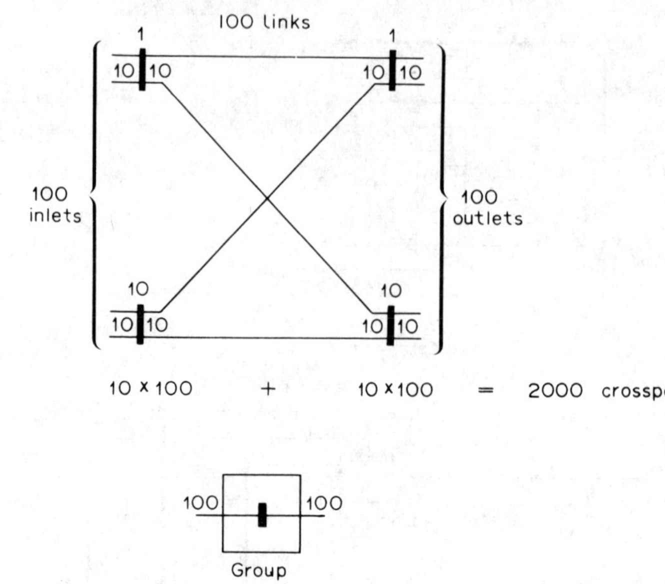

Fig. 4.14 Group selection (distribution) stage. The equivalent square matrix would require $100 \times 100 = 10\,000$ crosspoints. This uses two stage link trunking like Fig. 4.13 (4.3.7).

In large exchanges, the *concentration* of lines on to a smaller number of paths or *trunks* can be achieved as shown in Fig. 4.15.

A complete exchange switching network (Fig. 4.16a) then consists of 'line concentration' stages similar to Fig. 4.15 connected to 'call distribution' stages similar to Fig. 4.14. These give access to outlets arranged in groups serving different destinations or

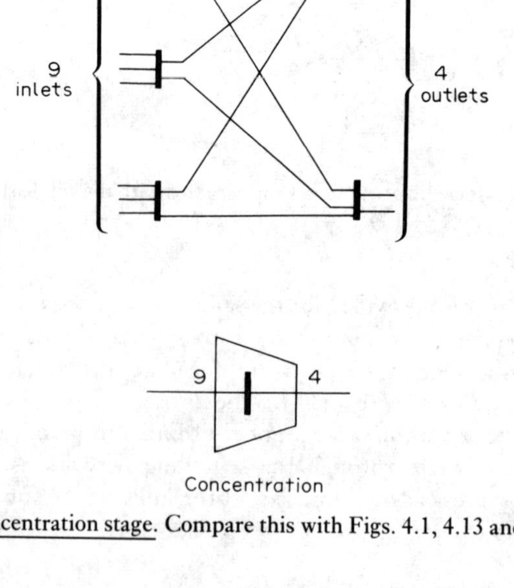

Fig. 4.15 Concentration stage. Compare this with Figs. 4.1, 4.13 and 4.14 (4.3.7).

routes. For those destinations which consist of subscribers' lines, a stage of *expansion* is provided which is arranged like a concentration stage in reverse.

The actual names given to the stages are not the same on all the systems. There may be more than one stage of each type and two functions may be combined on the same stage. In particular it is usual to provide the expansion at least in part by returning through the concentration stage. When the whole of the expansion is provided in this way the result is called a *folded network* (Fig. 4.16b).

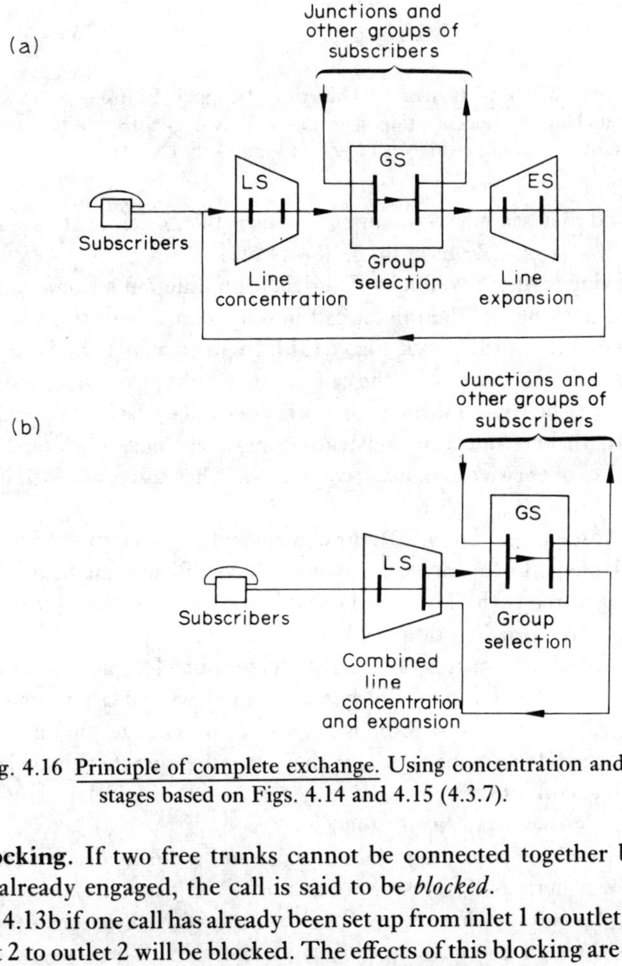

Fig. 4.16 Principle of complete exchange. Using concentration and distribution stages based on Figs. 4.14 and 4.15 (4.3.7).

4.3.8 Blocking. If two free trunks cannot be connected together because all suitable links are already engaged, the call is said to be *blocked*.

In Fig. 4.13b if one call has already been set up from inlet 1 to outlet 1, an attempted call from inlet 2 to outlet 2 will be blocked. The effects of this blocking are reduced in this case by the allocation of the groups of outlets to the three routes as shown. Instead of connecting the call to a destination in group 1 via outlet 1 (which is blocked), we can route it via one of the other outlets serving the same group (i.e. outlet 5 or 6). Notice that the outlet 2 is still available to calls entering at inlets 4, 5, 6, etc.

We have seen that the use of concentration is necessary to achieve an economic system and this will lead to some loss of calls due to all suitable paths or switches being engaged. Switch quantity calculations are based on providing a known acceptable probability of calls being lost due to such congestion (7.3.1). The effect of blocking is no different and can be taken into account in the same calculation.

In some systems, certain outlets may be permanently wired back to the inlet side of the network to permit a blocked call to seek a new path from a new inlet, e.g. if the outlet 9 of Fig. 4.13b were connected in this way to inlet 9, then the blocked call from inlet 2 can be

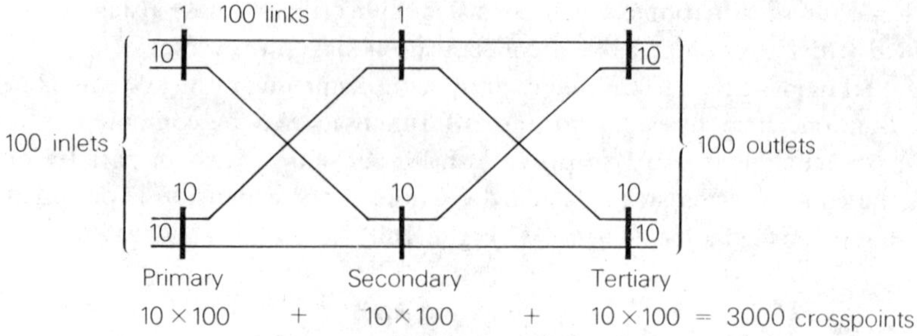

Fig. 4.17 Three-stage link trunking. This principle may be used to reduce blocking in concentration stages as well as in the distribution stage shown. Compare this with Figs 4.14 and 4.15. (4.3.8).

routed to outlet 9, and then via inlet 9 to outlet 2. This is the principle of *interaid* used in TXK 3 (**4.5.3**) and *planar transfer* used in TXE 4 (**4.6.2**).

The effect of blocking is further reduced if outlets from different switches are grouped for connexion to the next stage. The inlet 2 call in our example can still reach outlet 5 which is in the same group as outlet 2 (see Fig. 4.13b). In practice this implies a third stage of switches as in Fig. 4.17. In this case, if the call cannot reach the required outlet via one secondary switch, it may be routed through any other secondary having a free link to the required tertiary. In theory, one can construct completely non-blocking three-stage arrays, but this requires over twice as many secondary switches and is not used in practical commercial systems.

In electronic systems, e.g. TXE 4 (**4.6.2**), the control is able to examine the network and select from the available paths for each call the one which will give the next call the best chance of also finding a free path. This is achieved by *call-packing*, i.e. routing each call through the most heavily loaded section available to it.

4.3.9 Common control of matrix switches. A simplified typical arrangement is shown in Fig. 4.18. Each switching stage has associated with it a circuit capable of operating the required crosspoints in response to signals describing the inlet and outlet to be connected together. In crossbar systems these circuits operate the horizontal and vertical magnets in the correct sequence and may consist either of relays or electronics. In the present example they have been called 'markers'.

Registers are provided for recording dialled digits and a means of connecting them to the callers is shown here as a 'register connector'. A register is taken into use on each call only while the call is being set up. They are therefore fewer than the number of trunks provided through the exchange to carry the calls. The register connector must therefore provide some concentration. The point of connexion of the registers will also be at some point in the network at which some concentration has already been applied. The register connector function may be provided by matrix switches separate from the network, by separate relays, or by connecting the registers to the main network in the same way as junctions or lines.

To detect calling conditions and to indicate busy conditions, each subscriber's line needs a line circuit. This usually consists of two relays. During the call, a transmission bridge is required and is usually provided at a convenient point in the trunks between the switching stages. The markers need to be able to receive information relating to calling and called lines and therefore some connexion between them and the registers and line circuits is needed. This may include a central control as shown dotted in Fig. 4.18 especially in electronically controlled systems.

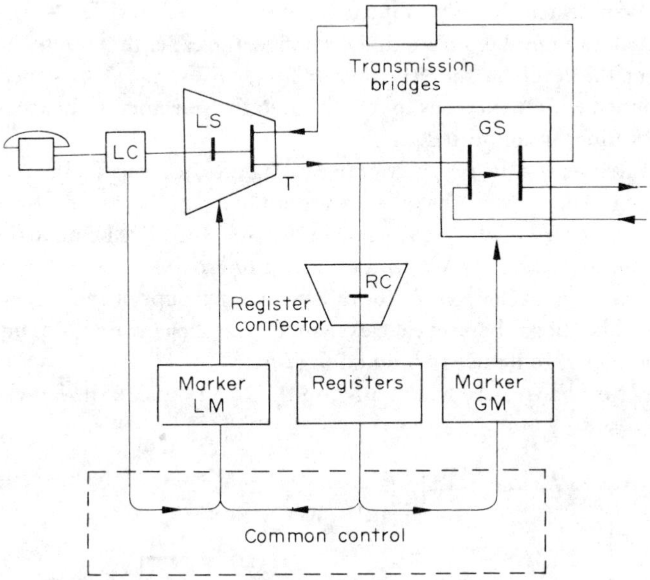

Fig. 4.18 Common control. The network is the same as Fig. 4.16. The control is typical of most crossbar and electronic exchanges but different names may be used for the various parts of the system (4.3.9).

The operation of Fig. 4.18 is as follows:

Calling loop is detected at *LC*.

LC signals to *LM* which line is calling.

LM sets *LS* and *RC* to connect the caller to a free register via a free trunk (*T*).

Dial Tone is returned to the caller from the register via *RC*, *T*, *LS* and *LC*.

Dial Pulses are received by the register from the caller via *LC*, *LS*, *T* and *RC*.

Called line identity is derived from the dialled digits and signalled to the markers.

GM sets *GS* to connect *T* to a free transmission bridge in the group serving the *LS* giving access to the wanted line. *LM* sets *LS* to connect this transmission bridge to the called line (if free).

Ringing and metering are controlled from the transmission bridge which also usually provides the holding conditions for the switches (e.g. crossbar vertical magnets).

Clearing signals are detected at the transmission bridge which causes the switches to release.

Now test yourself on the questions given on page 114.

PART 3 TYPICAL PRACTICAL SYSTEMS

4.4 Strowger

4.4.1 Non-director. In this version of the Strowger system, the subscribers' dial pulses control the selectors directly (4.3.4), a method known as *direct control*. No special equipment is provided to direct the call routing (4.4.2).

The basic trunking of a non-director exchange for local subscriber-to-subscriber or 'own-exchange' calls is shown in Fig. 4.7. Calls to other exchanges are provided for by connecting outgoing junctions at group selector levels. Incoming junctions from other exchanges are connected directly to incoming group selectors (Fig. 4.19).

To make a call to a number on another exchange, the caller first has to dial the 'routing' digits to select the level on the selectors in his own exchange to which the required junction is connected. He then has to dial through the selectors in the exchange at which the called subscriber is connected.

For a small area served by, say, four or five exchanges, it is relatively easy to allocate the junction levels in such a way that the same number may be dialled for any particular subscriber no matter from which exchange in the area the call is being dialled. A group of exchanges connected in this way is said to have a *linked numbering scheme.*

For larger groups of exchanges, or for calls between groups of exchanges, this becomes impracticable. The subscribers on each exchange must then have a different list of dialling codes (routing digits) to be used on junction calls.

Non-director exchanges are widely used in the U.K. outside the larger city centres.

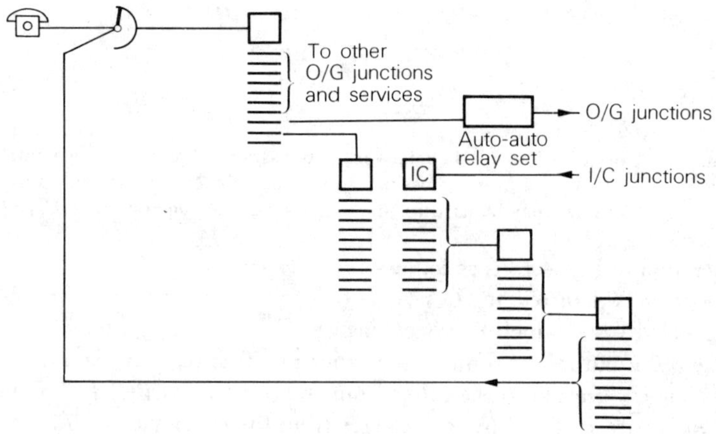

Fig. 4.19 Non-director trunking. Showing the method of connecting junctions. Compare this with Figs. 4.7 and 7.8 (4.4.1).

4.4.2 Director. A large city often has hundreds of exchanges, some less than a kilometre apart or even in the same building and with overlapping service areas. Under these conditions it would be unacceptable for callers to have to consult a different list of codes according to which exchange the telephone they happen to be using is connected. The need to keep the lists up to date would also severely restrict the periodic re-allocation of junction plant and selectors to meet changing patterns of demand.

To allow the same list of three-digit codes to be used anywhere in the area, director equipment is provided (Fig. 4.20) which translates the code dialled by the caller into the actual routing digits required to select the appropriate code selector levels. This is known as *indirect control.*

The code selector levels can be allocated to obtain the most economical use of selectors at each exchange independently of the dialling code allocated to each route. The number of second code (*C 2nd*) selectors to be provided can be minimized for example by taking off the heaviest traffic routes from the *C 1st* so that this traffic does not even reach the *C 2nd.* Generally, the junction routes to nearby exchanges will be more heavily loaded than those to the more distant exchanges in the area.

Director exchanges are in use in most of the larger cities in the U.K.

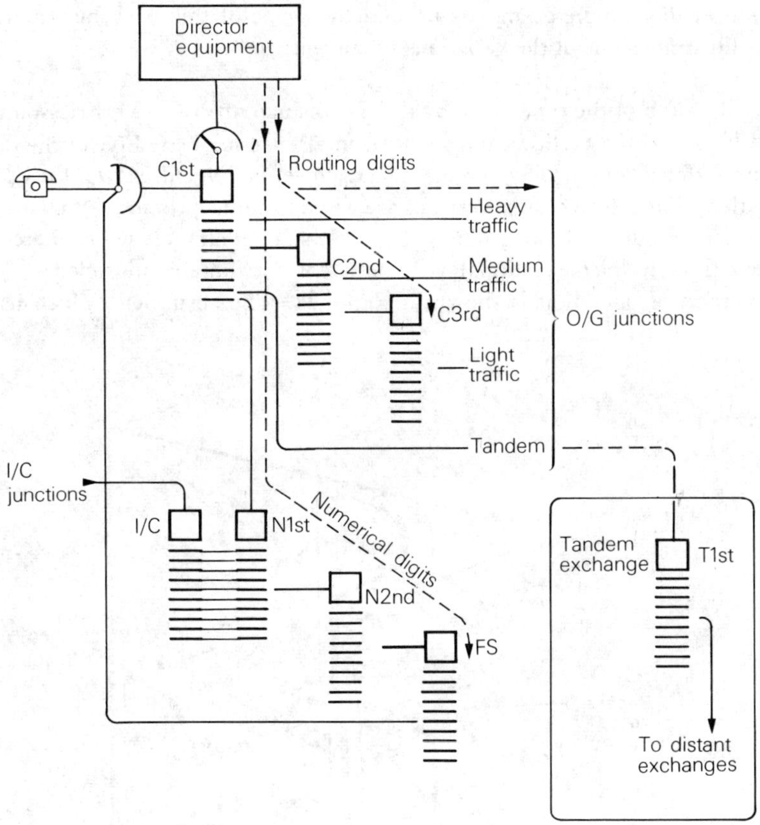

Fig. 4.20 Director exchange trunking. The director equipment may be either electromechanical or electronic. (4.4.2).

Referring to Fig. 4.20, the operation is as follows:

Caller lifts handset. The line uniselector seizes a free *C 1st* which seizes the director equipment from which dial tone is returned to the caller.

The code digits dialled are received by the director equipment which translates them and transmits up to six routing digits.

The numerical digits are next received by the director and stored until it has finished sending the routing digits. It then re-transmits the numerical digits without translation.

The routing digits route the call through the code selectors in the originating exchange (and through tandem selectors at an intermediate tandem exchange if required). All these selectors function as group selectors (4.3.4).

The numerical digits route the call at the destination exchange through the numerical selectors arranged as a four-digit exchange as in Fig. 4.7.

The director equipment releases when the last numerical digit has been re-transmitted and is available for use by other selectors.

4.5 Crossbar

4.5.1 Crossbar switch mechanisms. Most of the world's telephone switching equipment manufacturers make their own version of the crossbar switch mechanism. A description of the complete range is beyond the scope of this book but the following examples illustrate some of the variations in current use.

A typical switch of the type made by the Swedish company L. M. Ericsson is shown in Fig. 4.11. It has ten verticals and six horizontal selecting bars. Five of the horizontal bars control 10 sets of contacts per vertical. Each set is divided in two, thus providing 2×10 outlets. The inlets to these verticals are wired to contacts of the sixth bar which are used to select one out of the two groups of ten outlets. The complete switch therefore gives 200 outlets from 10 inlets (verticals) or 20 outlets if horizontally multipled.

The method of operation is shown in Fig. 4.21. The springsets, which have been

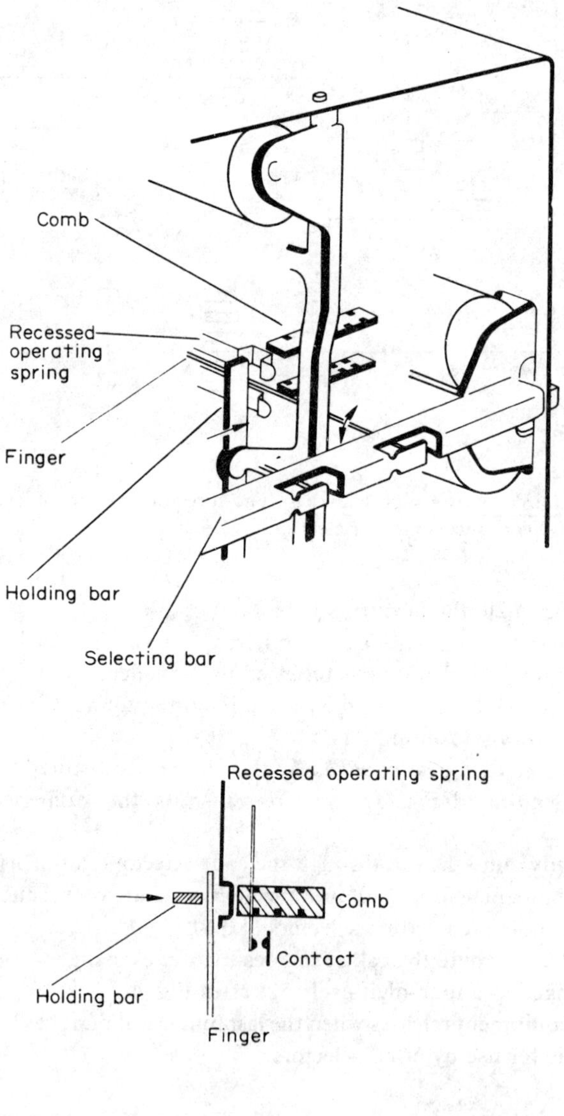

Fig. 4.21 Ericsson crossbar switch mechanism. For clarity the springsets have been omitted. A view of the complete switch is shown in Fig. 4.11 (4.5.1).

omitted from the diagram for clarity, are of the comb operated type (2.2.2). Associated with each springset is a recessed operating spring. When the selecting bar is unoperated, each finger rests in the space between a pair of springsets. The holding bar operating under this condition enters the recess and so does not move the comb to operate the springset. Operating the selecting bar causes the finger to cover the recess so that operation of the holding bar presses upon the finger to move the comb and operate the springset.

In another type of mechanism the recess for the selecting finger is provided in the form of a shoulder in the comb. A typical switch of this type is used on the ITT Pentaconta system, a U.K. version of which is made by STC for the British Post Office (system TXK3). This is a larger switch than the Ericsson one and has up to 22 verticals (known as *selectors*) and 14 horizontal bars.

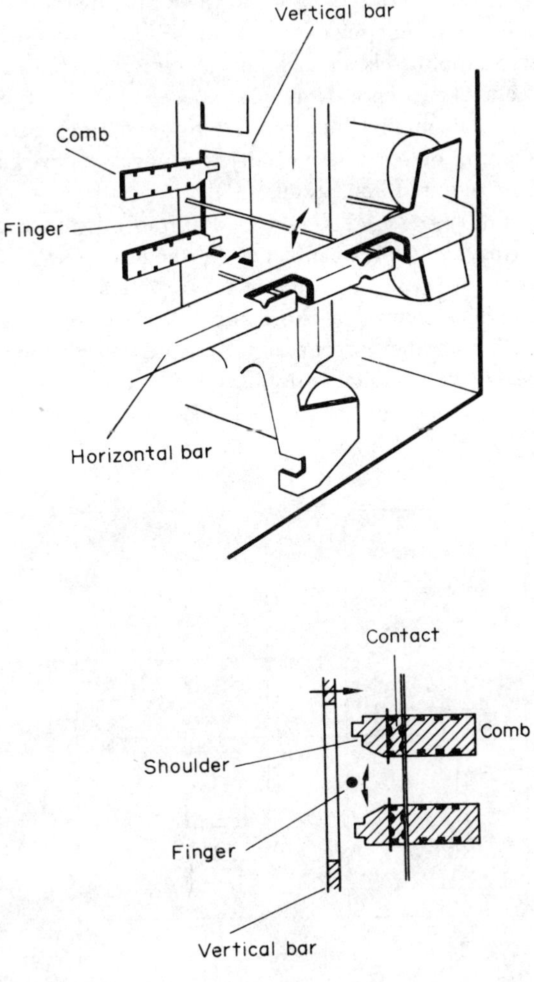

Fig. 4.22 Pentaconta crossbar switch mechanism. Compare this with Fig. 4.21 (4.5.1).

The method of operation is shown in Fig. 4.22. When the horizontal bar is unoperated, each finger rests in the space between a pair of combs. Openings in the vertical bar prevent it from engaging the combs if it is operated under this condition.

Operating the horizontal bar causes the finger to engage the shoulder of the comb so that operation of the vertical bar presses upon the finger to operate the springset.

The Plessey 5005 system (B.P.O. system TXK 1) uses a switch having 10 verticals (known as *bridges*) and 6 horizontal bars (known as *select* bars). The mechanism is generally similar to that shown in Fig. 4.21. In a typical application of the switch, a more complex switching arrangement of the type described above for the Ericsson switch is used to provide 28 outlets from 10 inlets.

4.5.2 British Post Office system TXK 1 (*Plessey* **5005**). In the U.K., TXK 1 is principally used in non-director areas (**4.4.1**).

The switching network using crossbar switches as described in **4.5.1** comprises a number of *line distributors* each of which provides the basic concentration and expansion (**4.3.7**) for 500 lines, and one or more *routers* which provide the traffic distribution (**4.3.7**) between the distributors. Junctions are also connected to the routers.

Fig. 4.23 shows a simplified block diagram of a typical TXK 1 exchange. The functions of the principal control elements are as follows:

The marker (MKR) contains 'hundreds', 'tens', and 'units' relays operated by signals from the router control on terminating calls to connect the marking lead by means of a 'contact tree' to the required line circuit (*LC*).

The transmission relay groups (TRG) provide a transmission bridge (**1.3.6**) and control the ringing, metering, and supervision on established calls (**3.5.1**). They also contain two connecting relays to give access to either of two registers.

The registers (REG) count and store the caller's dialled digits on relays for the transmission of called number information to the marker via the router control.

The router control (RC), which is duplicated for security, organises the sequence of operations for setting up one call at a time.

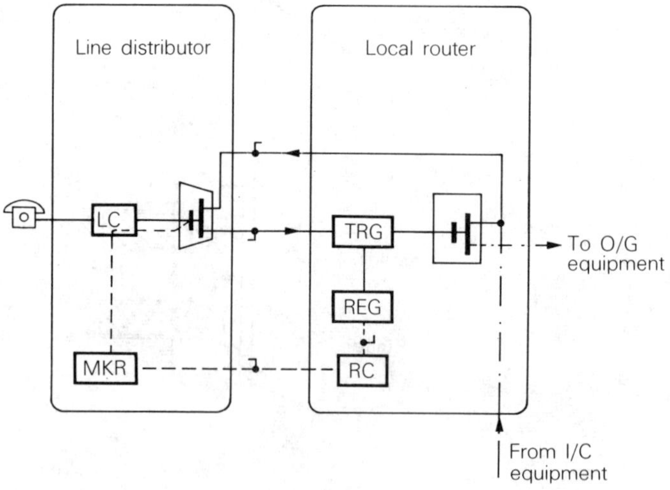

Fig. 4.23 Block diagram of TXK 1 exchange. Compare this with Figs. 4.16 and 4.24. (4.5.2).

The operation of the system on a local own-exchange call is as follows:

Calling loop causes the line circuit *LC* to mark the distributor switch outlet. Control relays associated with the *TRG* also mark an inlet connected to one free *TRG* having access to at least one free register. The combination of these two marking signals operates the crosspoints to connect the line to the *TRG* and its associated register.

Dial tone from the register is connected to the caller after it has received a class-of-service signal from the line circuit.

Dialled digits are counted and stored by the register which analyses the digits to determine whether the call is to be routed locally or to an outgoing junction.

For a local call, the full directory number is stored by the register and is then passed with the class-of-service information to the router control which transfers the called number to the marker of that distributor which contains the called line.

Busy or free conditions are detected at the router control over the marking lead through the marker and passed through the register to the *TRG*.

Connexion to the called line, if free, is established by the marker connecting a marking potential at the called outlet and the *TRG* connecting a marking potential at its inlet.

Ringing, tones, and metering as appropriate (3.5.1) are connected at the *TRG*. The register, router control, and marker are released for use on other calls.

Release at the end of the call is effected by the *TRG* removing the P-wire earth (6.4.1) to release the cut-off relay at the line circuit and the crossbar switch bridge magnets.

4.5.3 **British Post Office system TXK 3 (*ITT Pentaconta*).** In the U.K., TXK 3 is principally used in director areas (4.4.2).

The switching network using crossbar switches as described in 4.5.1 comprises a number of *line switching units* (*lsu*), each of which provides the basic concentration and expansion (4.3.7) for either 520 or 1040 lines, and one or more *group selection units* (*gsu*) which provide the traffic distribution between the *lsu*. Junctions are also connected to the *gsu*.

Fig. 4.24 shows a simplified block diagram of a typical TXK 3 exchange. The functions of the principal control elements are as follows:

The common relays (CR) are used only on originating calls and they control the choice of *lsu* inlet for connexion to the register junctors (*RJ*).

The markers and associated marking relays (MKR) extend marking signals to the crossbar switches during the selection process. The markers are duplicated for security.

The register junctors (RJ) are two-relay circuits for connecting a register to the caller during dialling and providing a through connexion to a *gsu* inlet when selection is complete.

The register finders (RF) are crossbar switches providing concentration from the junctors to the registers.

The registers (REG) count and store the caller's dialled digits and control the sequence of all operations during the setting up of a call.

Couplers are relay circuits used to provide one-at-a-time connexion between registers and translators and information paths.

The translator (XLTR) provides routing and charging information to the register.

The local feed junctors (LFJ) provide a transmission bridge (1.3.6) and control the ringing, metering, and supervision on an established call (3.5.1).

The information paths (IP) each consists of a pair of 20-wire channels over which the data for controlling a call are signalled rapidly between the registers and the markers.

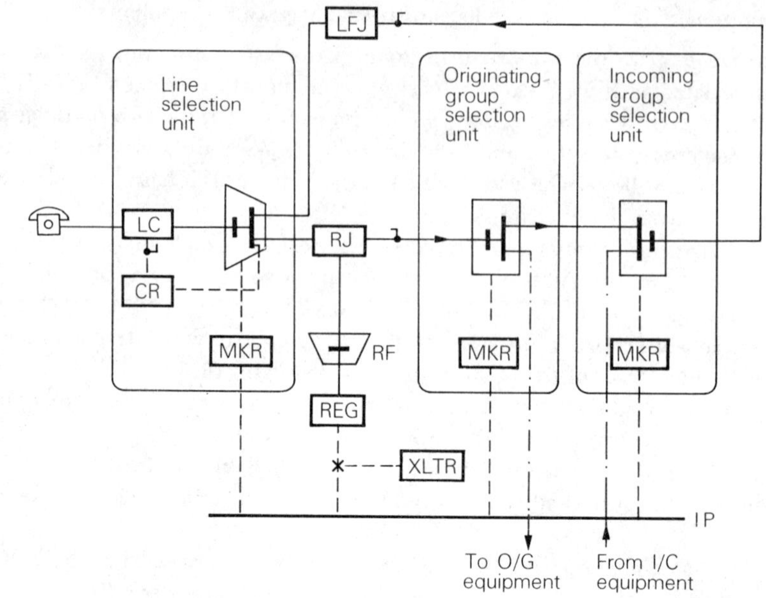

Fig. 4.24 Block diagram of TXK 3 exchange. Compare this with Figs. 4.16 and 4.23. (4.5.3).

When required, a marker seizes an *IP* and signals over the speech path to the register, telling it which path it has seized. The *IP* is released as soon as the necessary data have been transferred.

The operation of the system on a local own-exchange call is as follows:

Calling loop causes the line circuit *LC* to extend a calling signal to the common relays which pass a marking forward to all accessible register junctors having access to a free register. Control relays select one free register and one of the marked junctors and operate the register selector crosspoints. This causes the marker to identify and mark the *lsu* outlet of the calling line and signal its class-of-service over the *IP* to the selected register. The register marks the *lsu* inlet via *RF* to operate the *lsu* crosspoints to connect the calling line to the register. The common relays and marker release.

Dial tone is returned to the caller from the register via the *lsu*.

Dialled digits are counted and stored by the register which transmits the initial digits via the coupler to the translator. The number of digits sent to the translator depends on the area numbering scheme but is typically three, e.g. for a director area (**4.4.2**). The translator sends back to the coupler a two-digit identification of the first stage routing required. A signal from the register via the register junctor causes the *gsu* marker to use the *IP* to obtain the routing data from the coupler, enabling it to select a free outlet in the required group.

The marker seizes the *IP* again to advise the register that a link to the incoming *gsu* has been seized. The marker then releases for use on other calls.

On a local call, the selection process is repeated for the incoming *gsu* to connect a free local feed junctor.

Connexion to the called subscriber, if free, through the *lsu* is similar except that no translation is required and the register passes the last three digits of the dialled number over the *IP* to the *lsu* marker to select the called line.

Busy or free conditions are detected by the appropriate marker at each stage of the selection process and signalled back over the *IP* to the register.

Ringing, tones, metering, and release are controlled by the *LFJ* in a similar way to the *TRG* in the TXK 1 system (4.5.2).

4.6 Reed electronic

4.6.1 British Post Office system TXE 2 (*Plessey Pentex*). In the U.K., TXE 2 is used for the smaller local exchanges (typically less than 3000 lines) where it is gradually supplanting the Strowger and Crossbar systems.

The basic switch unit is a 5 × 5 matrix of four-contact reed relays (2.7.2). These are grouped to form switches of various sizes, e.g. 25 × 25, 5 × 15, etc. The complete switching network consists of four stages of such switches *A, B, C,* and *D*. The *A, B,* and *C* switches provide concentration (4.3.7) on originating calls and expansion on terminating calls. The *D* switches provide traffic distribution on terminating calls only, including incoming junction calls. No separate distribution stage is provided for outgoing junctions, traffic studies having shown that it is more economical to provide additional *C* switch outlets for this purpose.

Fig. 4.25 shows a simplified block diagram of a typical TXE 2 exchange. The functions of the principal control elements are as follows:

The call control is duplicated for security and organises the sequence of events for setting up calls through the network. Contained within the call control are markers (M) for marking lines or supervisory relay sets to which a connexion is required.

Switch selectors and access relays (S) select and mark free paths through the switch network under instruction from the call control.

Registers (REG) store the called and calling numbers and co-operate with the call control during the setting up of a call.

The supervisories (SUPY) provide a transmission bridge (1.3.6) and control the ringing, metering, and supervision on established calls (3.5.1). They also provide access to registers via the register switch. Separate groups of supervisories are provided for outgoing junctions, incoming junctions, and own-exchange calls.

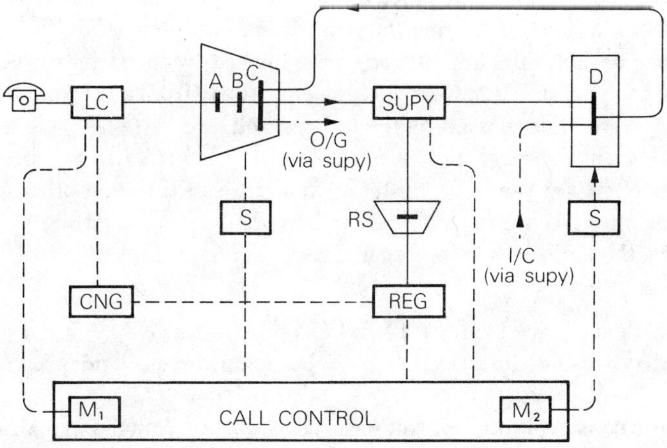

Fig. 4.25 Block diagram of TXE 2 exchange. Compare this with Figs. 4.16 and 4.26. (4.6.1).

The register switch (RS) is a reed relay matrix providing concentration from the supervisories to the registers.

The calling number generator (CNG) generates the calling subscriber's number and passes it to a free register for storage.

The operation of the system on a local own-exchange call is as follows:

The calling loop causes the line circuit *LC* to send a pulse to the *CNG* which electronically generates the directory number (*DN*) of the calling line. A register selector circuit associated with the *CNG* allocates a free register and the calling line *DN* is stored in the register.

The register applies to the call control for selection of a free supervisory. For exchanges at which most calls are junction calls, an outgoing supervisory on the most used route would be the first choice for this initial allocation. If most of the calls are locally terminated, the own-exchange supervisories would be the first choice. If no first choice supervisories are available, a free supervisory is selected in another group.

The call control marks the *C* switch at the chosen supervisory and uses the calling number information from the register to mark the calling line. The switch selectors select and mark a path through the *C, B,* and *A* switches in that order from the supervisory to the calling line. The register switch is marked from the supervisory and from the register to operate the appropriate crosspoint.

A class-of-service field associated with the line marker (M_I) generates a class-of-service signal which is passed to the register.

Dial tone is returned to the caller by the register when the class-of-service signal is received.

Dialled digits are received and stored by the register which analyses the number dialled to determine the routing required. If an outgoing junction supervisory has been connected, some or all of the initial digits may be repeated over the junction as well as being analysed locally.

On a local call, in cases where an outgoing supervisory has been allocated initially, the register releases the connexion as soon as it identifies that a local call is required (usually after the first digit). From its stored calling number information it resets the connexion, this time using an own-exchange supervisory, an operation which takes only 50 ms and is easily completed within the interdigital pause (3.3.4).

Busy or free conditions at the called line are detected by the call control via its marker which marks the called line using the dialled number information from the register.

Connexion to the called line, if free, is established by the switch selectors setting a path through the *D, C, B* and *A* switches respectively from a marking potential at the supervisory to the line marked by the call control.

Ringing, tones, metering, and release as appropriate (3.5.1) are provided at the supervisory. The register and register switch release for use on other calls. The call control was only taken into use for short periods during the setting up of the call as required by the register, serving other registers in between.

4.6.2 British Post Office system TXE 4 (*STC*). In the U.K., TXE 4 is used for the larger (typically over 2000 lines) exchanges in both non-director and director areas (**4.4.1, 4.4.2**).

The switches consist of four-contact reed relays (2.7.2) arranged in matrices of various sizes according to their place in the network, e.g. 10×8, 18×12, etc.

A single *universal* switching network of four stages (*A, B, C,* and *D*) is employed with no separate switching for such purposes as register connexion. The network is completely *folded* (4.3.7), all external connexions being made at the *A* stage and all paths through it using all stages, i.e. *A–B–C–D–C–B–A*.

The traffic concentration and expansion (4.3.7) is provided by the *A*, *B*, and *C* switches with distribution at the *D* stage. The dimensions of the *A* switch matrices can be chosen to suit the traffic requirements of the various terminations such as lines, junctions, and registers.

The network consists of a number of *switching units* containing the same *even number* of standard *sub-units*. The number of units can be increased as required to provide additional lines or traffic capacity.

All like-numbered sub-units on the exchange are grouped to form a *plane*. The number of planes, usually six or eight, remains unchanged throughout the life of the exchange. Each plane has its own marking control so that the effect of individual failures is limited. All lines and junctions have access to all planes at their *A* switch. This method of dividing a network up into separately controlled planes is called *sectionalization*.

Fig. 4.26 shows a simplified block diagram of a typical TXE 4 exchange. The functions of the principal control elements are as follows:

The cyclic stores (CS) hold semi-permanent data, e.g. directory number, class-of-service, etc. for all terminations, and they also scan on a regular basis to determine the

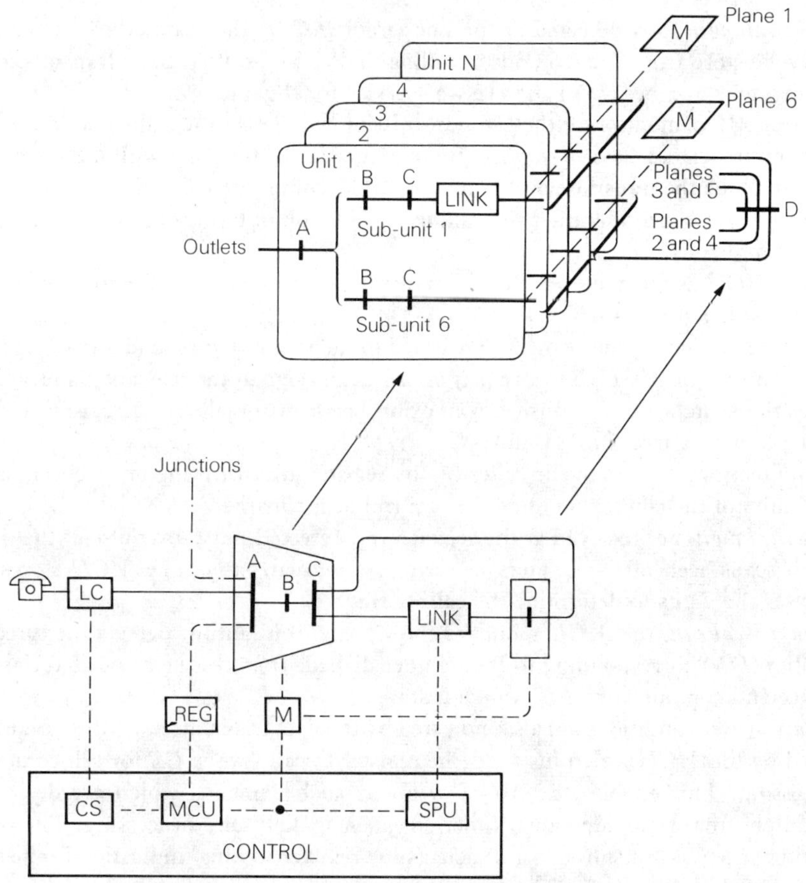

Fig. 4.26 Block diagram of TXE 4 exchange. Compare this with Figs. 4.16 and 4.25. (4.6.2).

current state, e.g. busy or free, of each. The data relating to the terminations it serves is continuously transmitted by each *CS* over a 'highway' to all main control units (*MCU*). Each *CS* rack serves up to 480 lines in addition to junctions and other circuits.

The main control units (MCU) are stored-program units which control the setting up of all calls. A minimum of three per exchange is provided, the number increasing according to the traffic capacity required up to a maximum of 20.

The registers (REG) receive and transmit dialled information as required under the control of an *MCU*, each of which provides the storage and sequencing of operations for up to 36 registers.

The supervisory processing units (SPU) are common-control units, each serving the bridge links and junction circuits on a single (odd-numbered) plane, for which they provide storage and control the sequencing of operations.

The link circuits provide a transmission bridge (**1.3.6**) for local calls and apply the ringing, metering, and supervision on established calls under the control of *SPU*. Separate *through-links* without bridges and not controlled by *SPU* are used for junction calls and register connexions.

The markers and interrogators (M) identify free paths through the network on request from the *MCU* and apply marking potentials to operate the crosspoints.

The operation of the system on a local call is as follows:

The calling loop is detected at the line circuit *LC* by the scanner associated with the cyclic store *CS*. The *CS* adds a 'calling' indication to its regular transmission of the directory number (*DN*) and class-of-service for the line.

A free *MCU* monitoring the *CS* transmission will detect the calling condition and store the associated class-of-service. By reference to the time slot within the scanning cycle at which the transmission occurs, the *MCU* computes and stores the equipment number (*EN*) which identifies the calling termination in terms of switching unit and outlet numbers.

The *MCU* again monitors the *CS* transmission to identify a free one of its own registers and stores its *EN*.

The *MCU* signals the two *EN* concerned to the interrogators to identify free paths in all planes. The *MCU* chooses a pair of planes and signals the relevant plane markers to set the switches to establish a connexion between the allocated register and the calling line via a free through-link.

Dial tone is returned to the caller by the register after carrying out a check on the continuity of the connexion under the control of the marker.

Dialled digits are received at the register and detected pulse-by-pulse at the *MCU* which scans each of its registers in turn on a regular basis. The *MCU* stores and analyses the digits to determine the call routing.

On a local call, the *MCU* monitors the *CS* highway until it detects the directory number (*DN*) corresponding to the number dialled. It stores the associated class-of-service and computes the *EN*, which it stores.

Busy or free conditions are also indicated by the data transmitted by the *CS* and are stored by the *MCU* which instructs the register to call via the *CS* for call connexion processing. This enables the *MCU* to queue such demands which are dealt with separately from time dependent functions such as dial pulse processing. The *MCU* monitors the *CS* transmission and detects the register 'calling' indication in the same way as a line 'calling' indication.

Connexion to the called line, if free, is established via a bridge-link by the markers and interrogators under the control of the *MCU*. The register is then instructed by the

MCU to release itself and the switches connecting it to the caller.

Ringing, ringing tone, metering, and release are provided at the bridge link under the control of the *SPU*.

Busy or NU tone as required are provided by setting up a separate connexion to special tone circuits connected to *A* switch outlets.

PART 4 SUMMARY

4.7.1 Manual exchanges

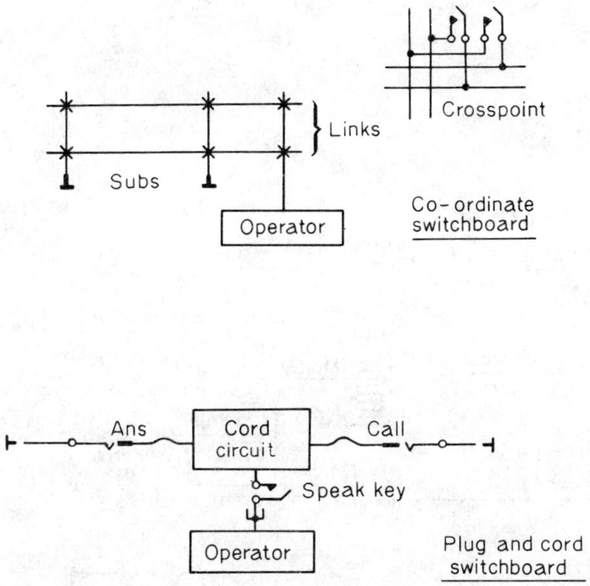

Fig. 4.27 Manual exchange principles—summary.

4.7.2 Strowger system principles

Mechanisms: reverse-acting, simple interrupter, not suitable for two-motion selectors.

Forward acting, needs toggle interrupter or interacting relay.

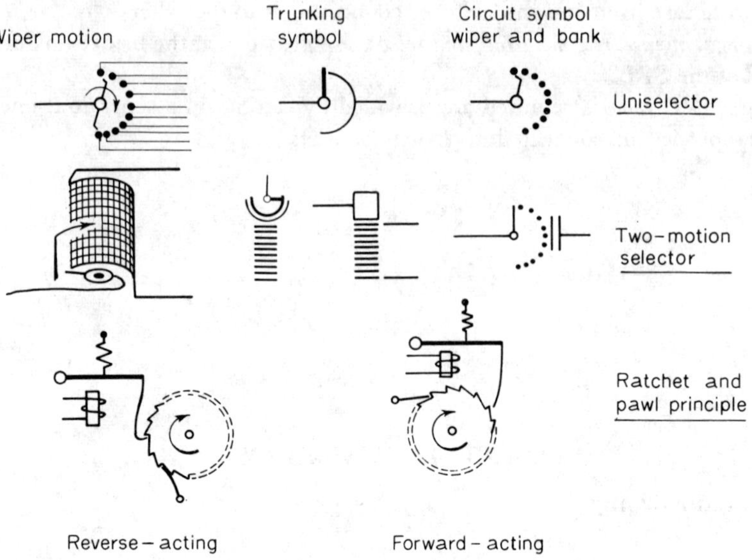

Fig. 4.28 Basic selector mechanisms—summary

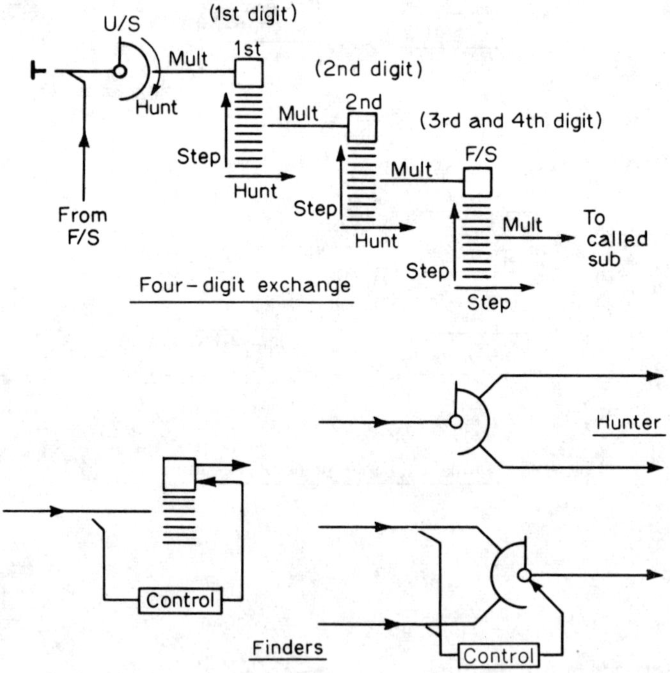

Fig. 4.29 Strowger trunking principle—summary.

4.7.3 Common control automatic exchanges

Advantages of auto: reduced cost per call, faster switching, uniform service, privacy, more traffic for same staff.

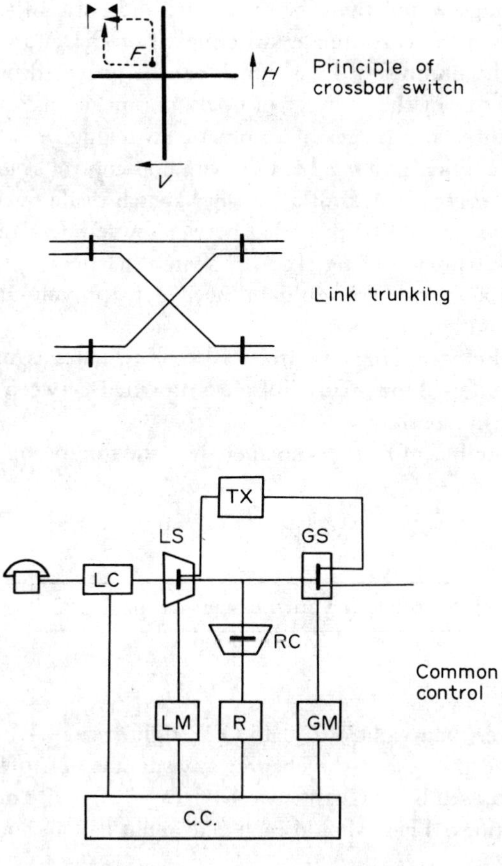

Fig. 4.30 Common control—summary.

Now turn to page 114 and test yourself on the questions there before leaving this chapter.

PART 5 SELF-TEACHING NOTES

4.8.1 First read **4.1.1** to **4.3**.

4.8.2 Now try the following twelve questions.

Q.1 How many *crosspoints* are required on a co-ordinate switchboard for ten subscribers and one operator's telephone with three links (connecting circuits)?

Q.2 What is the maximum number of calls the switchboard described in Q.1 can have in progress at one time?

Q.3 How many plugs would there be on a plug and cord switchboard having the same capacity as the co-ordinate switchboard of Q.1 and Q.2?

Q.4 What limits the number of lines which can be provided on a manual switchboard irrespective of the number of operators' positions?

Q.5 Write down three advantages of automatic switching.

Q.6 What is the function of a *marker* in a common control system?

Q.7 Why are the *selecting fingers* of a crossbar switch made flexible?

Q.8 How many links would be provided between nine inlets and nine outlets by a two-stage link trunking using six 3×3 switch matrices?

Q.9 How many 4×4 switches would be needed to provide 16 links between 16 inlets and 16 outlets?

Q.10 A two-stage link trunking of fifteen 10×20 switches is used to connect 200 inlets to 50 outlets. How many links are provided between the two stages?

Q.11 Define the term *blocking*.

Q.12 Why is the number of *registers* smaller than the number of trunks through the exchange?

Now check your answers on page 262.

4.8.3 If you got fewer than eight of Q.1 to Q.12 right, read **4.1.1** to **4.3** again. Then go on to read the rest of the chapter anyway, i.e. **4.4** to **4.7.2**.

4.8.4 A few days after studying Chapter 4 (**4.1.1** to **4.7.2**), write out complete answers to the next two questions. These should each take about half an hour.

Q.13 Explain what is meant by the term *switch matrix*.
With the aid of suitable sketches, show how such matrices can be interconnected to provide the switching network for a complete exchange.

Q.14 With the aid of simple sketches describe the principle of operation of a crossbar switch.

Now check your answers on page 262.

5 · STROWGER SYSTEM—SELECTOR MECHANISMS

For self-teaching notes, see page 138

PART 1 INTRODUCTION

5.1.1 Why Strowger? As we saw in Chapter 4, many different automatic exchange systems are in use of which Strowger's was the first. The Strowger system is still very widely used and existing exchanges of this type will remain in service for some years yet. The standard 2000-type equipment of the British Post Office is of this type, for example. The need to interwork with these older exchanges has a profound influence on the design of the newer types of exchange.

Strowger is one of the more straightforward systems to understand, but illustrates most of the essential principles of automatic switching systems in general.

5.1.2 What is a selector mechanism? The switching networks of automatic exchange systems are of two main types. Some are basically non-mechanical, such as reed-relay or electronic crosspoint systems, but the rest employ electro-mechanical devices such as crossbar or rotary switches. We use the term *selector mechanism* to describe the mechanical working parts of these switches. In practical selector circuits, the functions of the mechanism are controlled by a group of relays, as we shall see in the next chapter.

Typical switches used in modern Strowger exchanges are shown in Figs 5.13 and 5.14.

5.1.3 In this chapter, you will learn first of all the principles on which the selector mechanisms operate. At the end of the chapter are details of some typical practical mechanisms.

The circuit arrangements for controlling these selectors are dealt with in Chapter 6.

PART 2 PRINCIPLES OF STROWGER SELECTORS

5.2 Uniselector principles

5.2.1 Uniselector mechanisms. The types of uniselector usually employed for telephone and telegraph switching have reverse-acting mechanisms. The reason for this is that both the construction of the mechanism and the provision of interrupter contacts is generally simpler than on switches with forward-acting mechanisms.

Figure 5.1a shows a typical self-driving arrangement and Fig. 5.1b shows the general arrangement for pulse-controlled stepping.

Pulse-controlled stepping. Each 'break' pulse from the subscriber's dial, or other pulsing device, releases relay *A* for the duration of the pulse. Each time relay *A* releases, it operates the magnet at contact *A*1. In practice, relays *B* and *CD* are also required as described in **3.6.1**, **6.3.1** and **6.3.2**.

Self-drive operation. A break contact (*dm*) operated by the armature of the selector

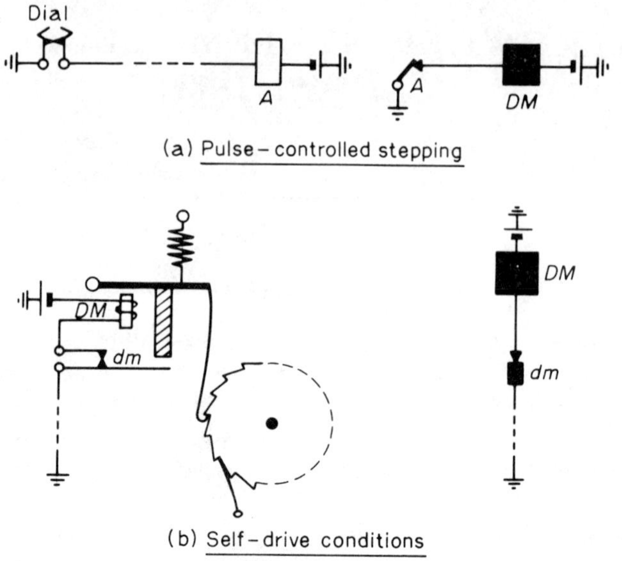

Fig. 5.1 Pulse-controlled and self-drive conditions. Compare (a) with Fig. 3.19 and (b) with Fig. 4.6 (5.2.2).

magnet is connected in series with the magnet. Each time the magnet operates, this *interrupter* contact opens the magnet circuit just as the pawl drops into the next tooth. This releases the magnet and steps the wipers. As the armature returns to normal, the contact closes again and re-connects the magnet. The force of the restoring spring is such that the armature returns to normal so quickly that it completes its movement of the wipers before the current has time to rise to a value which will re-operate it. The magnet then re-operates and moves the pawl into the next tooth and the cycle is repeated.

The chief difficulty with interrupter contacts is to ensure that they do not open so soon during the operating stroke of the armature that they disconnect it before it has moved the pawl far enough to step the wipers. At the same time, they must not close again in the return stroke so soon as to prevent the armature restoring before commencing the next operation.

The use of reverse-action makes it possible to arrange for the pawl to step between $1\frac{1}{4}$ and $1\frac{1}{3}$ ratchet teeth. The interrupter contacts are then adjusted to open as the pawl drops into the next tooth. This would not be possible if a forward-acting mechanism were used, because the pawl must stop when it reaches the end of the tooth to avoid stepping the wipers more than one step.

On the reverse stroke, the contacts can be allowed to close earlier in the stroke than this, because the restoring spring is already exerting its greatest force at the start of the stroke. It is therefore able to complete the restoration of the armature in spite of the effect of the current in the magnet coil, which takes time to reach its full va ue after the contacts have closed.

In both types of mechanism, the pawl must be held against the ratchet during stepping to prevent it jumping out of the tooth. This is very easily achieved on a reverse-acting mechanism by attaching the pawl to the armature by means of a flat spring (pawl spring), the tension of which presses the pawl to the ratchet. Such a spring would tend to bow outwards if used on a forward-acting mechanism.

In principle, two types of stop are required to limit the movement of the armature and pawl. The exact form of these stops varies considerably from one design to another but the main features are illustrated by the typical arrangement shown in Fig. 5.2.

The movement of the armature on its operating stroke is limited by an armature front-stop. This function may be performed by the magnet pole face in some designs. Adjustment of the armature travel is then achieved by varying the position of the magnet or by varying the position of the armature pivot or knife-edge. The travel must be adjusted to ensure that the pawl moves just far enough to engage with the next tooth and operate the interrupter contact with no risk of its stepping over two teeth.

The return movement of the armature is limited by an *armature back-stop*, the position of which must be adjusted so that the wipers are properly positioned on the bank contacts at the end of the stroke, In addition to this, a *pawl back-stop* is provided against which the pawl wedges so that it is held firmly on the rachet wheel in its normal position. This prevents the ratchet and wiper assembly from continuing to move under its own momentum after the armature and pawl have come to rest at the end of the return stroke. In some designs, these two back-stop functions may be combined in one component.

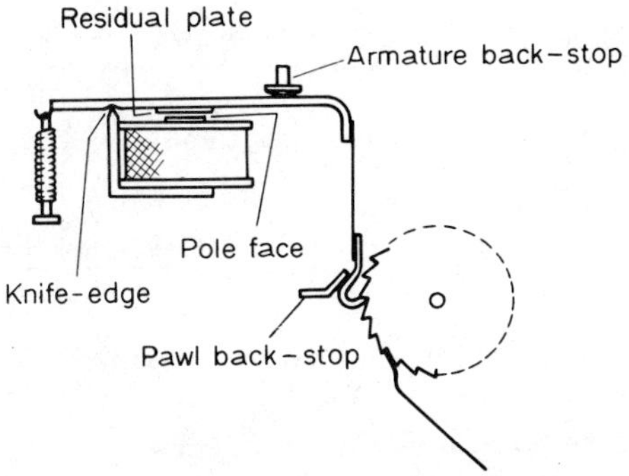

Fig. 5.2 Basic uniselector mechanism. Compare this with Figs 4.6a and 5.1 (5.2.1).

The pitch of the teeth on the ratchet wheel depends, of course, on the number of contacts on the bank. The angle between successive teeth must be the same as the angle between successive bank contacts. For a 25 outlet uniselector, for example, the twenty-five contacts are equally spaced round a semi-circular arc. The ratchet wheel therefore has twice this number of teeth (fifty teeth) equally spaced round the complete circumference.

5.2.2 Uniselector banks and wipers. In modern designs, each arc consists of a set of contacts moulded into one plastic insulator. As many of these arcs are then clamped together as are required to form the complete bank.

In the earlier types, the banks are constructed by clamping the set of contacts of each arc in place between flat insulators and metal spacers.

These two types of construction are shown in Fig. 5.3.

For some circuit applications, it may be necessary to connect a series of contacts together on certain arcs of the switch. This is sometimes done by constructing these arcs with a solid segment extending over the section concerned. This is shown in Fig. 5.13b, where a 25 outlet switch has single contacts on the arc illustrated in positions 1 and 12, and solid segments over positions 2 to 11 and 13 to 25. The solid segments actually consist of two separate contacts side by side which are connected together by the wipers.

Uniselector wipers are usually in the form of flat springs shaped at the ends to slide smoothly over the surface of the bank contacts. The wipers are in pairs, one tensioned against each side of the contacts to provide good contact pressure. The wipers used for ordinary switching purposes are so shaped that they move off one contact before touching the next, thus providing a 'break-before-make' action like the change-over

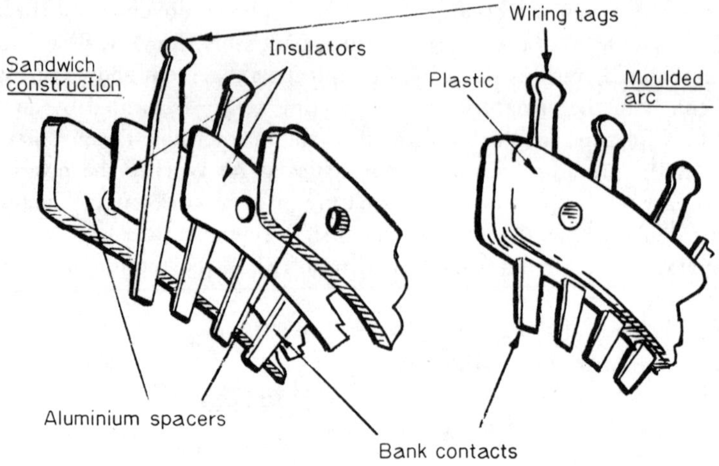

Fig. 5.3 Uniselector bank construction (5.2.2).

contact action of a relay. Such wipers are called *non-bridging* to distinguish them from those which have elongated tips to provide a 'make-before-break' or *bridging* action. Both types are shown in Fig. 5.4.

Bridging wipers are used on those arcs of the switch which are required to carry current while the wipers are actually moving. If non-bridging wipers were used, they would break the current on each step resulting in sparking between the wiper and bank contact which would damage the wiper tips.

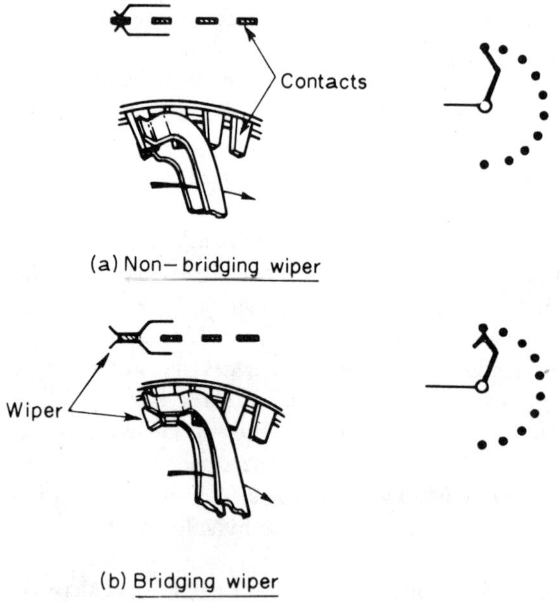

Fig. 5.4 Bridging and non-bridging wipers. The bank symbols here show ten outlets only. This is often used as an abbreviated symbol for larger uniselectors; the full number of outlets are only shown when necessary (5.2.2).

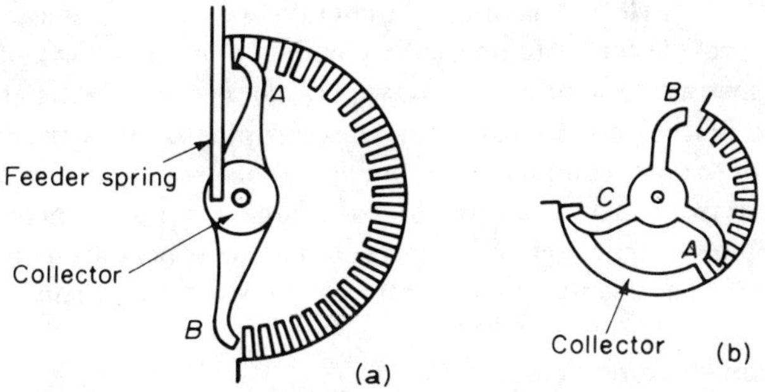

Fig. 5.5 Uniselector wiper feed. The fixed feeder spring type (a) is used on the larger uniselectors and the fixed collector type (b) is used on the smaller uniselectors. Type (b) permits closer spacing of the arcs but restricts the outlet availability (5.2.2).

The wipers can be stepped right round the bank to reach the point at which they first started. The circuit connexions cannot, therefore, be made by wiring directly on to the wipers. A sliding contact is provided to make contact with the wipers. This consists either of a fixed feeder spring with a *collector* ring attached to the wiper and rotating with it, or a fixed collector segment as part of the bank with an extra wiper to function as a feeder spring. The principle of these two systems is illustrated in Fig. 5.5.

In one case a double-ended wiper is used. One end enters the bank at the first contact as the other end leaves the bank at the last contact. There is thus always one end of the wiper in the bank and one end out of the bank.

In the other case, a three part wiper is used and the contacts are arranged around a third of a circle (120°) instead of around a semi-circle (180°). There is thus one part of the wiper in the contact section of the bank at a time. Of the other two parts of the wiper, one is outside the bank and ineffective at this time while the other is in contact with the collector segment which also occupies one third of an arc. (If the contacts occupied only a quarter of a circle, four wiper arms would be needed.)

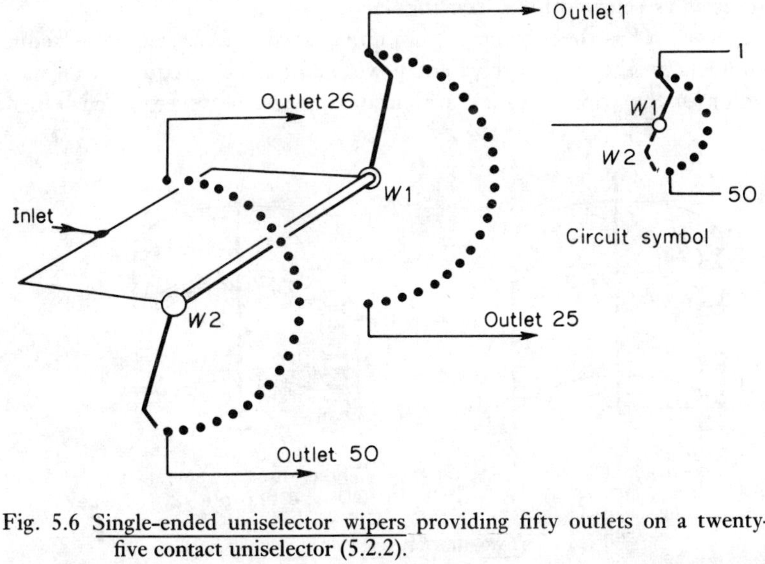

Fig. 5.6 Single-ended uniselector wipers providing fifty outlets on a twenty-five contact uniselector (5.2.2).

In the case of a uniselector having, say, twenty-five contacts in a semi-circular arc, it is possible to provide for switching to fifty outlets. This is achieved by using single-ended wipers instead of double-ended wipers connected alternately as shown in Fig. 5.6. As the first wiper of a pair completes stepping over its twenty-five contacts, the second one of the pair commences stepping over its own set of twenty-five contacts. The first of these two wipers does not enter the bank again until the second wiper has completed stepping over its set of contacts. The two wipers are connected together so that they provide access to a total of fifty contacts between them.

5.3 Two-motion selector principles

5.3.1 Vertical and rotary stepping mechanisms. The vertical and rotary stepping are carried out by separate forward-acting ratchet and pawl mechanisms. The two ratchets are, of course, both attached to the wiper assembly.

Reverse-acting mechanisms are not used on two-motion selectors for two reasons:

(*a*) In reverse-acting mechanisms, the pawl is held firmly engaged with the tooth of the ratchet when at rest. The friction between the pawl and tooth of one ratchet would therefore interfere with the movement of the other ratchet if used on a two-motion selector.

(*b*) Vertical stepping is a straight line motion and it is clearly impossible to return the wipers to normal by stepping them on until they come back to where they started as we can on a uniselector. A restoring spring is therefore used to return the wipers to normal at the end of a call. If a reverse-acting mechanism were used, the pawl could not easily be lifted clear of the ratchet during release and correctly engaged in the tooth again on completion of the release.

In practice, there are two release systems used in different types of selector. One uses a third electromagnet to lift the vertical and rotary detents clear of the ratchet teeth causing the wipers to return to normal over their original stepping path. They therefore release by first rotating and then dropping vertically.

The other method is to allow the rotary mechanism to self-drive to the end of the bank where the wiper assembly is allowed to drop vertically and then rotate below the contacts of level 1 to return to normal without retracing its path through the bank. This is sometimes referred to as 'rectangular' or 'square' release.

This last system has the advantage that no separate release magnet is required and that all contacts on the level receive equal wear and are all frequently cleaned by the sliding action of the wipers. There are circumstances, however, when it is desirable

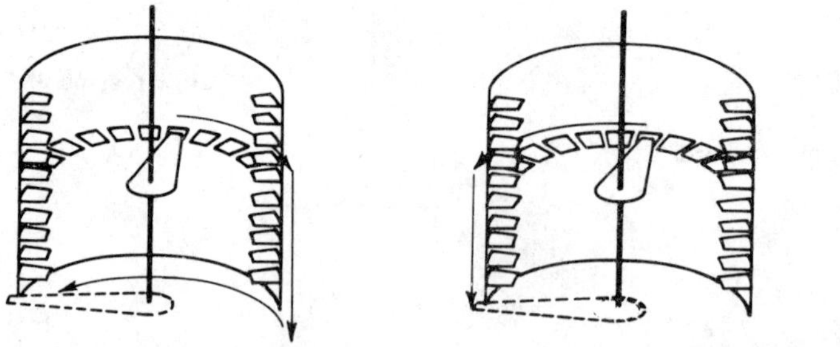

Fig. 5.7 Two-motion selector release (5.3.1).

to be able to restore a selector to normal from the first outlet of a level within the period of the interdigital pause. This is done to 'absorb' digits which are sometimes included in a subscriber's number to provide a uniform numbering scheme but are not required for switching. Only the separate release magnet can restore the wipers to normal fast enough for this function.

There are basically three ways in which the vertical and rotary ratchet teeth can be combined to form one unit. These are shown in Fig. 5.8.

The earliest method used consists of one solid ratchet cut with two sets of teeth of sufficient length to be aligned with their pawls and detents in any position. The total length of this combined ratchet is at least twice the vertical movement required at the wipers and the mechanism is consequently larger than is necessary.

To reduce the size of the mechanism, either the vertical or the rotary teeth must be shortened.

If the vertical teeth are made short as in Fig. 5.8b, a separate support must be provided for the wiper assembly once rotary stepping commences. This is the purpose of the comb plate and disk which, being separate from the vertical ratchet, must be adjusted in relation to the ratchet to ensure alignment of the wipers in the bank.

If it is the rotary ratchet teeth that are made short as in Fig. 5.8c, the vertical ratchet and the wiper assembly must be free to slide relative to the rotary ratchet in a vertical direction. A spline or slot in the rotary ratchet carriage ensures that it carries the wipers with it during rotary stepping.

The forward movement of the armature on its operating stroke is limited by the magnet pole face which thus functions as an *armature front-stop*. The forward

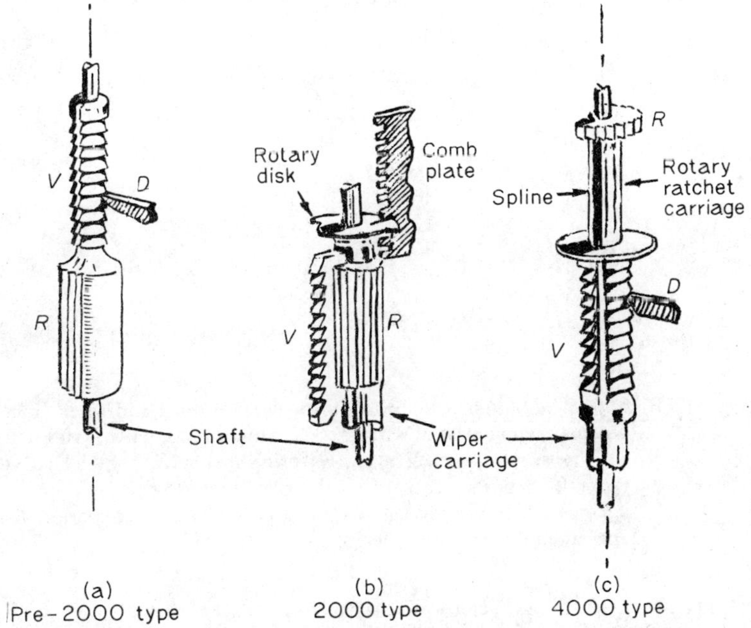

Fig. 5.8 Two-motion selector ratchets. In each case V indicates the vertical ratchet teeth. On types (a) and (c), D is the stationary detent which supports the weight during rotary release. The position of the detents and comb plate relative to the ratchets corresponds to the wipers stepped to the fourth contact of the third level (5.3.1).

movement of the pawl, however, is limited by a separate *pawl front-stop*, the position of which is adjusted to ensure that the wipers are moved into the correct position on the bank contact on each step. The pawl wedges between the pawl front-stop and the ratchet tooth to arrest the motion of the ratchet at the end of each step.

In its normal (unoperated) position, the pawl rests on the *pawl guide*, being held there by the *pawl spring*. The position of the pawl guide is adjusted so that the pawl is clear of the ratchet tooth when at normal but moves correctly into the root of the tooth at the commencement of its forward movement on operation of the armature. In some designs, the pawl itself rests on the pawl guide while in other designs an extension of the pawl or the pawl arm rests on the guide. The principle is the same either way.

The pawl is not directly attached to the pawl spring as it usually is on reverse-acting mechanisms on uniselectors (**5.2.1**). Instead it is part of a pawl arm which is pivotted at the end of the armature. The pawl spring bears on the pawl arm to hold the pawl against the ratchet during stepping.

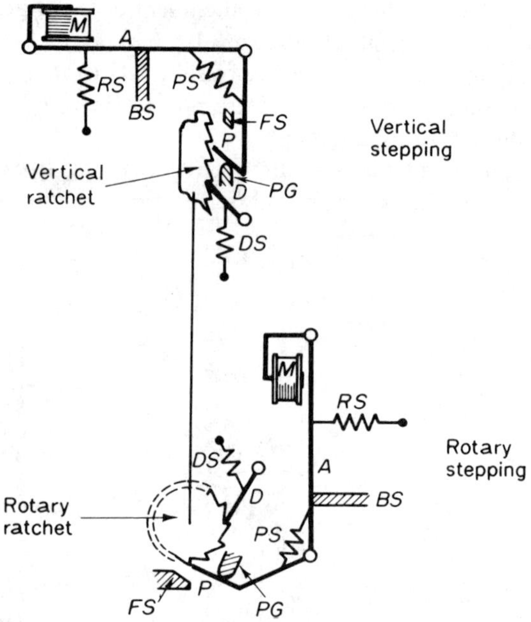

Fig. 5.9 Basic two-motion selector mechanism. The vertical and rotary mechanisms each consist of a magnet *M*, armature *A*, pawl *P* and detent *D*. The armatures each have a restoring spring (*RS*) and a back stop (*BS*). The pawls are held on the pawl guides (*PG*) by their pawl springs (*PS*) and have a front stop (*FS*). The detents are held in the ratchet teeth by the detent springs (*DS*) (5.3.1).

5.3.2 Interrupter contacts (two-motion selectors). As we saw in **5.2.1** when considering the reverse-acting mechanism of a uniselector, it is essential that the contact which is used to disconnect a magnet to provide self-drive conditions must operate very late in the operate stroke of the armature. A simple break contact operated directly by the armature of a forward-acting mechanism would require very critical adjustment to meet this condition and yet open sufficiently by the time the pawl completed its forward movement.

One solution to this difficulty is to use a 'make' action interrupter contact to operate a relay (usually designated *G*). A 'break' contact of this relay is then connected in series with the selector magnet. The operate and release lags of the relay delay the disconnexion and subsequent re-connexion of the magnet.

The method used on the later types of switch is to provide a *toggle action interrupter* contact which does away with the need for a separate interacting relay. This consists of a pivoted contact operating arm linked to a spring. The function of the spring is to hold the arm in either of its two stable positions. The interrupter contacts are held open by the arm under the action of this spring in one of the two positions but not in the other position.

The arm is moved from side to side by a *striker* which consists of a U-shaped extension of the armature. The striker is in contact with the operating arm during only part of the operate and release strokes of the armature.

During operation of the armature, one side of the striker engages the contact operating arm which turns on its pivot. The spring tension opposes the movement of the arm until it passes the mid-point of its movement. The spring then aids the movement of the arm, completing the rest of its travel independently of the armature and striker. The arm is thus able to complete its movement and open the contact after the armature has reached the end of its travel.

As the armature restores, the other leg of the striker engages the contact operating arm at a late point in the return stroke. Again the arm is able to complete its movement independently of the armature, closing the contact after the armature has completed its return stroke.

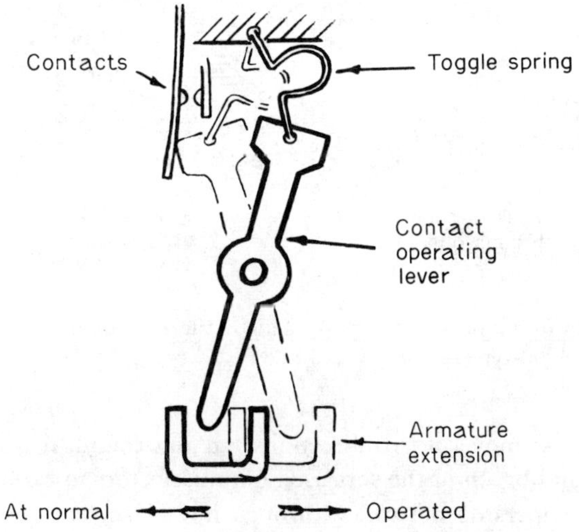

Fig. 5.10 Principle of toggle interrupter (5.3.2).

5.3.3 Wipers and banks (two-motion selectors). A complete bank consists of one or more sections, each having ten levels. Each level has two rows of contacts insulated from each other and each having eleven contacts. Some versions have only a single row of contacts per level.

The moulding technique used on uniselector banks presents difficulties when applied to double rows of contacts so the contacts are clamped in place between flat insulators and spacers. The insulators may be moulded in plastic with recesses to hold the contacts so providing an even surface for the wipers.

In addition to the main contact banks a separate *vertical marking bank* may be fitted. This consists of a set of contacts arranged in two vertical columns of eleven contacts each. These correspond to the ten levels of the main bank plus one contact in the vertical 'normal' position. The wipers for the vertical marking bank are lifted by the mechanism during vertical stepping along with the main wipers, but do not rotate with them during rotary stepping.

The vertical marking bank is useful when a selector is required to provide different facilities on different levels. An example of this on some types of exchange is the provision of different call charging or metering rates on different routes. The vertical marking bank contacts are connected to meter pulse supplies at the appropriate rate for the routes served at each level.

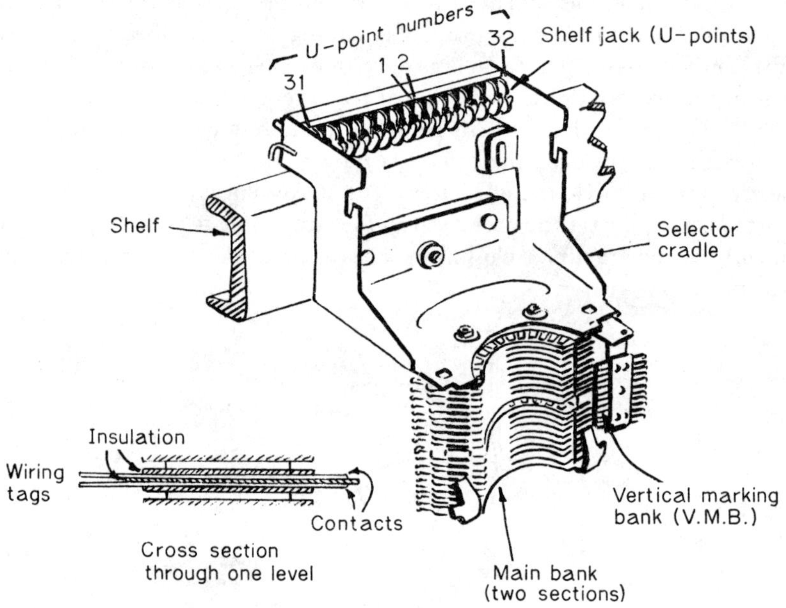

Fig. 5.11 Two-motion selector bank. This is the type used for 2000-type and 4000-type selectors (5.3.3).

Because the wipers move in two directions, up and round, it is difficult to use any form of collector and brush of the sort used on uniselectors to make connexions to the wipers. Since the wipers do not rotate through more than one revolution as the uniselector wipers do, it is possible to make direct connexions to the wipers, however, and flexible wiper cords are used for this purpose. A separate single cord is used for each wiper, employing a similar tinsel-on-fabric construction to the plug cords on a manual switchboard.

The wipers are in pairs in order to make contact with the two rows of contacts in each level. It is therefore necessary to adjust them so that the two wipers of a pair have sufficient clearance between their tips for them not to touch one another as they move from contact to contact, so allowing for some bounce. The clearance must not be so large that either of the wipers fails to touch the bank contact or to exert sufficient contact pressure on it (see also Fig. 5.15).

Bridging wipers are not provided on two-motion selectors. You may remember that they were required on uniselectors on those arcs carrying current while the switch is moving its wipers. This usually applies when self-driving the uniselector to normal, by connecting the supply to the magnet through all contacts of the arc, except the 'home' contact (see Chapter 6—Circuits). In two-motion selectors, this type of operation is performed by means of an off-normal contact (see 5.3.4), so bridging wipers are never required.

5.3.4 Mechanically operated springsets. These are sets of contacts basically the same as relay springsets. Instead of having their own magnets and armatures, however, each springset is operated when the wipers reach some pre-determined position or positions. The operation of the springsets is achieved by means of suitable cams attached to the wiper carriage or shaft.

The main types of mechanically operated springsets are known as the *N*, *NP*, *NR* and *S* springs, the first two being operated by the vertical movement of the wiper carriage and the second two by the rotary movement.

The N springs are the *vertical off-normal* contacts. These function as soon as the wipers take the first vertical step and do not restore until wipers return to normal again.

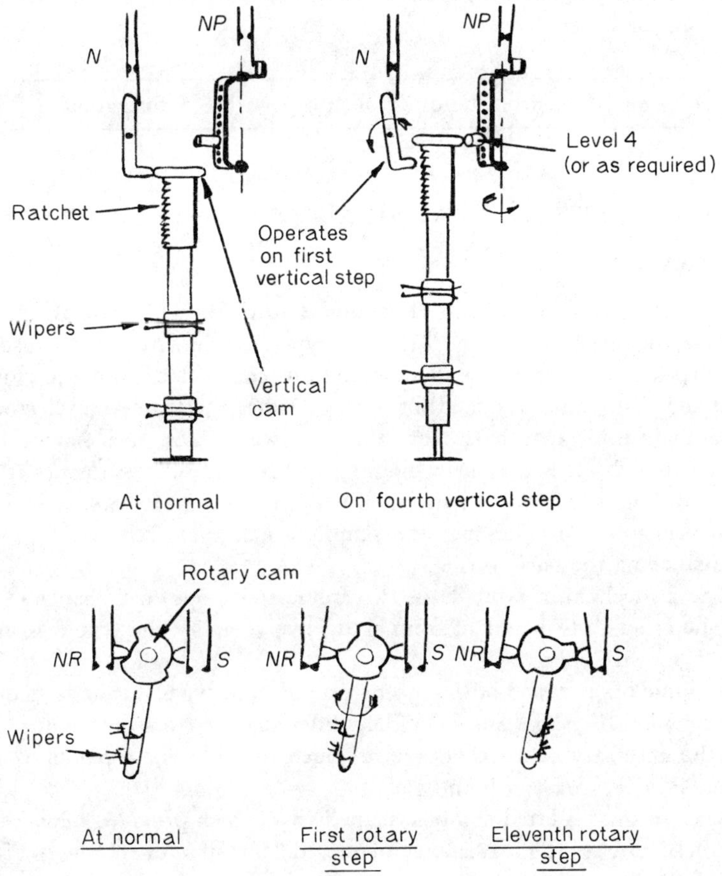

Fig. 5.12 Selector mechanically operated contacts. *N* is the vertical off-normal, *NP* is the normal post, *NR* is the rotary off-normal and *S* is the eleventh step contact unit (5.3.4).

In practice, this is usually achieved by a vertical cam holding the lever springs of each contact under tension when the wipers are at normal. As the cam moves off-normal, the contacts operate under their own spring tension.

The NP springs or 'normal post' springs can be arranged to operate on any desired level or levels. The contact operating lever has a hole in it for each level and a stud can be fitted in each level on which the springset is required to operate. Each time the vertical cam reaches a level for which a stud has been fitted, it engages the stud, causing the lever to move and operate the springset.

The purpose of these contacts is similar to the vertical marking bank.

Sometimes, two independent sets of *NP* springs with two independent operating levers and sets of studs may be fitted to a selector. They are then designated as *NPA* and *NPB* springsets.

The NR springs are the *rotary off-normal* contacts. They are operated by the rotary cam on the first rotary step and remain operated the whole time the wipers are in the bank.

The S springs are the *eleventh step* contacts and they are operated by the rotary cam on the eleventh rotary step only. This is the last rotary position and in the case of a group selector is only reached when all the earlier contacts are busy. The eleventh contact is not used as a working outlet and so the *S* springs are used for connecting 'busy tone' to the caller.

Test yourself on the questions on page 138 before going on.

PART 3 TYPICAL PRACTICAL SWITCHING APPARATUS

5.4 Typical selectors

5.4.1 25-point uniselectors (B.P.O. types 2 and 3). The general appearance of these uniselectors is shown in Fig. 5.13. The type 2, is the larger and is used for most general purposes. The type 3 is a more lightly constructed and therefore cheaper version of the same thing designed for use as a subscriber's line uniselector.

The bank assembly may have from three to twelve arcs, each having twenty-five contacts. One of the arcs may be a 'homing' arc having solid segments as illustrated in Fig. 5.13b. The wiper brushes are mounted on the contact bank assembly and the rest of the mechanism forms a separate assembly which can be removed for maintenance without disturbing the bank wiring.

The type 2 mechanism is made in two frame sizes, one to fit banks of up to five arcs and the other to fit banks of from six to twelve arcs. The type 3 is made in the narrower frame size only.

The principle of operation of the mechanism of both types is exactly as described in (5.2) and shown in Figs. 5.1 and 5.2. They differ slightly from one another in the way in which the necessary adjustments are provided for. The main points of adjustment are illustrated in Fig. 5.13 for both types.

The electromagnet is fitted in a box-shaped yoke which provides a good, low reluctance, magnetic circuit. The resistance of the coil for 50 V operation is 75 Ω for the type 2, and 100 Ω for the type 3.

A residual plate of stainless steel is fixed to the armature to ensure a uniform release in the same way as the residual stud on a relay armature.

Wipers and contacts are made of nickel-silver. The wipers can be provided in any combination of bridging, non-bridging, single-ended or double-ended types as required.

The bank contact assembly is mounted either on spring clips or rubber supports to reduce the effects of vibration.

The position of the wipers round the bank is indicated by a pointer fixed to the frame beside the *index wheel* attached to the wiper spindle at the opposite end to the ratchet wheel. The pointer can be set in either of two positions. If the uniselector is used in

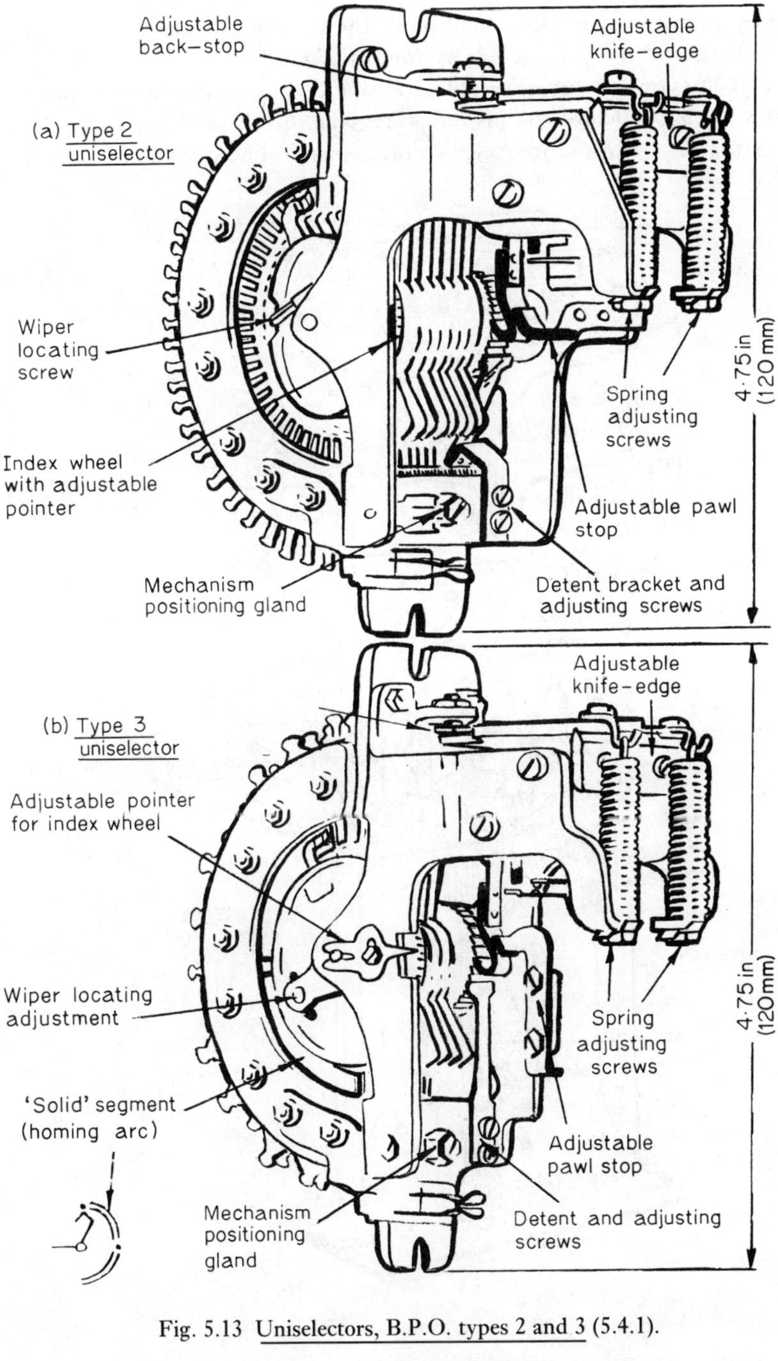

Fig. 5.13 Uniselectors, B.P.O. types 2 and 3 (5.4.1).

a circuit where it restores its wipers to the normal position at the end of the call (homing), the pointer is positioned to indicate the first bank contact as 0, i.e. the bank contacts number 0 to 24 or 0 to 49 depending on whether double-ended or single-ended wipers are used. If used in a circuit where the wipers remain on the contact last used at the end of each call (non-homing), the pointer is set in its other position, the first contact is then numbered 1, i.e. the bank contacts number 1 to 25 or 1 to 50.

The self-drive stepping speed is about 60 steps per second (50 rev/min to 100 rev/min depending on the number of wipers and the adjustment of the mechanism). Under pulse-controlled stepping conditions, the magnet operate time is about 20 ms (8 ms to 30 ms) and the magnet release time is about 8 ms (3 ms to 12 ms).

5.4.2 Pre-2000 type two-motion selector. Until the introduction of the 2000 type mechanism in 1936, each of the manufacturers supplying the British Post Office had their own design of two-motion selector mechanism. These followed the same basic

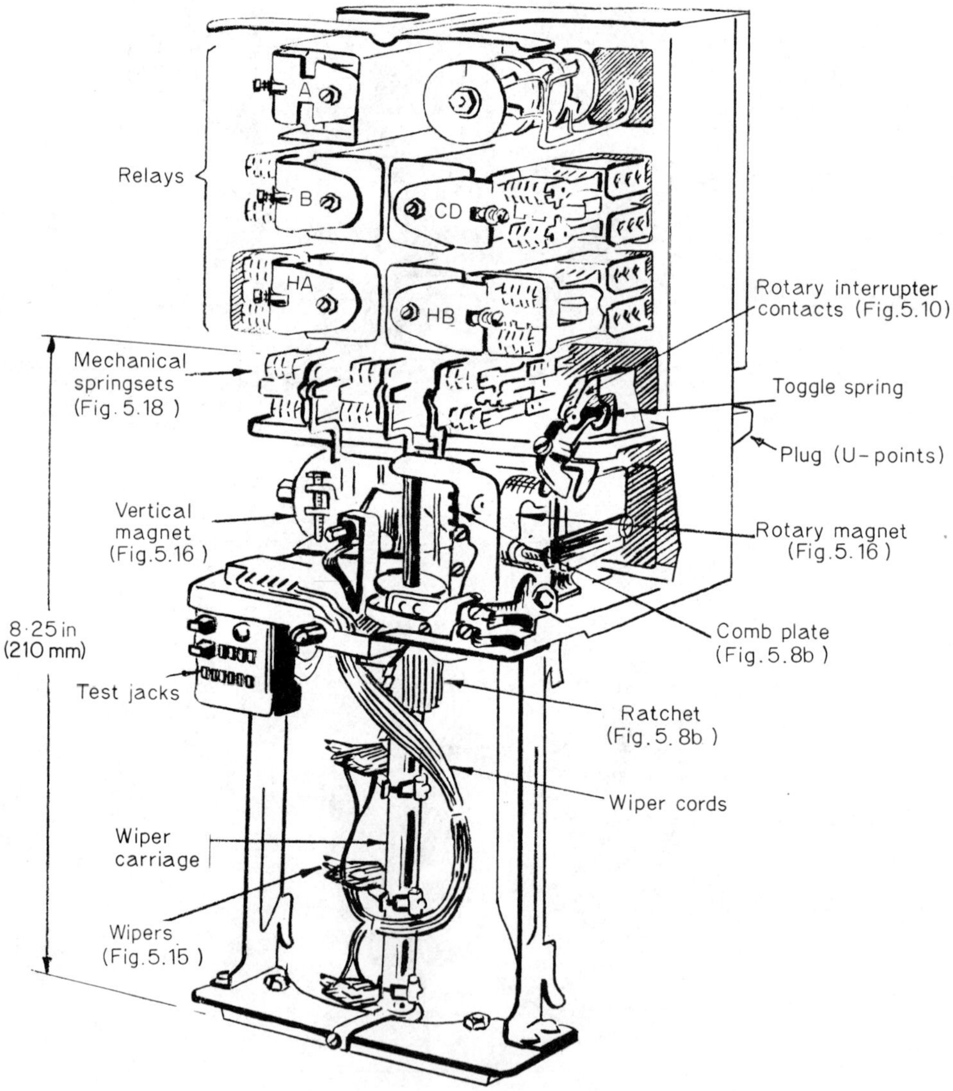

Fig. 5.14 2000-type group selector (5.4.3).

pattern but differed in detail, particularly with regard to the mechanism used to operate the mechanical springsets.

In spite of their differences, all of these early types of selector became known as 'pre-2000' types on the introduction of the 2000 type selector.

The pre-2000 type mechanisms were all of the three-magnet type with the vertical ratchet above the rotary ratchet as in Fig. 5.8a. They are similar to the Strowger selectors of American manufacture.

5.4.3 2000 type two-motion selector. This was adopted by the British Post Office in 1936 to supersede the earlier types. The same design of switch is made by each of the British manufacturers supplying the Post Office.

It is a two-magnet mechanism with the vertical ratchet alongside the rotary ratchet as shown in Fig. 5.8b. It is consequently smaller than the pre-2000 type selectors.

The wipers are fixed to a *wiper carriage* which slides on a stationary shaft. The shaft is supported at its lower end to prevent any 'whip'. This permits the use of more contact bank sections than was possible with the pre-2000 type. These earlier types were limited to three banks whereas as many as twelve have been used on a 2000 type mechanism.

The general appearance of a group selector employing a 2000 type mechanism is shown in Fig. 5.14.

The frame is a one-piece die-casting of an alloy of aluminium and silicon in the form of an upper and a lower platform joined by two members which support the magnets and the shaft. The upper platform accommodates the mechanically operated springsets and the lower platform the detents, stops and adjusting screws.

The bank contact assembly is of the type illustrated in Fig. 5.11 and is fixed to the rack by means of a bracket which also serves as a cradle into which the selector mechanism is jacked. The vertical marking bank, when provided, is attached to the same cradle on a vertical spindle or pivot so that it can be swung clear of the wipers to allow the selector to be removed from the cradle.

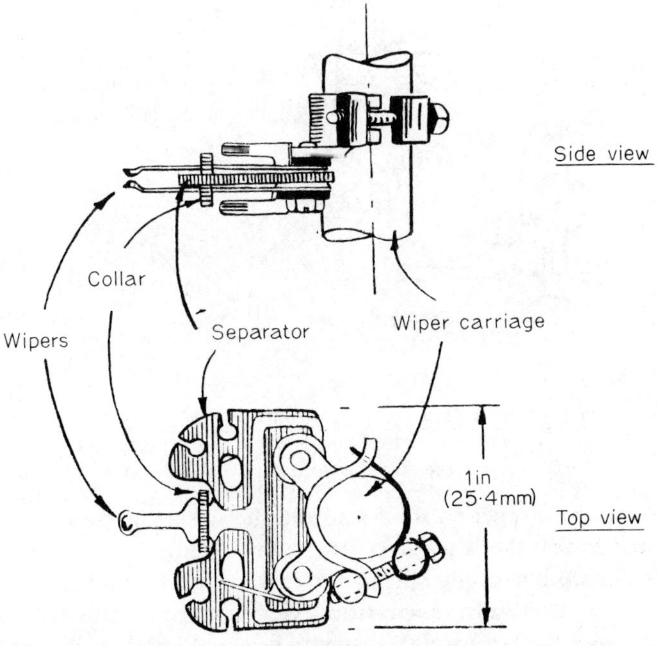

Fig. 5.15 Selector wipers, 2000-type and 4000-type. (5.4.3).

The wipers are made of nickel-silver and are provided in pairs, each having a spoon shaped tip to form a domed contact surface. The two wipers are held parallel by a fibre collar, and are attached to opposite sides of a separator in the form of a plate of insulating material. The separator provides an anchorage for the wiper cords and is fixed to a bracket which is held on to the wiper carriage by means of a spring steel clamp as shown in Fig. 5.15.

The vertical and rotary magnet assemblies are identical. Each consists of a single coil wound on a cast iron core having a large cross-sectional area to give a low reluctance magnetic circuit. The high resistivity of cast iron reduces eddy current effects which could otherwise 'slug' the magnet (see Chapter 4—Relays).

The vertical stepping mechanism is shown in Fig. 5.15 and its operation is basically as already described for a forward-acting mechanism in **4.3.3** and for two-motion selectors in general in **5.3.1**.

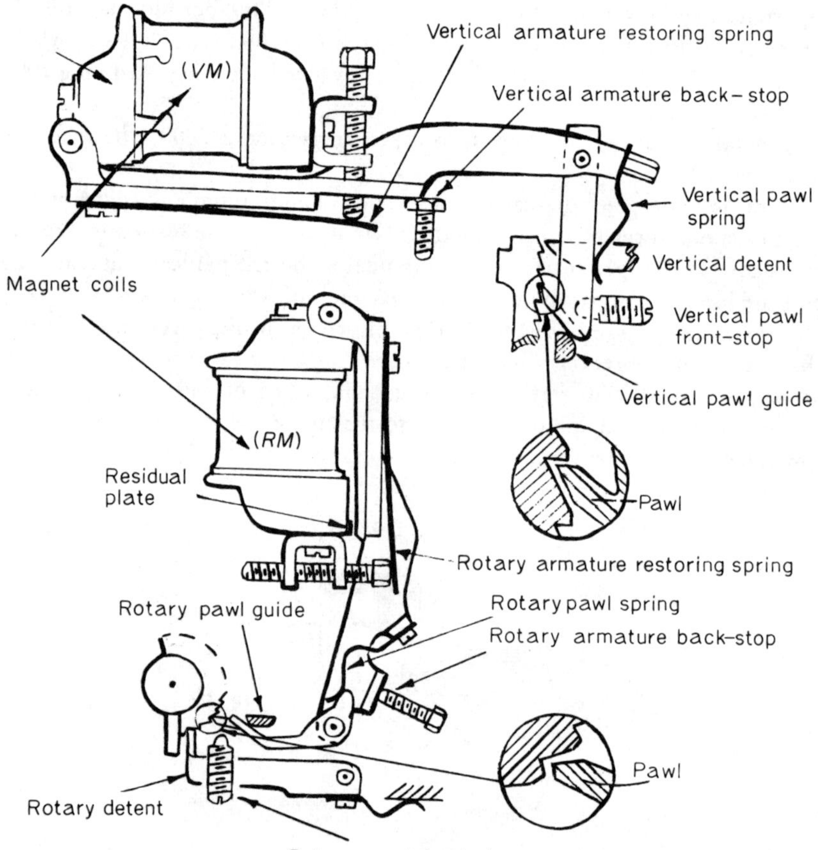

Fig. 5.16 Vertical and rotary stepping mechanism, 2000-type. This shows a side view of the vertical mechanism and a plan view of the rotary mechanism. Compare this with Figs 5.8, 5.9 and 5.14 (5.4.3).

The position of the magnet on the frame and the setting of the vertical pawl front-stop are adjusted so that the latter arrests the vertical movement of the wiper carriage just before the armature has fully operated to make contact with the magnet pole face. This is, of course, an easier adjustment than trying to make the two operations coincide exactly. It also causes the armature extension to become slightly bowed as the armature completes its stroke. When the current is disconnected to release the magnet, the armature straightens itself imparting a 'flick' action which assists the action of the restoring spring to give a faster release.

At the end of vertical stepping, the ratchet is held in position by the vertical detent with the rotary disk opposite one of the notches in the comb plate.

When rotary stepping commences, the vertical ratchet moves out of engagement with the vertical detent and the rotary disk engages the notch in the comb plate. The weight of the wiper carriage is then supported by the comb plate as the wipers engage the bank contacts.

Apart from the physical shape of the pawl and ratchet, the rotary stepping mechanism is very similar to the vertical stepping mechanism. Its operation is as already described for the vertical mechanism.

To restore the wipers to the normal position, they must first be stepped to the twelfth rotary step. At this position, the wipers are clear of the last contact of the bank. The rotary pawl, operating for a further step, finds no tooth with which to engage and therefore moves over under the action of the pawl spring to come to rest with its locking projection held against part of the comb plate provided for the purpose. The rotary disk is also clear of the comb plate notch on the twelfth rotary step.

The pawl is unable to complete its stroke so that the interrupter contacts are not opened and the magnet remains energized.

The disengagement of the rotary disk allows the wiper carriage to fall. The restoring force is provided by a helical spring contained within the wiper carriage and fixed at one end to the shaft and at the other end to the inside of the carriage.

At the 'normal' level, the rotary ratchet disengages from the rotary detent and the rotary disk is opposite a notch in the comb plate. The restoring spring is thus able to rotate the carriage back to the normal position, the wipers passing underneath the first contact level. The movement is arrested at the normal position by the rotary cam engaging with the comb plate.

Interrupter contacts of the 'toggle' type described in **5.3.2** can be provided on the vertical or rotary magnets as required.

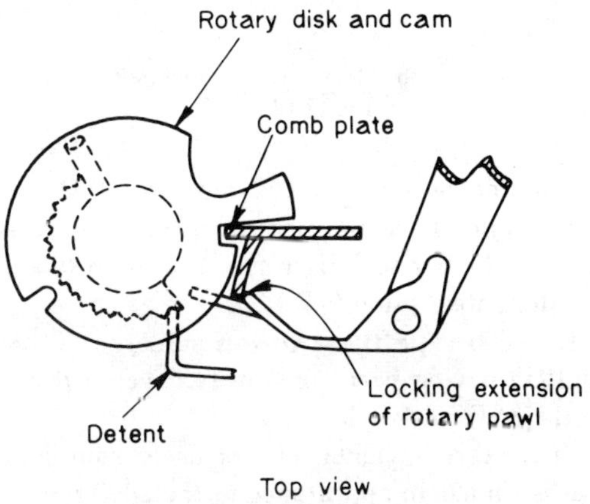

Fig. 5.17 Release of 2000-type selector. Conditions on twelfth rotary step and during vertical restoration. Compare this with Fig. 5.16b (5.4.3).

The operating levers for the mechanical springsets are pivotted between bearing pins, a fixed one for each lever in the upper platform of the frame, and an adjustable one for each lever in the lower platform.

A single cam plate mounted on the top of the ratchet functions as both a vertical and a rotary cam for the operating levers controlling the *N*, *NR* and *S* springs. A second vertical cam is provided for the *NP* springsets having separate steps for the *NPA* and *NPB* springset operating levers.

Each springset assembly consists of a pair of springsets of the same type as those of the B.P.O. 600 type relay mounted on a bracket with an auxiliary armature engaging both springsets. The operating lever bears on the auxiliary armature to operate the contacts. The *NP* springset assembly has no auxiliary armature. This allows the *NPA* and *NPB* springs to be operated by separate operating levers.

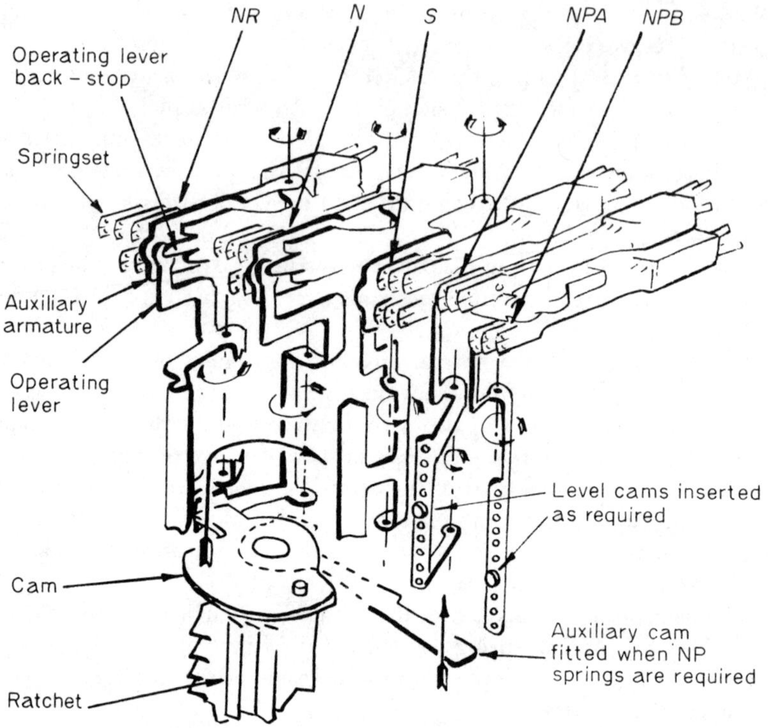

Fig. 5.18 Mechanically operated springsets, 2000-type. Compare this with Fig. 5.12 (5.4.3).

The magnets have a resistance of 50 Ω and the self-drive speed is about 35 steps per second. In some cases, a special low resistance (25 Ω) magnet has been used to obtain a higher self-drive speed of some 46 steps per second to provide a faster release for digit absorption (5.3.1).

5.4.4 4000 type two-motion selector. This was adopted by the British Post Office in 1958 having been previously developed by the General Electric Co. Ltd (G.E.C.) as their SE 50 switch.

It is a three-magnet selector using a similar release method to the pre-2000 selectors but is similar in appearance to the 2000 type. It is interchangeable with the 2000 type as regards rack mountings and banks.

The frame casting is similar to that of the 2000 type mechanism but it has separate vertical members for the two stepping-magnet assemblies and a pillar at the front for the various adjusting screws. Most of the adjustments are controlled by nuts and screws, bending adjustments having been practically eliminated. The lower end of the wiper shaft is supported by a second casting attached to the underside of the frame.

The ratchets are of the type shown in Fig. 5.8c. The rotary ratchet carriage is free only to rotate on the fixed wiper shaft. The wiper carriage to which the vertical ratchet is attached is free to slide vertically over the rotary ratchet carriage.

The bank contact assembly is identical to that of the 2000 type selectors. The same wipers are also used for both types of mechanism.

To simplify maintenance and to permit the use of mass production techniques during manufacture, the magnet, pawl and armature of the vertical and rotary mechanisms form two self-contained sub-assemblies. Each can be adjusted away from the

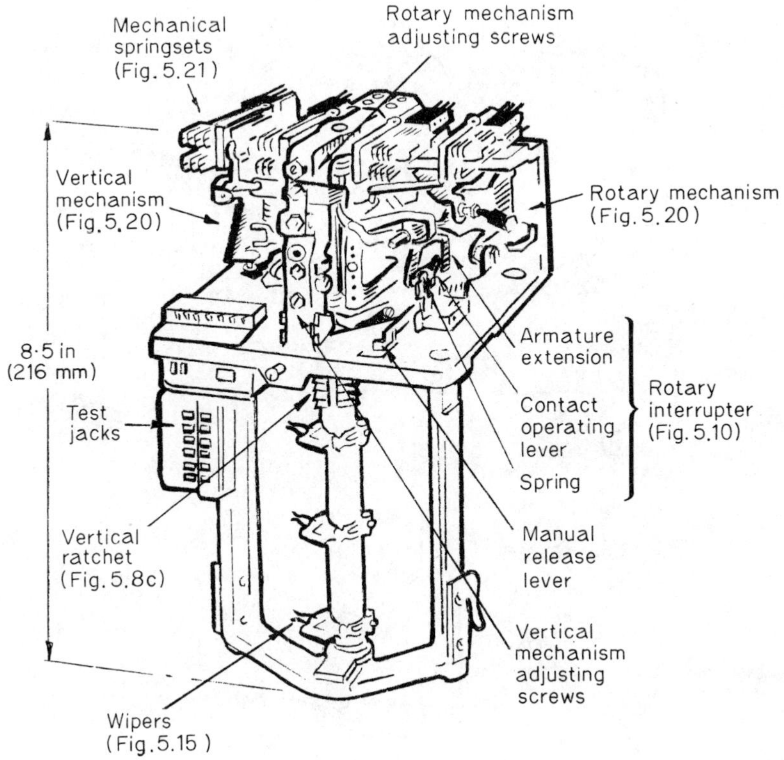

Fig. 5.19 Selector mechanism, B.P.O. 4000 type (5.4.4).

selector and can then be fitted to the selector without upsetting its own or any other adjustments. Interrupter contacts, when provided, form part of the vertical or rotary magnet sub-assembly.

The magnets have box-shaped yokes similar to those of the uniselectors (**5.4.1**).

The vertical stepping mechanism is shown in Fig. 5.14 and its operation is similar to that of the 2000 type selector (**5.5.2**). The pawl is positioned, however, at a greater distance from the tooth in the idle state to allow sufficient clearance for the return motion of the ratchet during release. The pawl guide is in the form of an eccentric (i.e. off-centre) nut on the front column of the frame which engages the pawl arm extension. The eccentricity of the nut permits adjustment of the position of its working face.

During vertical stepping, the weight of the wiper carriage is supported by the vertical detent at each step. This detent is mounted on a pivot and is held against the ratchet teeth by a spring. A second fixed or stationary detent is also provided but during vertical stepping this is located in a vertical slot in the ratchet and is, therefore, ineffective. On the first rotary step, however, it engages the circular part of the vertical ratchet, the detent teeth, to support the weight of the carriage.

The short ratchet teeth engaged by the vertical pawl are straight, not circular like the detent teeth. This provides a greater bearing surface for the pawl and therefore less wear.

The rotary stepping mechanism is similar to the vertical mechanism but the pawl guide in this case bears directly on the pawl arm. The operation is otherwise the same for both mechanisms.

A release magnet *ZM* is located between the vertical and rotary magnets at the rear

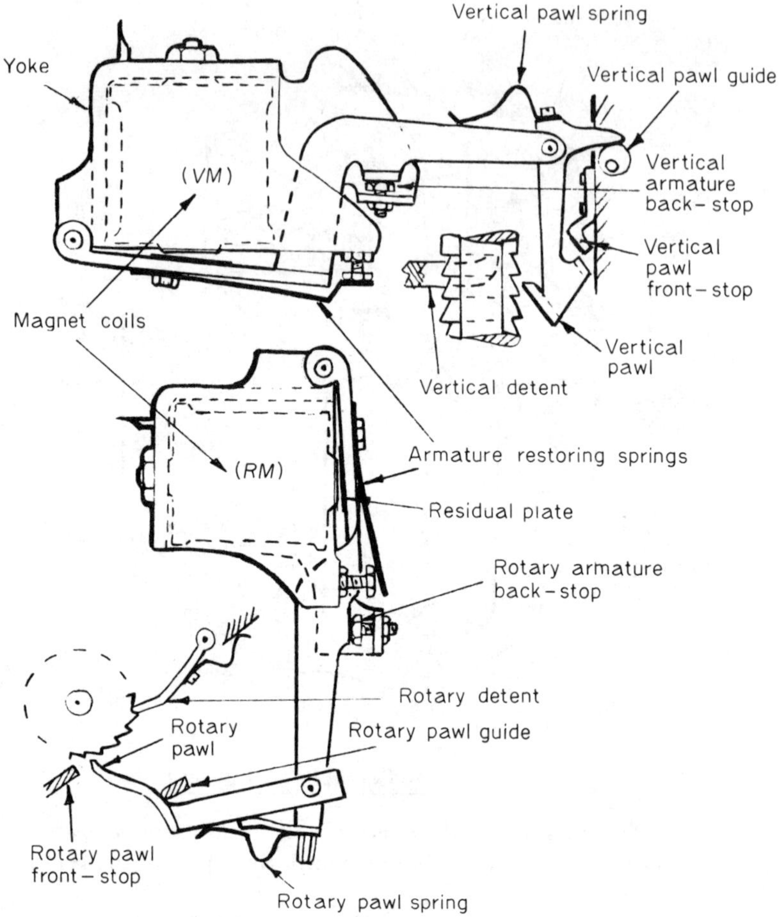

Fig. 5.20 Vertical and rotary stepping mechanism, 4000-type. Compare this with Figs. 5.8, 5.9, 5.14 and 5.16 (5.4.4).

of the selector. The release magnet armature is linked to the vertical and rotary detents so that when it operates both detents are lifted clear of their respective ratchets. The helical restoring spring inside the wiper carriage then returns the wipers to the normal position. During the rotary part of the release, the carriage is supported by the stationary (vertical) detent until the wipers have rotated clear of the bank. The stationary detent is then in line with the slot as it was during vertical stepping and the wipers are then restored vertically.

An extension of the release magnet armature is provided at the front of the switch for manual release of the mechanism. An auxiliary magnet can be provided which disengages only the rotary detent.

Interrupter contacts of the toggle type can be provided. These are similar in principle to those of the 2000 type but are of different construction, as shown in Fig. 5.19.

The rotary ratchet carries a rotary cam. The *NR* and *S* springsets are positioned adjacent to this cam so that it bears directly on to their auxiliary armatures without the need for separate operating levers.

The vertical ratchet carries a vertical cam for the *N* and *NP* springsets. The *N*

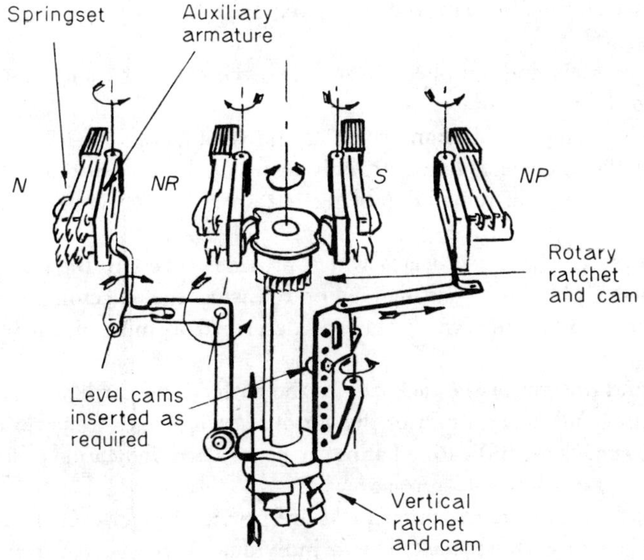

Fig. 5.21 Mechanically operated springsets, 4000 type. Compare this with Fig. 5.12. A second *NP* operating lever bearing on the same vertical cam can be provided when separate *NPA* and *NPB* actions are required (5.4.4).

springset assembly has an auxiliary armature operated by a linkage of horizontally pivotted operating levers. The *NPA* and *NPB* springs are operated by vertically pivotted levers similar to the 2000 type. These are shown in Fig. 5.21.

The performance of the selector is similar to that of the 2000 type.

5.5 Selector racks

In the exchange, the selectors are mounted on steel frameworks called racks. As can be seen from Fig. 4.10, there are quite large numbers of each type of selector in even a medium-sized exchange. The selectors of each type occupy several racks approximately 10 ft 6 in high and 4 ft 6 in wide. These racks are stood side by side to form a suite of racks and are held in place by a framework suspended from the ceiling of the apparatus room.

The selectors are mounted on horizontal members known as shelves, all the selectors on a shelf having their bank contacts connected as one multiple. The shelf multiples so formed may also be connected together in pairs to form a larger multiple. The multiple wiring is taken to terminals where further interconnexion of several multiples can be carried out (see Chapter 7—Grading).

On subscribers' line uniselector racks, the uniselectors and their control relays are fitted to the rack separately and the rack wiring connected directly to them. They cannot be removed without first disconnecting the wiring.

On two-motion selector racks, the banks are mounted on the shelves together with the selector cradles and selector shelf jacks. The complete selector consists of a mechanism attached to a relay mounting plate large enough to accommodate all the control relays required by that particular type of selector. This relay plate is pierced with a pattern of holes suitable for the relays as illustrated in Chapter 2 (Fig. 2.18). Attached to the rear of the plate is a plug assembly (Fig. 5.14) which engages with the shelf jack (Fig. 5.11) when the selector is positioned in its cradle. This arrangement permits the selector to be removed when required for servicing without disturbing the rack wiring.

The selector inlets and supplies are wired to the shelf jack and the selector outlet connexions to the bank contacts.

Each shelf is equipped with bank contacts and shelf jacks for ten selectors. To allow for growth of the exchange, however, only some of these positions will be occupied by selectors initially, the remainder being left spare for selectors to be added later as required.

The selectors each have a test jack which provides access to the selector circuit for testing purposes, plugs for busying out or releasing the selector and an indicating lamp if required. The function of these is described in more detail in Chapter 6—Circuits.

The left-hand upright of the rack carries the battery and earth supply bus-bars and fuses; the right-hand upright carries the various services such as alarm control relays. At the top of the rack are indicating lamps giving an alarm indication of fault conditions such as failure of a selector to release.

In addition to selector racks, other racks are provided in an exchange for mounting relay circuits not directly associated with individual selectors. An example of this is the signalling circuits used on junctions. These are equipped as relay sets which are similar to the relay mounting plates of selectors and fit similar shelf jacks but without contact banks or multiple wiring.

A typical group selector rack is shown in Fig. 5.22.

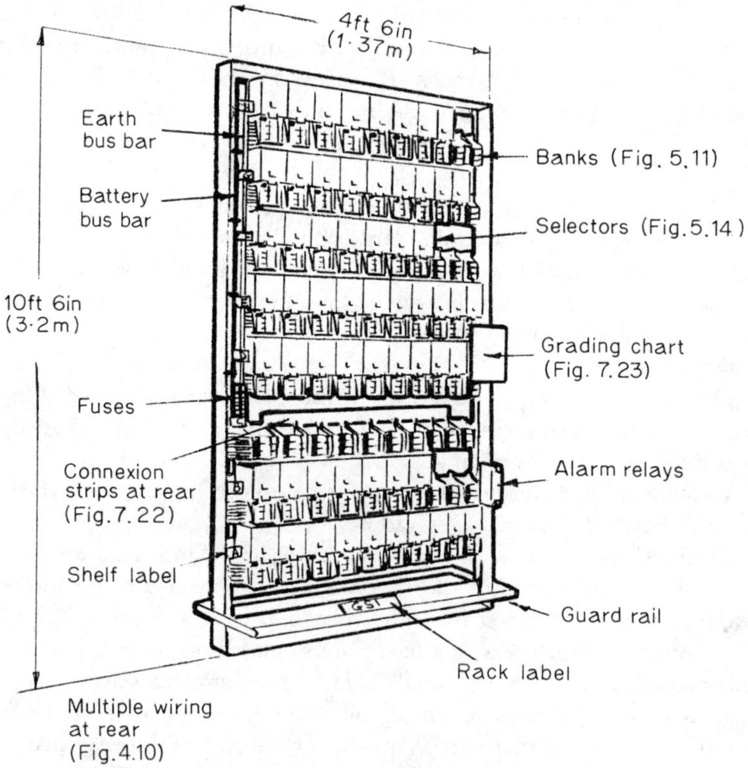

Fig. 5.22 Group selector rack, 2000-type. The empty banks allow for growth by the addition of selectors later (5.5).

PART 4 SUMMARY

5.6 Practical selector mechanisms

Two-motion selectors: pre-2000, 3-magnet, not standardized.
2000 type, 2-magnet, rotary release.
4000 type, 3-magnet, *ZM* release.

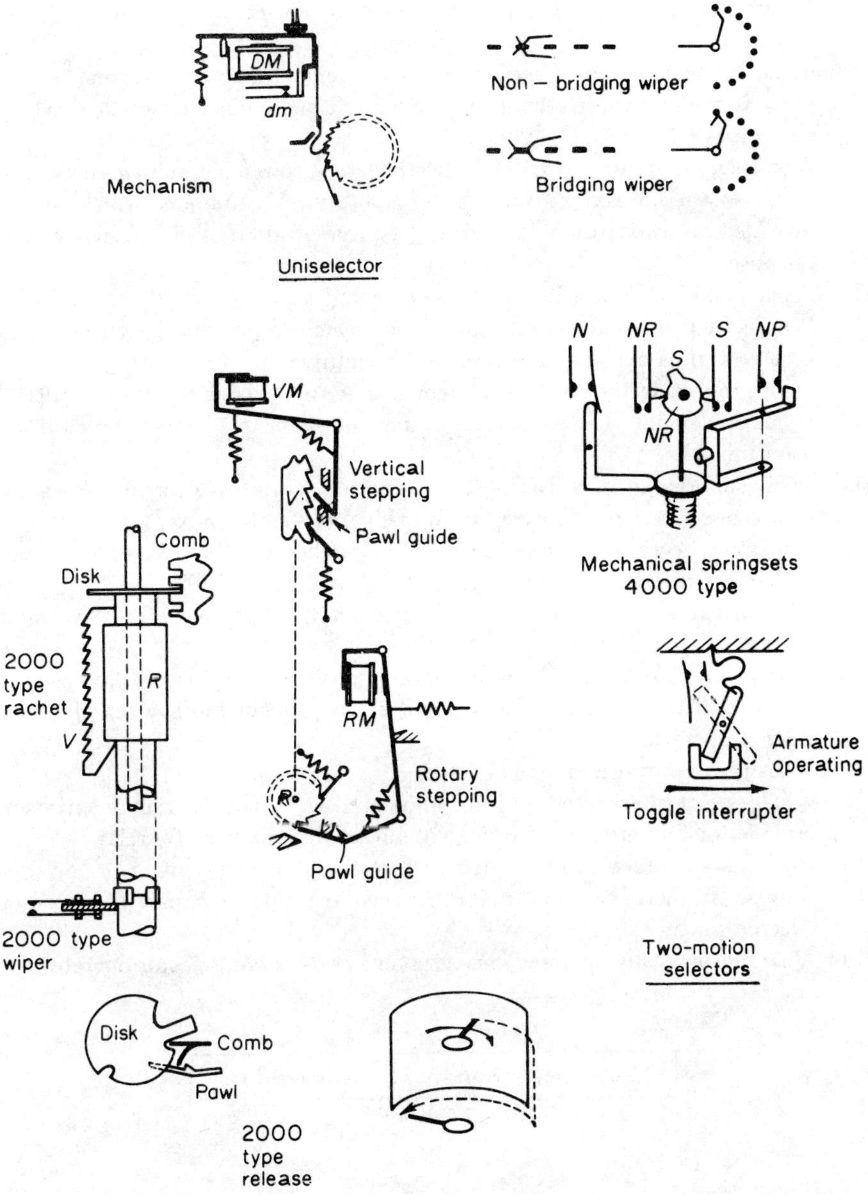

Fig. 5.23 Practical selectors—summary.

Now try the questions on page 139 before going on to the next chapter.

PART 5 SELF-TEACHING NOTES

5.7.1 Read **5.1.1** to **5.3.4** and revise **4.3.3** and **4.3.4**, which deal with the trunking principles and selector multiples.

5.7.2 After reading **5.1.1** to **5.3.4**, try the following questions.

Q.1 How many 'levels' has the bank of a two-motion selector?

Q.2 What is the name given to the component of a selector mechanism which acts on the ratchet tooth to step the wipers?

Q.3 What is the function of the detent in a reverse-acting mechanism?

Q.4 How is the interrupter contact of a uniselector connected to give self-drive conditions? (Sketch or describe.)

Q.5 How many ranks of two-motion selectors are required in a four-digit exchange?

Q.6 What is the maximum number of subscribers' numbers which could be provided on a four-digit exchange? (Assume all digits to be available for this purpose.)

Q.7 What is the difference between a hunter and a finder?

Q.8 Why is each two-motion selector *not* connected to just one selector on each of its levels, thus avoiding the need for a multiple and for hunting?

Q.9 On a group selector level, which contact, if any, carries the fewest calls?

Q.10 What advantage has a reverse-acting mechanism over a forward-acting mechanism?

Q.11 What name is given to the type of uniselector wiper which provides a make-before-break action in moving from one contact to the next?

Q.12 How many teeth would be needed on the ratchet wheel of a uniselector having ten outlets in a bank occupying one third of a circle (120°)?

Q.13 Give one reason for *not* using a reverse-acting mechanism for a two-motion selector?

Q.14 What would you adjust the position of on a two-motion selector to ensure that the vertical pawl moves into the root of the ratchet tooth when the vertical magnet operates?

Q.15 What is the purpose of a pawl spring?

Q.16 What type of interrupter contact can be used on forward-acting mechanisms to provide for self-drive conditions without an 'interacting' (*G*) relay?

Q.17 How many contacts are provided on a vertical marking bank?

Q.18 Why must there be a clearance between the tips of a pair of two-motion selector wipers?

Q.19 What mechanically operated springset on a two-motion selector operates on the first rotary step?

Now check your answers on page 263.

5.7.3 You should have answered fifteen out of these nineteen questions correctly. If you scored less than this, read **5.1.1** to **5.3.4** again and try them once more.

With fifteen or more questions correct, read the rest of the chapter (**5.4.1** to **5.6**).

5.7.4 The whole of Chapter 5 should have been studied before attempting the following questions.

Write complete answers to these questions spending no more than half an hour on each.

Check each answer on page 263 before starting on the next question.

Q.20 Describe:

(*a*) a forward-acting stepping mechanism,

(*b*) a reverse-acting stepping mechanism.

Illustrate your answers with sketches adequate to show details of the ratchet, pawl and detent in each case. (C & G 1964)

Q.21 Sketch and describe the construction of the mechanical arrangements provided on a two-motion selector for operating contacts when the wipers are:

(*a*) off-normal,

(*b*) raised to a particular level, and

(*c*) stepped in to any level.

What other mechanically operated springsets may be fitted to the mechanism and under what circumstances do they operate? (C & G 1963)

Q.22 Sketch and briefly describe the general construction of a 220 point bank for a two-motion selector.

What is the relationship between the layout of the bank contacts and the tooth spacing of the rotary and vertical ratchets of the selector mechanism? (C & G 1959)

6 · STROWGER SYSTEM—SELECTOR CIRCUITS

For self-teaching notes, see page 171.

PART 1 INTRODUCTION

6.1.1 Relay control of selectors. The selector mechanisms described in the last chapter can only perform the simplest of operations on their own. All they can do in fact is to move their wipers step by step as instructed by suitable pulses of current to the appropriate magnets.

Relays, of course, are even more limited individually but they can be connected together in groups to form circuits capable of quite complex sequences of operations. Circuits of this type are used for controlling the stepping, hunting and releasing functions of selectors.

In the first automatic exchange, installed by Almon Strowger's nephew Walter Strowger at La Porte in 1892, each subscriber was connected to his selector at the exchange by five wires. Three of these wires were for the three keys on which the hundreds, tens and units digits respectively of the required number were tapped out; another wire was for speaking over; and the fifth wire was for releasing the selector. Earth return was used for the speaking and signalling currents.

On modern systems, the subscriber is connected to the exchange by only two wires, without earth return, for all speaking and signalling purposes. Instead of the subscriber deciding which magnet to operate on the selector, a relay circuit is provided to do it. Also as we saw in the last chapter, to avoid the need for each subscriber to have his own selectors, relay circuits are provided to enable each selector to choose a disengaged selector in the next rank to receive the next digit—a process we have called 'hunting'.

If the selector mechanism is regarded as replacing the plugs and cords of the manual exchange, then the relays replace the operator's thought processes while setting up a call.

If all this makes automatic switching sound like a very complicated way of doing what appears to be the simple job of switching a call, you must remember that in an automatic exchange, the apparatus has to make all the decisions that the operators are employed to carry out in a manual exchange. On a local call, for example, these decisions are briefly as follows:

See who wants a call.
Find a free cord circuit.
Find out what number is required.
Decide that the call is a local call.
See if the required line is already engaged.
If it is engaged, tell the caller.
If it is free, ring and wait for an answer.
Charge the caller for the call.
Wait for the call to finish before disconnecting it.

6.1.2 In this chapter, we shall first examine the circuits in terms of what are called circuit elements. These are little bits of the complete circuits, the idea being that each element is concerned only with one specific function, (e.g. one of those listed above) making it that much easier to follow. By this means, you will learn enough of the principles of operation of such circuits for you to be able to work out for yourself the operation of any selector circuits you may meet in future.

To show you how the elements go together to make complete circuits, a number of typical selector circuits are given at the end of the chapter.

The circuit symbols for relays have already been given in Chapter 2 and those for selector mechanisms in Chapter 5. The diagrams in this chapter consist entirely of symbols of this sort and it is important that they be understood. To help you, all the symbols used here are shown in Appendix 5 so that you may refer to them when in doubt.

It is also necessary for you to be familiar with the detached contact system of diagrams as described in **2.4**.

PART 2 CIRCUIT ELEMENTS

6.2 Line signalling

The signals used on automatic exchange subscribers' lines have already been covered in Chapter 3. From the calling subscriber to the exchange, there is a loop for calling, 'break' pulses for dialling and disconnexion for clearing. The ringing signal from the exchange to the called line is a low frequency alternating current (e.g. $16\frac{2}{3}$ Hz, see **1.3.3**) and information on the progress of the call is returned to the caller in the form of audible tones. From the called line to the exchange, the answer signal is a loop and the clearing signal a disconnexion. The connexions of the line relays for these signals are shown for each of the selectors in a connexion in Fig. 6.1.

The initial calling signal is received on relay *LR* in the subscriber's line circuit. Contacts of this relay cause the uniselector to commence hunting for a free first selector under the control of a second relay *K* of which only the contacts have been shown in Fig. 6.1. When the uniselector finds a free first selector, this relay *K* operates to disconnect *LR* and extend the calling signal through to the selector's line relay *A*.

Relay *LR* is thus only connected to the line at a time when the subscriber is neither sending nor receiving speech, tones or pulses. It is not, therefore, necessary to go to the expense of providing two equal windings on it to maintain the line in a balanced condition (see **3.3.3**).

The first selector *A* relay, however, is used as a transformer to connect Dial Tone to the calling subscriber and must have balanced windings and nickel-iron sleeves (see **2.2.8** and **3.3.7**). This relay also receives the dial pulses for the first digit and its contact repeats them to the selector vertical magnet. To keep the distortion to a minimum, relay *A* has an isthmus armature (see **2.2.7**).

At the end of the first digit (pulse train), the first selector hunts for a free outlet to a second selector. When it finds a free second selector, relay *CD* (relay *H* in some designs) operates and disconnects its own *A* relay and extends the calling loop through to the second selector *A* relay. If the first selector fails to find a free second selector, its *A* relay is used as a transformer to connect Busy Tone to the caller.

The circuit of the second selector is the same as the first selector except that it does not provide Dial Tone.

The final two digits are received on the final selector *A* relay which repeats the pulses of one train to the vertical magnet and those of the last train to the rotary

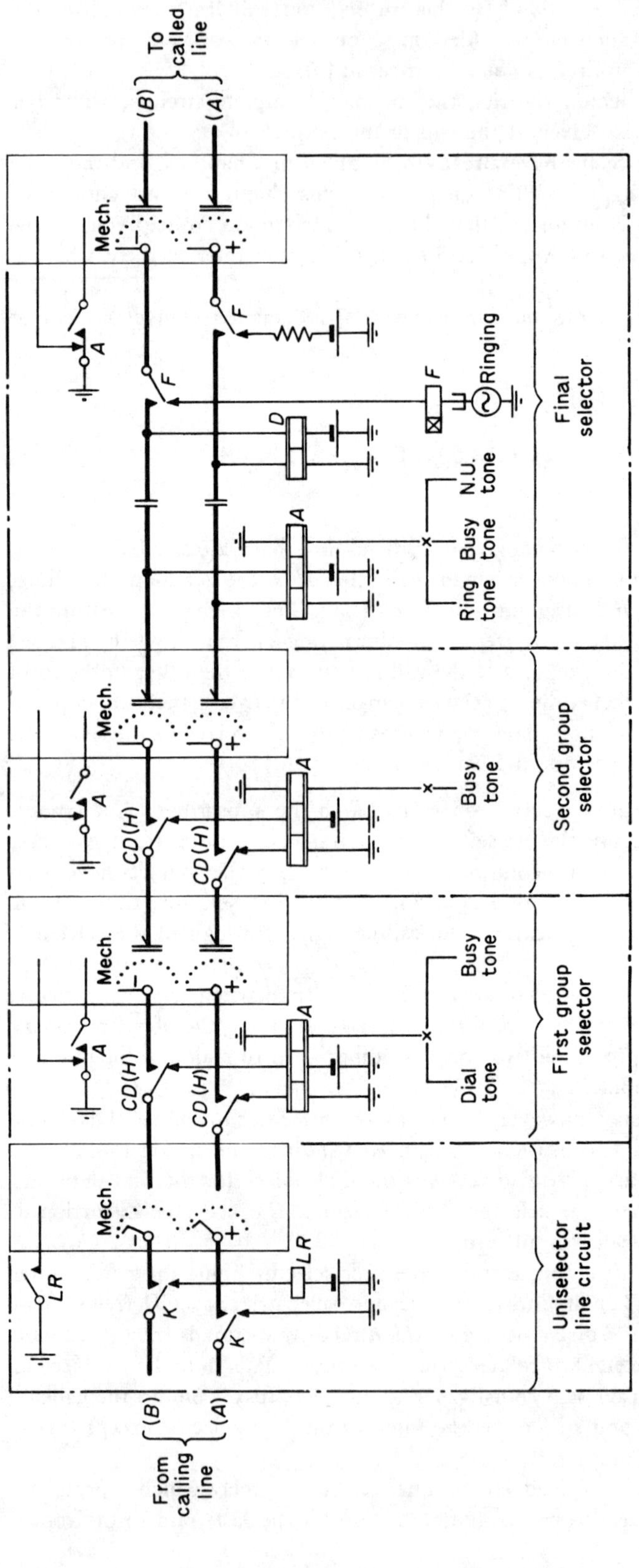

Fig. 6.1 Line signalling conditions. The K contacts shown in the line circuit operate when the uniselector finds a free first selector. The CD (or H) contacts shown in the group selectors operate when the selectors find a free outlet in the required level (6.2).

magnet under the control of certain other relays not shown in Fig. 6.1. The called line is 'tested' to determine whether it is free and the appropriate tone returned to the caller via relay A in the final selector.

Ringing is applied to the called line, if free, via relay F as described in **3.5.1**. When the called subscriber answers, looping the line, relay F operates and relay D is then connected to the called line. Relay D operates to control the operation of the calling subscriber's meter to charge him for the call.

Relays A and D in the final selector form a transmission bridge (see **1.3.6**).

When the called subscriber clears, disconnecting the calling loop, relay A in the final selector releases and, through other relays not shown in Fig. 6.1, causes the selectors to be restored to normal.

The following sections are concerned with how these line signals and line signalling relays together with other relays in the selector circuits cause the selectors to step on the dial pulses, to self-drive and to hunt for free outlets and finally to release.

6.3 Pulse-controlled stepping circuits

The basic principle of pulse-controlled stepping is shown in Fig. 5.1a. It can be applied to any type of ratchet operated switch of either the two-motion or uniselector type. In the simpler type of Strowger exchange, however, only the two-motion selectors employ pulse-controlled stepping. It will be sufficient, therefore, for our present purposes to consider only the circuit elements for selectors of this type.

6.3.1 ***B*** **relay.** Dial pulses are received from the line in the form of 'breaks' or disconnexions of the line current. The selector magnet must operate once for each break, so we need to connect it to a break contact of the line relay A. The magnet thus operates each time relay A releases on a disconnexion of the line current.

In this simple form, the magnet would be permanently energized when the circuit is idle, relay A being unoperated under this condition. A second relay B as shown in Fig. 6.2 prevents this.

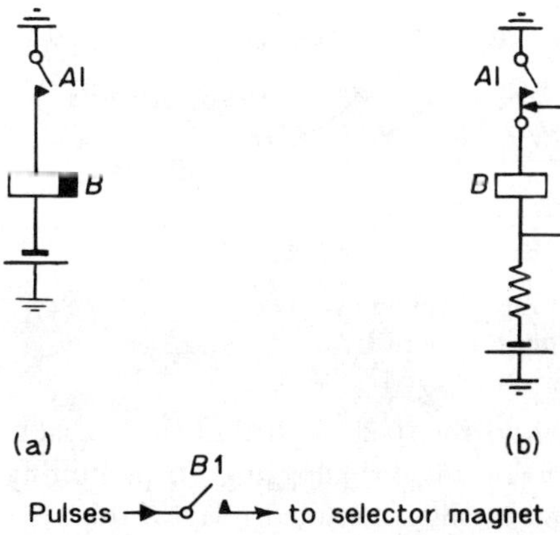

Fig. 6.2 Principle of B relay. The slugged relay used mainly on pre-2000-type circuits is shown at (a) and the short circuit or (stehlic) relay, used in most modern circuits, is shown at (b) (6.3.1).

When the circuit is first taken into use, the calling loop operates relay A which operates relay B at contact $A1$. Relay B is made slow to release so that it does not release during the short time that relay A is released on each break pulse. If the subscriber clears, however, the loop is disconnected for a much longer period and relay A is released for long enough to release B as well. A break contact of relay B can, therefore, be used to initiate release of the selector. Relay B thus also distinguishes between dial pulses, during which it does not release, and clearing signals, during which it does release even though both types of signal consist of disconnexions (**3.3.4**).

The variation of the magnetic flux in the B relay is shown graphically in Fig. 6.3.

If the current is disconnected for long enough, the relay releases when the flux decays to its release value (see also **2.2.4**). This is shown in Fig. 6.3a.

If the current is disconnected for one pulse as in Fig. 6.3b, the flux builds up again during the following 'make' pulse. It has not time, however, fully to regain its original value before the commencement of the second break pulse. During this second pulse, the flux therefore falls to a lower level than it did during the first pulse.

On a continuous pulse train, the flux variation settles about a steady value after a few pulses as shown in Fig. 6.3c.

To ensure that the minimum flux *D* is still sufficient to hold the relay operated even after a large number of pulses, the release lag of the relay under static conditions as shown in Fig. 6.3a must be considerably longer than the duration of a single break pulse. In practice, a release lag of 250 ms minimum is required of a slugged relay of the type shown in Fig. 6.2a.

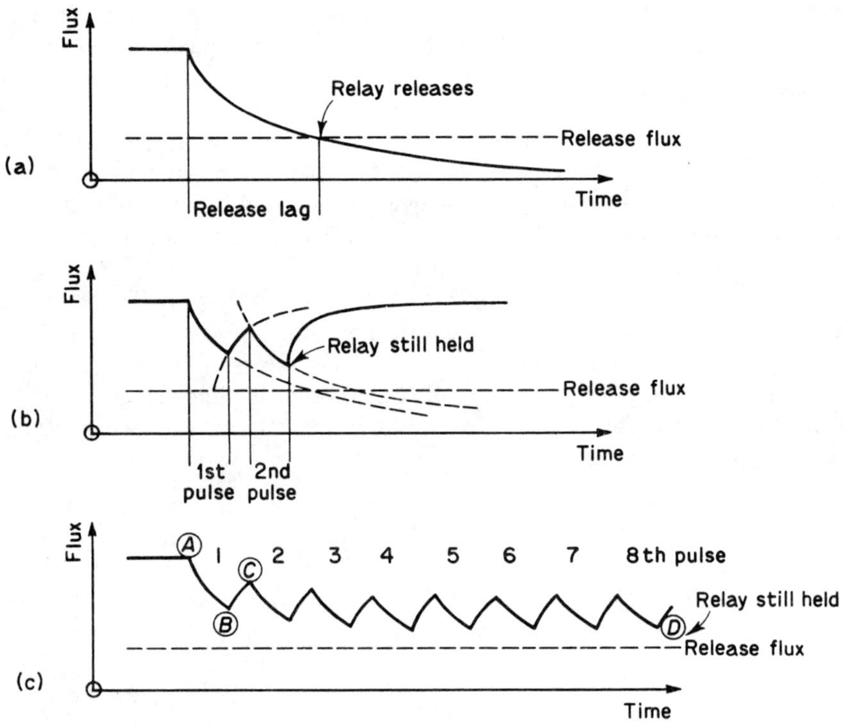

Fig. 6.3 B relay flux. (a) relay releasing, (b) relay holding over two pulses and (c) relay holding on continuous pulse train (6.3.1).

An advantage of the short circuit relay shown in Fig. 6.2b is that all the available winding space is used both for operating the relay and for 'slugging' it, permitting greater spring loads than with the slugged relay. A short circuited relay of this type is sometimes referred to as a 'stehlic' *B* relay.

An examination of Fig. 6.3c will show that a faster (i.e. steeper) rise of flux (*BC*) will permit the use of a higher value of release flux without the relay releasing on long pulse trains. Consequently the relay does not need such a long release lag and can, therefore, carry more contacts. This condition is also met by the use of a relay without a slug as in Fig. 6.2b. The release lag is produced by connecting a short circuit across the relay to provide a path in which the back e.m.f. can circulate a current (**2.2.6**).

In the case of the slugged relay, although the use of a heel-end slug has negligible effect on the operate lag of the relay ordinarily, it does delay the build-up of flux in a relay which has already operated. This is because the armature is in the operated position giving a low reluctance air-gap and therefore the same magnetic circuit as when the relay is releasing (**2.2.5**).

The short circuit type of *B* relay has additional advantages when used in conjunction with a selector magnet as we shall see later (**6.3.3**).

6.3.2 ***CD* relay.** In addition to relays *A* and *B*, a third relay is required in a pulse-controlled stepping circuit. This is designated '*CD*' (the use of the designation '*C*' is avoided for relays because it is used for capacitors).

The function of the *CD* relay is to determine when the dial pulse train has ended. Its contents are used both to prevent the selector magnet from being stepped by further trains of pulses and to cause the selector to perform its next set of operations. In the case of a group selector, this means commencing to hunt for a free outlet.

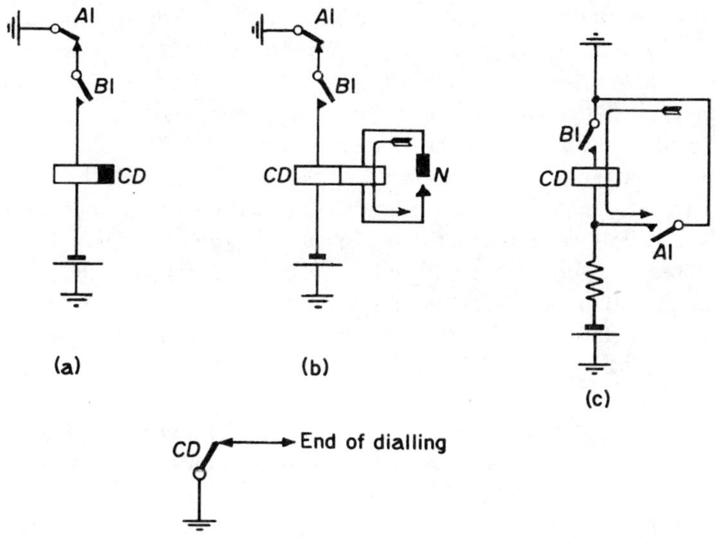

Fig. 6.4 Principle of *CD* relay. The slow release can be obtained by means of a slug as in (a), a short circuit second winding as in (b) or a short circuit during the 'make' (*A* operated) pulses as in (c) (6.3.2).

The principle of operation of a *CD* relay is shown in Fig. 6.4. In the examples shown there, *CD* operates during the first break pulse of the train and is provided with a slow release feature so that it will not release during the short period for which relay *A* is operated on the make pulses (i.e. between the break pulses).

In practical selector circuits, relay *CD* operates on initial seizure of the circuit (when both *A* and *B* first operate) rather than on the first dial pulse. This is known as pre-operating relay *CD* and some typical circuits employing pre-operated *CD* relays are described in **6.3.3**.

If a *CD* relay is not pre-operated, it must be only lightly loaded (i.e. carry few springs) to prevent its releasing during the first make period because the flux has only had the duration of one break pulse in which to build up. In the pre-operated case, however, it has already reached its full value long before the first dial pulse commences.

6.3.3 Selector vertical stepping circuit. A group selector is only required to respond to one digit. This single train of pulses is used to step the wipers vertically and the wipers then self-drive to hunt for a free outlet.

To control the vertical stepping, the selector needs all three of the relays we have discussed. Relay *A* receives and inverts the pulses. Relay *B* discriminates between dialling and release signals and prevents permanent energization of the selector magnet. Relay *CD* determines when vertical stepping shall cease and rotary hunting commence.

The selector line signalling relay *A* is required to operate on relatively small currents in series with the resistance of the line wires and is therefore provided with only one contact unit. This single contact has to be used for controlling relays *B* and *CD* as well as pulsing the magnet.

For a slugged *B* relay, this simply means connecting *B* to the make of a change-over and the *VM* magnet circuit to the break action. This is shown in Fig. 6.5a.

The arrangement for the short circuit *B* relay is different as shown in Fig. 6.5b. Here the *VM* magnet circuit replaces the resistor previously shown for the battery connexion in Fig. 6.2. The resistance of the relay is chosen so that when *A* is operated, the current in the magnet in series with the coil of relay *B* is too low to hold the magnet operated. When *A* releases the magnet operates to current from earth via contact *B*1 and the break portion of contact *A*1.

A further advantage of the short circuit *B* relay compared with the slugged version is that in Fig. 6.5a contact *A*1 interrupts the magnet circuit so that the back e.m.f. from the highly inductive magnet causes sparking which damages the contact. In Fig. 6.5b, when *A* operates there is still a current path through the magnet via the coil of relay *B* without the current jumping any contact gap as a spark. The back e.m.f. of the magnet assists the start of current in the coil of *B* on each make pulse so that relays of this type will function satisfactorily with release lags as short as 150 ms.

The pre-operate circuit of *CD* in each case is via the higher resistance winding of relay *CD*. Once the selector has taken one vertical step on the first pulse of the train,

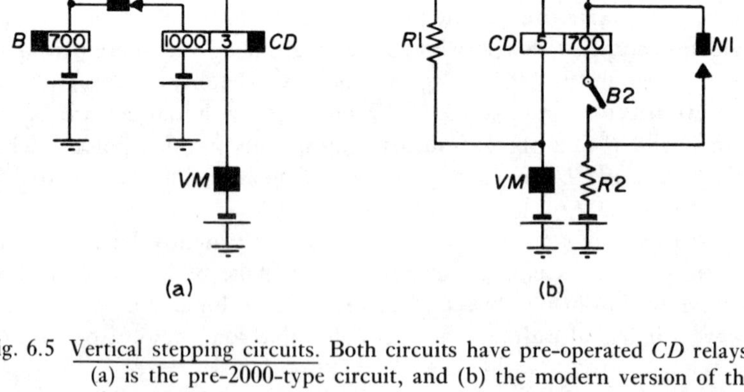

Fig. 6.5 Vertical stepping circuits. Both circuits have pre-operated *CD* relays. (a) is the pre-2000-type circuit, and (b) the modern version of the element (6.3.3).

*N*1 contact operates so that the pre-operating path is ineffective. Relay *CD* is then only energized on the break pulses (i.e. when *A* releases).

The use of a make-before-break (K-action) contact on relay *A* (contact *A*1) gives the maximum length pulse to both *B* and *CD* without the loss which would occur due to transit time of a simple change-over.

The operation of the vertical stepping circuit of Fig. 6.5b as used in modern selector circuits may now be summarized as follows:

Seizure: Loop operates *A*.
*A*1 operates *B* via *R*1. *B*2 operates *CD* on 700 Ω coil.
Dialling: *A* releases once for each break pulse.
Each release of *A* operates the magnet once.
On first vertical step *N*1 operates.
*N*1 short circuits 700 Ω coil of *CD*, making *CD* slow to release.
Break pulses energize *CD*, make pulses energize *B*. *B* and *CD* remain held.
End of digit: *A* re-operates at end of each pulse. *A* therefore operates at end of last pulse.
*A*1 holds *B*.
*A*1 releases *CD*.
*CD*1 disconnects the magnet operating path.

6.3.4 Selector vertical and rotary stepping circuit. A final selector must respond to two trains of pulses, one stepping the wipers vertically and the other one stepping them in a rotary direction. The vertical stepping is basically the same as that already described for the group selector circuits. The line relay *A* receives the dial pulses and controls the stepping of the vertical magnet. Relays *B* and *CD* are provided with slow release features so that both hold during the pulse train. Relay *CD* releases at the end of the train of pulses.

Once *CD* has released at the end of the first of the two trains of pulses, another relay *E* is brought into use. Its function is to prepare the circuit for the next digit and to arrange for it to be directed to the rotary magnet instead of to the vertical magnet. It is designated *E* rather than *D* because the relay providing transmission bridge facilities to the called subscriber is called *D* (see Fig. 6.1).

The basic circuit for relay *E* is shown in Fig. 6.6.

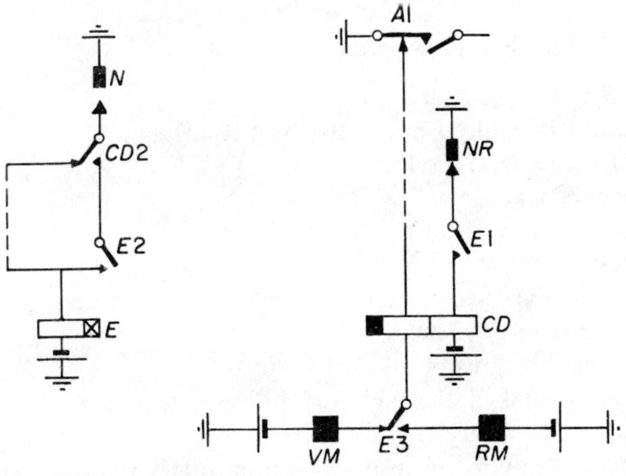

Fig. 6.6 Principle of final selector *E* relay. The *CD* relay shown here is as shown in Fig. 6.5a. Compare this with Fig. 6.7 (6.3.4).

When *CD* releases at the end of the first of the two pulse trains, *CD*2 operates *E*. (*N* contact prevents *E* from being operated before the first digit.)

*E*1 now re-operates *CD* in readiness for the second digit while *E*2 holds *E* to prevent it from being released by the operation of *CD*2. A further contact, *E*3 switches the pulsing circuit over from the vertical magnet to the rotary magnet. The circuit is arranged so that the functions performed by *N* contacts during vertical stepping are performed by *NR* contacts during rotary stepping.

At the end of the second and final pulse train, *CD* again releases. Other contacts of *CD*, not shown here start the next sequence of operations—testing the line for busy or free conditions.

The combination of relays *B*, *CD* and *E* as used in modern final selector circuits is shown in Fig. 6.7.

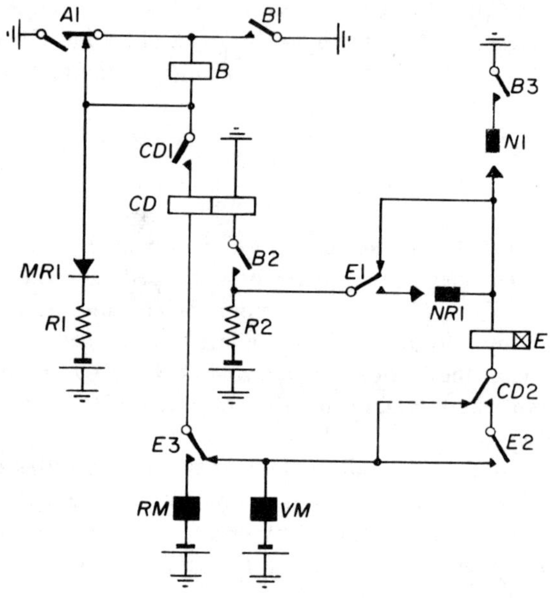

Fig. 6.7 Circuit for vertical and rotary stepping as used in 2000-type final selectors (6.3.4).

The operation of this element is as follows:

Seizure: *A*, *B* and *CD* operate.
1st digit: *B* and *CD* hold, *A* pulses the vertical magnet.
At the end of the digit, *CD* releases.
B remains operated.
*CD*2 releasing, operates *E*.
*E*1 re-operates *CD*.
The short circuit is applied to relay *CD* during vertical stepping by *N*1 contact.
2nd (final) digit : *B* and *CD* hold.
A pulses the rotary magnet.

CD releases at the end of the digit and prepares the next stage of the circuit operation.

The short circuit is applied to relay *CD* during rotary stepping by *NR*1 contact.

Relay *E* has a front-end slug, making it slow to operate by delaying the movement of the armature until the flux has built up to nearly its full value (**2.2.5**). This ensures that the relay is sufficiently fluxed before its contacts operate relay *CD* to prevent it releasing during the transit time of contact *CD*2.

6.4 P-wire and rotary hunting circuits

6.4.1 Use of the 'private' wire (P-wire). Rotary hunting means that the wipers of a switch self-drive until they reach an outlet connected to a free selector in the next rank. By a 'free' selector is meant one which is not already engaged on a call, having been taken into use by another switch in the same multiple as the one we are considering. On reaching such a free outlet, the wipers stop and the calling line is connected through as in Fig. 6.1.

The circuit of the hunting switch must inspect each outlet in turn to see if it is free. This process is known as *outlet testing*.

The testing process must not be carried out on the line or 'speech' wires (− and + wires), because this would interfere with calls already in progress on the busy outlets. A third wire is therefore provided on all connexions through the exchange, the potential of which indicates the condition of the circuit. This avoids intrusion on calls in progress and is called the private wire, usually abbreviated to '*P*-wire'.

A very commonly used system is for all busy circuits to have their *P*-wires at earth potential and all free circuits to have a disconnexion or absence of earth on the *P*-wire. The outlet testing then consists of testing for an 'absence-of-earth' condition, as illustrated in Fig. 6.8.

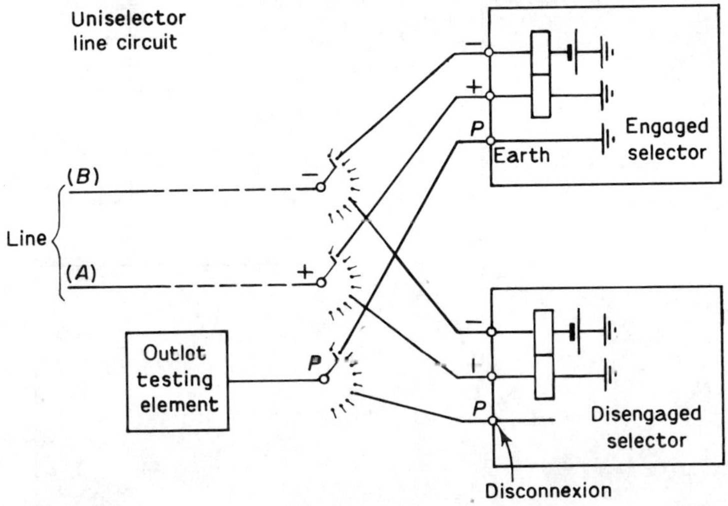

Fig. 6.8 Private wire (*P*-wire) illustrating absence-of-earth testing (6.4.1).

There are certain advantages to using a battery potential as a definite indication of a free *P*-wire instead of the more negative indication of 'no earth'. Systems of this sort are known as battery testing **(6.4.4)**.

The reason for the earth on the busy *P*-wires in both absence-of-earth and battery testing systems is that once the calling loop has been switched through to the next circuit, an alternative holding condition is necessary. This consists of holding the switching relays to the *P*-wire earth as shown in Fig. 6.9. These are the relays whose contacts are shown in the − and + wires in Fig. 6.1.

When the caller clears at the end of the call, the final selector removes the *P*-wire

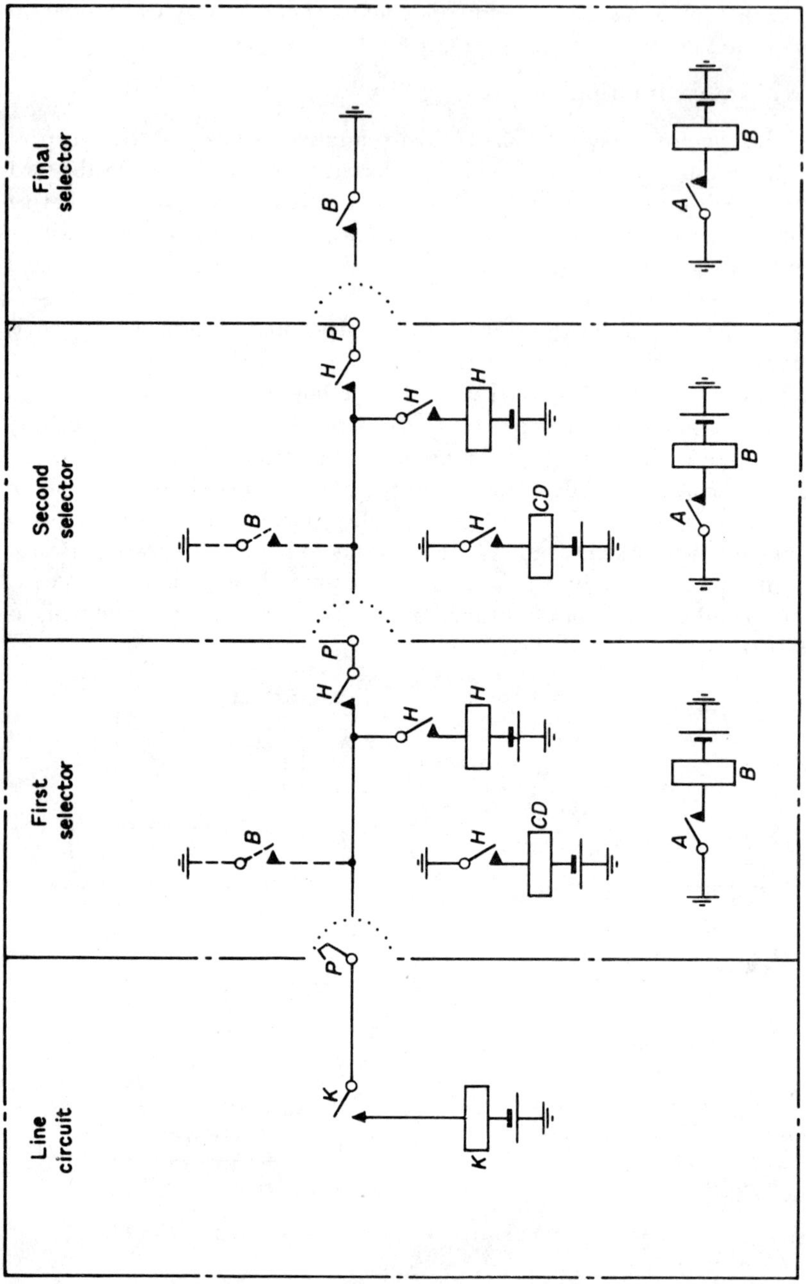

Fig. 6.9 P-wire holding conditions. The contacts of these relays in the line wires are shown in Fig. 6.1 (6.4.1).

earth to release the *H* or *K* relays of the other selectors and these in turn release their selector mechanisms (**6.5**).

The method illustrated in Fig. 6.9 is described as *backward holding*. The transmission bridge which provides the holding earth in this case is in the last circuit in the connexion. In some exchanges, such as director exchanges, the transmission bridge and holding earth are at the first selector and *forward holding* is used to subsequent selectors.

6.4.2 Uniselector absence-of-earth testing. The self-drive conditions for a uniselector using a reverse-acting mechanism have already been discussed (**4.3.3** and

Fig. 5.1). A combined self-drive and outlet testing circuit suitable for uniselectors is shown in Fig. 6.10, and the operation is as follows:

Start: *LR* operates to calling loop as in Fig. 6.1.

*LR*1 contact operates *DM* magnet via first bank contact, uniselector *P* wiper and *K*1 contact.

Relay *K* is short-circuited by earth on both sides of its coil (via *P* and *K*1) and so does not operate

The operation of *DM* causes the interrupter contact *dm* to open, releasing *DM*. The release of *DM* steps the wipers one step (reverse-acting mechanism **4.3.3**).

Busy outlets: These are indicated by earth on the *P*-wire.

On such contacts, the magnet *DM* operates to the *P*-wire earth via the *P* wiper. *K* is still prevented from operating by having earth on each side, one from *LR*1,

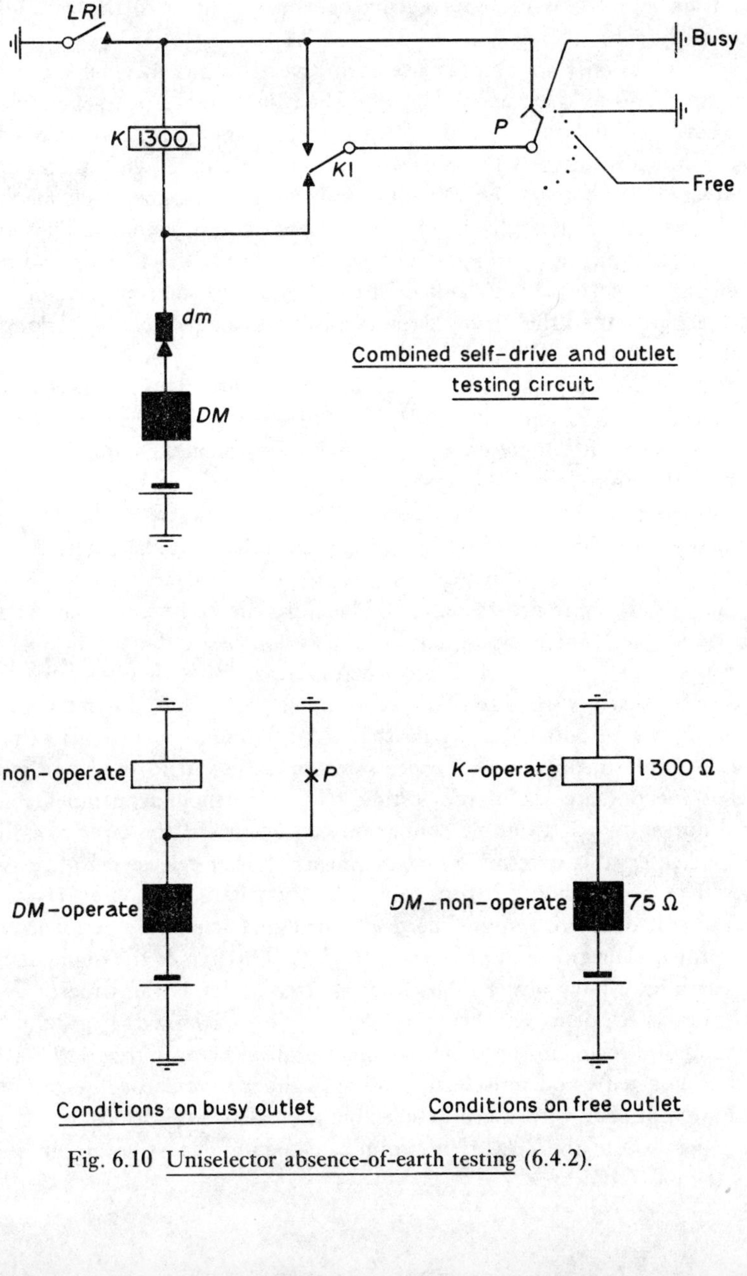

Fig. 6.10 Uniselector absence-of-earth testing (6.4.2).

the other one from *P*. *DM* opens *dm* as before and the wipers step. The wipers thus step off each busy outlet so that the switch self-drives over all busy outlets.

Free outlet: This is indicated by the absence of any earth on the *P*-wire to operate the magnet so the wipers stop.

There is now only one earth connexion to relay *K*, at *LR*1 contact. Relay *K* therefore now operates in series with the magnet *DM*. *K*1 contact connects earth from *LR*1 to busy the outlet to other selectors until the holding earth is returned.

The resistance of *K* is such that *DM* does not operate in series with it.

Contacts of *K* as shown in Fig. 6.1 connect the loop through to the next (group) selector, releasing *LR*. A holding earth is meanwhile returned over the *P*-wire by the group selector to hold *K* via *K*1 (operated) and to busy the outlet.

A bridging wiper is used because a non-bridging wiper would disconnect the short circuit from relay *K* as it stepped from one busy contact to the next allowing *K* to operate at the wrong time.

This circuit is only suitable for use with reverse-acting mechanisms. It is a feature of such mechanisms that the wipers step when the magnet is de-energized and therefore no current is flowing when the *P* wiper steps from a busy contact to a free contact. If this circuit were used with a forward-acting mechanism, current would be flowing in the magnet circuit via the *P* wiper as it stepped because such mechanisms step when the magnet is energized. The movement of the wiper would thus break the magnet current causing sparking which would damage the wiper tip and bank contact.

6.4.3 Group selector absence-of-earth testing. Two-motion selectors use forward-acting mechanisms so the testing circuit must be arranged to avoid taking the magnet current through the wiper-to-bank contact (**6.4.2**).

On the pre-2000 type selectors, the interrupters do not have a toggle action and an interacting relay *G* is required (**5.3.2**). The current for this relay can be taken through the wiper because it is only energized while the interrupter contact is operated. The current is therefore broken by the interrupter contact at each step as the magnet releases in preparation for the next step. Relay *G* is not necessary for selectors such as the 2000 type which have toggle interrupter contacts (**5.3.2**). The circuit shown in Fig. 6.11 is therefore used instead.

In this circuit, *H* is pre-operated well before the commencement of hunting and is arranged to hold on busy outlets but release on free outlets. The reason for doing it this way round is that heavily loaded relays release faster than they operate and outlet testing relays usually require a lot of contacts. We want the relay to perform its function on a free outlet quickly so that it will connect an earth to the *P* wiper as soon as possible to prevent any other selector testing in to the same outlet. In Fig. 6.10 it is the operate lag of the testing relay *K*, which determines the unguarded interval during which a double connexion can occur. In the circuit of Fig. 6.11, the unguarded interval is determined by the rather shorter release lag of relay *H*, assisted by the effect of the induced 'back' e.m.f. of the rotary magnet *RM*.

This circuit does not require the use of bridging wipers, which it may be recalled are not provided on two-motion selectors (**5.3.3**). This avoids the momentary connexion of the earth on a busy outlet to an adjacent free outlet as the wiper moves from one to the other as happens with bridging wipers. This effect of the bridging wiper could cause another selector in the same multiple to miss the only free outlet available to it. This does not matter on uniselectors because the switch can come round again for a second attempt, but with a group selector, the call would be lost.

The operation of the 2000 type group selector outlet testing circuit shown in Fig. 6.11 is then as follows:

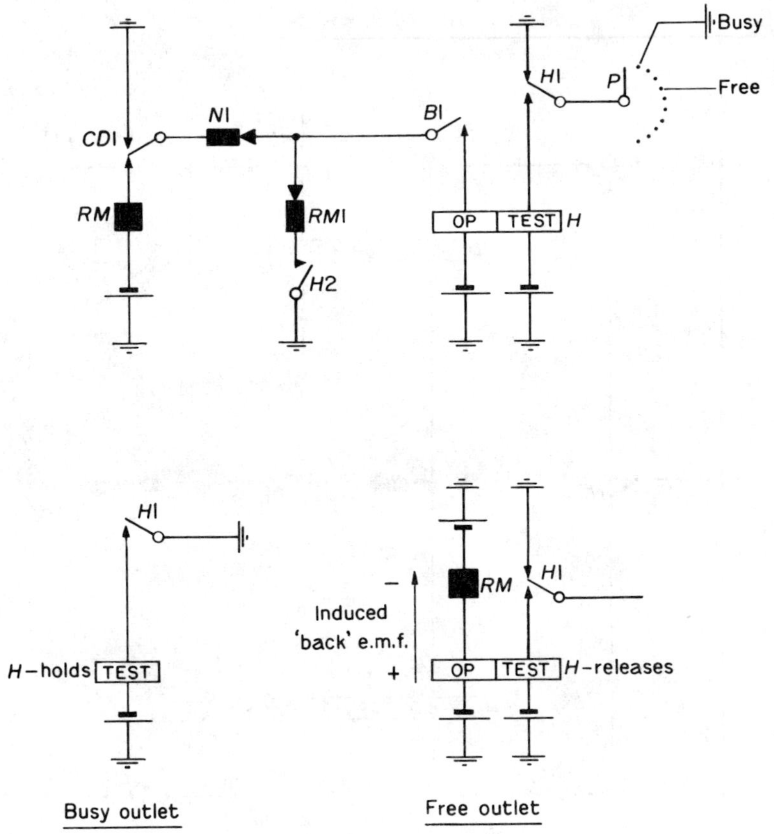

Fig. 6.11 2000-type circuit. Absence-of-earth testing for two-motion selectors with toggle type interrupters. Relays *B* and *CD* are as shown in Fig. 6.5b (6.4.3).

Before dialling: Relays *B* and *CD* operate as in **6.3.3**.

On the first vertical step, *N*1 operates *H*. (A contact of *H* may then be used to short circuit *CD* in place of the *N* contact shown in Fig. 6.7b.)

End of dialling: Relay *CD* releases as in **6.3.3**.

A self-drive circuit is connected for the rotary magnet *RM* via *H*2, interrupter contact *RM*1, *N*1 and *CD*1. *H* is held via *H*2 and *B*1.

RM operates, stepping the wipers.

First outlet busy: As the wipers step on to the outlet, *RM*1 contacts open disconnecting both *RM* and the operate winding of relay *H*. *RM* releases.

H holds on its 'test' winding to the *P*-wire earth via *H*1 and the *P* wiper. The release of *RM* closes contact *RM*1 which re-operates the magnet *RM* to step the wipers. The wipers thus self-drive over all busy outlets.

Free outlet found: *RM*1 contacts open as the wipers step on to the outlet as before. *RM*1 releases *RM* again but now also releases *H* because there is no earth to hold it on its 'test' winding.

Contact *H*2 prevents any further operation of *RM*, thus stopping the wipers on the free outlet. *H*1 connects earth to busy the outlet.

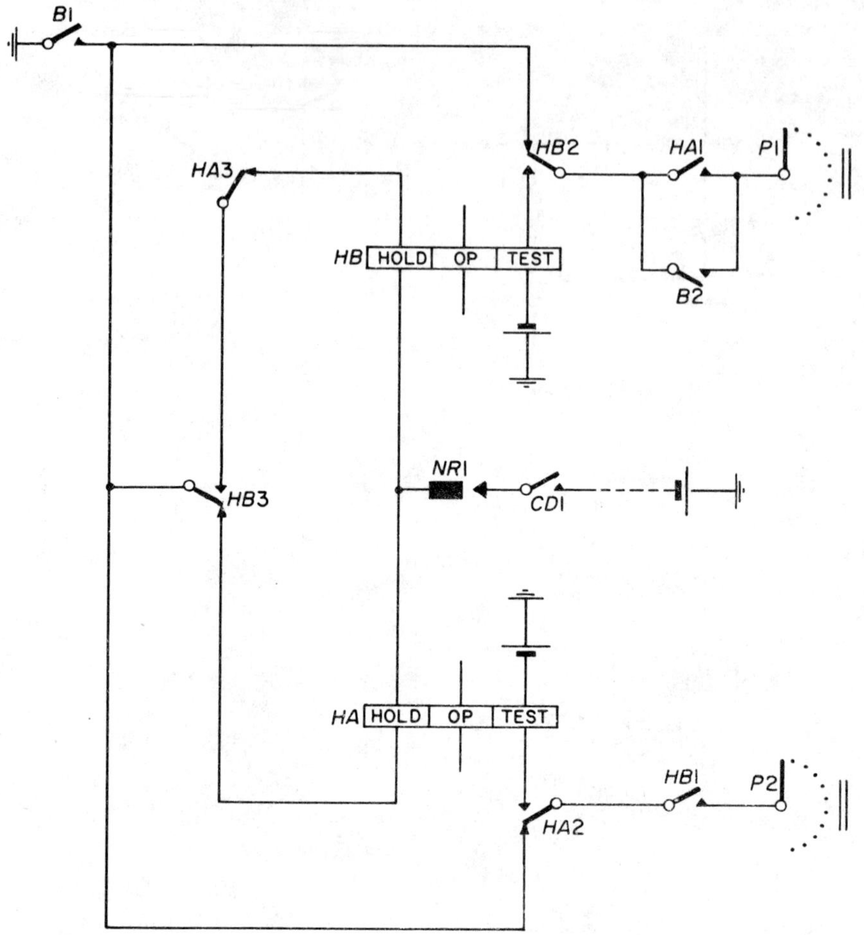

Fig. 6.12 Absence-of-earth dual testing. Principle of 2000-type, 200-outlet group selector testing (6.4.4).

When the wiper steps off the busy outlet, it disconnects the current in the 'test' winding of relay *H*. No sparking occurs, however, because the operate winding maintains the flux in the relay.

When *RM*1 disconnects *RM* and *H* on a free outlet, the back e.m.f. induced in *RM* exceeds that in the coil of relay *H*. Its direction is such as to tend to keep the current flowing in the magnet *RM* and therefore tends to reverse the current in *H* thus hastening its release. This reduces the unguarded interval on the *P*-wire of the selected outlet.

In any circuit using a pre-operated testing relay, it is necessary to have another relay which operates on a free outlet to switch the calling line through. It is convenient to use *CD* for this purpose. This is easily arranged by using selector *NR* contacts to connect the switching contacts of *CD* during vertical stepping only after rotary stepping and the pulsing circuit contacts of *CD* are completed, as shown in Fig. 6.20.

6.4.4 Absence-of-earth dual (200 outlet) testing. We saw earlier (4.3.3) that the double row of contacts on each level of a two-motion selector bank can be used to provide an outlet availability of 20 to give a total of 200 outlets on the selector. With this arrangement it is necessary to test both the upper and the lower contact at each rotary step, and to step the wipers on to the next contact only if both outlets are engaged. This means using two testing relays as shown in Fig. 6.12.

The dual testing circuit shown has pre-operated testing relays *HA* and *HB* similar to the single testing relay *H* of Fig. 6.11. Relay *HB* tests the *P*1 (lower contact) and *HA* tests the *P*2 outlet (upper contact).

Either relay releasing, on finding a free outlet, is arranged to operate *CD* again, and this provides a holding circuit for the other one of the two relays.

If both outlets are free, *HA* and *HB* both release and the operation of *CD* then re-operates *HA*.

Relay *HA* then holds to the holding earth returned over the *P*1 wire, or *HB* holds via the *P*2 wire if that is the free outlet.

6.4.5 Battery testing. The advantage of absence-of-earth testing is that no special provision needs to be made to indicate that a circuit is free: the absence of the holding earth which is always present on engaged circuits is sufficient.

There are three important disadvantages to absence-of-earth testing which can be overcome by using battery testing. These are as follows:

(*a*) Double connexions can occur due to two selectors testing in to the same outlet at the same time, or one of them testing in during the operate or release lag of the other one's testing relay (**6.4.3**).

(*b*) A disconnected *P*-wire due to some fault condition such as a dirty contact will appear as a free outlet.

(*c*) The holding earth must be disconnected for a short period at the end of a call in order to release the switching relays (Fig. 6.9). During this time, any of the selectors in the connexion could be taken into use on another call before they have restored to normal. The new call would go to the wrong number.

These effects result in a certain number of calls failing on systems using absence-of-earth testing. However, provided the switching relays are fast, giving only short unguarded intervals, the result is not serious.

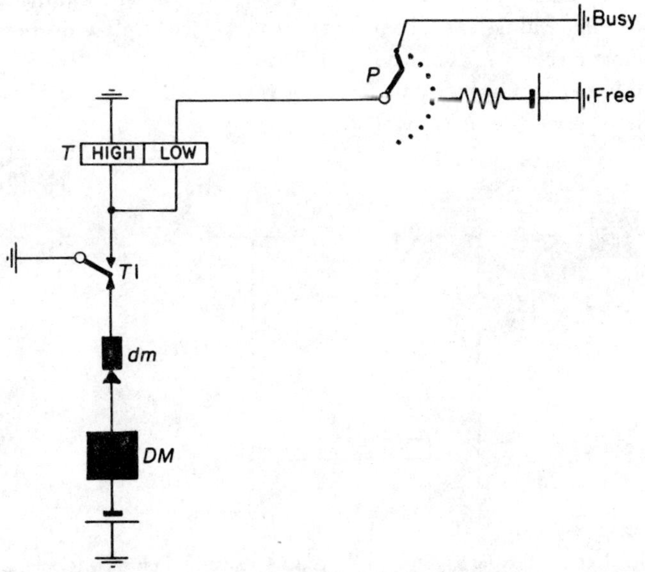

Fig. 6.13 Battery testing principle (6.4.5).

In systems where the switching apparatus represents only a small part of the cost of the whole system, it is often considered worth the additional expense of providing battery testing. It is for this reason that Telex exchanges use battery testing throughout.

Battery testing is also used for common apparatus taken into use for a short time on each call as in certain circuits on the 'director' system.

The principle of battery testing is illustrated by the circuit of Fig. 6.13.

The testing relay *T* in Fig. 6.13 can only operate when the *P* wiper encounters a battery potential. On free outlets, the battery is connected through a resistance low enough for sufficient current to flow to operate the relay over both its windings in series. As soon as the relay operates, its contact *T*1 disconnects the selector magnet self-drive circuit. The same contact also short circuits the high resistance winding of relay *T*, lowering the potential at the bank contact. This lower potential is insufficient to operate another testing relay and so effectively guards the outlet until a holding earth is provided.

The testing relay must operate quickly in order to stop the selector before it steps the wipers on. This restricts the contact load which the testing relay can carry and necessitates the use of another relay for switching.

A typical circuit including this switching relay is shown in Fig. 6.14, the operation of which is as follows:

Start: *LR* operates to the calling signal.

*LR*1 connects the self-drive circuit via *T*1, *H*1 and *dm*.

Relay *T* is connected to the *P* wiper on both coils in series via *LR*1, *MR*1 and *LR*2.

Free outlet: Relay *T* operates to 50 V battery potential via 550 Ω resistance.

*T*1 disconnects *DM* stopping the drive, the resistance of *H* being too great for *DM* to operate in series with it.

H operates.

With *T*1 operated, the 7 Ω coil of *T* and the 550 Ω resistor form a potential divider giving a *P*-wire potential of about 3 V. This potential is too low to operate the *T* relay of any other selector testing the same outlet, thus guarding the outlet.

Switching: Contacts of *H*, not shown here, connect the calling line through, releasing *LR*. The holding earth returned on the *P*-wire in place of the 550 Ω battery condition, short circuits *T* causing it to release.

This completes a holding circuit for relay *H* to the holding earth via *T*1, *H*1 and *R*1 and to battery via the magnet *DM*.

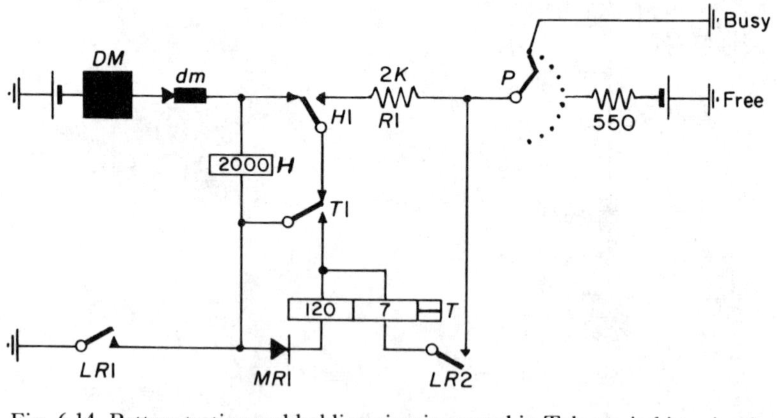

Fig. 6.14 Battery testing and holding circuit as used in Telex switching circuits. Although illustrated here for a uniselector line circuit, the same basic arrangement is used on group selectors (6.4.5).

Let us now see how a circuit of this sort overcomes the disadvantages stated above as applying only to absence-of-earth testing.

If two selectors test in to one outlet together, both testing relays draw their current through the same 550 Ω resistor and both operate. Each now short circuits its high resistance (120 Ω) winding. With both relays now drawing current on their low resistance (7 Ω) 'hold' windings from the same 550 Ω resistance, neither receives enough current to hold. Both release and self-drive and testing re-commences. In practice,

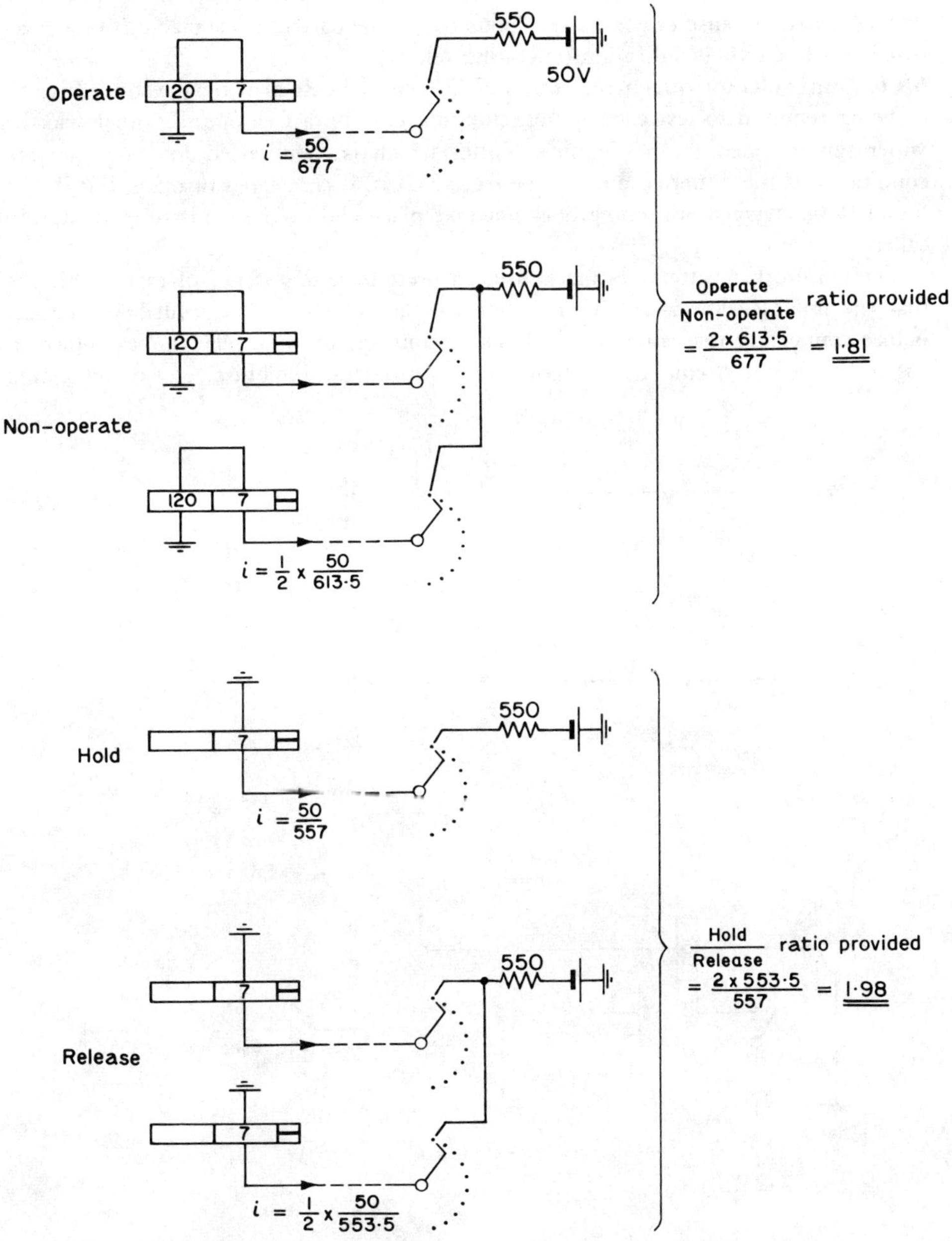

Fig. 6.15 Battery testing conditions. The use of a rectifier in series with the testing relay increases both ratios in practice (see *MR*1 in Fig. 6.14) (6.4.5).

it is found that no two testing relays are sufficiently alike for them to continue in this way in step so that neither manages to seize any of the outlets before the other.

These conditions are illustrated in Fig. 6.15, from which it can be seen that the ratio of the hold currents between the case of one relay and the double connexion case is greater than the corresponding ratio for the operate currents. This is why the testing relays are designed to release rather than non-operate under simultaneous testing conditions.

There is no possibility of a selector testing in to a disconnexion due to a fault or during release because only a battery potential of the correct resistance can be recognized as a free circuit indication (operating relay *T*).

6.4.6 Final selector outlet testing. This is a special case of battery testing. Instead of being required to test each outlet, stopping only on a free one, the final selector wipers are stepped to the required outlet which is then tested for busy or free conditions. If the outlet is found to be free, the call is connected through, but if it is found to be engaged, switching does not take place and busy tone is returned to the caller.

The reason that battery testing is used in preference to absence-of-earth testing is that the final selector is required to operate the *K* relay of the called line circuit before connecting the call through. This disconnects the line relay *LR* as shown in Fig. 6.1. The line circuit *K* relay, however, forms part of the uniselector outlet testing

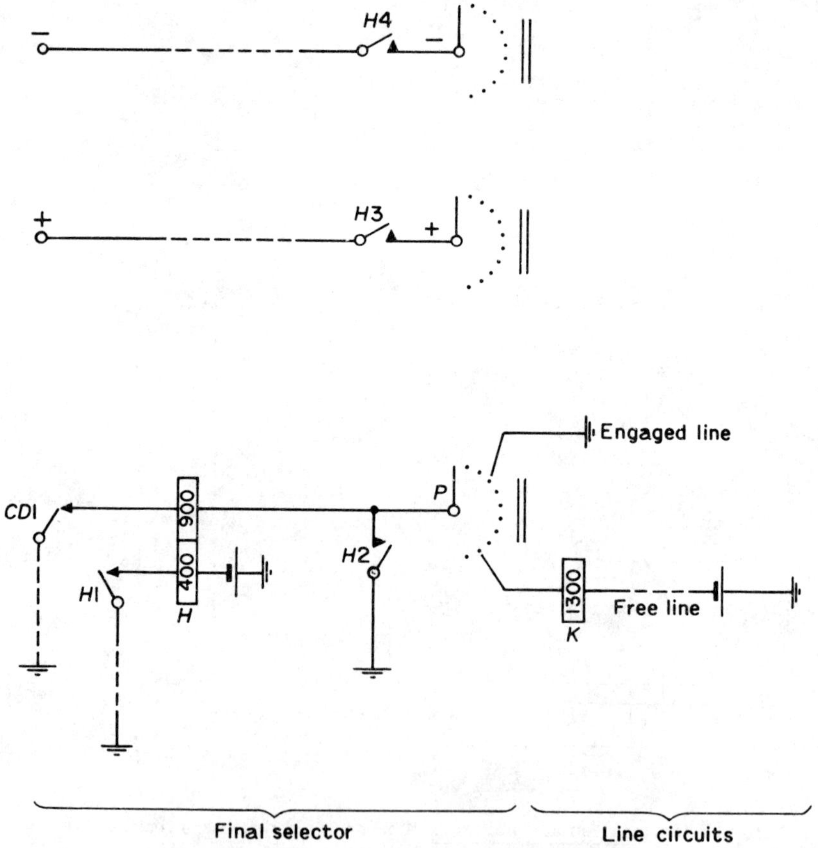

Fig. 6.16 Final selector outlet testing (6.4.6).

circuit and is connected to battery potential as shown in Fig. 6.10. An earth-connected testing relay in the final selector thus provides a very convenient way of testing for the *K* relay battery potential and also operating relay *K*.

A simple testing element used in final selectors is shown in Fig. 6.16 the operation of which is briefly as follows:

> *CD* releases at the end of the final pulse train (**6.3.4**). If the line is engaged, there will be an earth potential on its *P*-wire at the final selector outlet and *H* does not operate. If the line is free, the *P*-wire condition is a battery potential via the line circuit *K* relay. On the release of *CD*, *H* now operates via *CD*1 earth in series with *K* which also operates. *H*1 holds *H* and *H*2 applies an earth to busy the *P*-wire to other final selectors. *H*3 and *H*4 connect the call to the called line.
>
> In the complete final selector circuit, *H* is connected to the *P* wiper, for a short time only. This 'fleeting test' period is timed by the release lag of relay *E* (**6.3.4**). If *H* does not operate by the end of this period, the final selector is arranged to return Busy Tone to the caller and no connexion is made to the called line.

Most modern circuits test the *P*-wire for earth potential before connecting *H* to test for the battery potential. Such circuits are arranged to return Number Unobtainable tone (N.U.) if neither earth nor battery potential is found on the *P*-wire. It is thus only necessary to leave the *P*-wire disconnected on spare numbers. With the simple circuit of Fig. 6.16, it is necessary to connect special N.U. tone circuits to the spare number outlets.

The complete final selector circuit shown in Fig. 6.21 tests for both battery and earth in this way.

6.4.7 P-wire metering. A ratchet operated meter connected at each line circuit records the calls made from each line. The meter readings are used for calculating subscribers' bills. The meters are pulse-operated, and step the number indicated one unit for each pulse. The meter pulses are generated by the selector circuits or by metering circuits associated with the selectors. The meter is only operated on calls on which the called party answers and may operate one or more times during the call depending on the charge appropriate to the call being made.

The dependence of the meter pulses on the called subscriber's answer signal makes it necessary to control the generation of the pulses at whichever circuit provides the transmission bridge facilities (**6.2**).

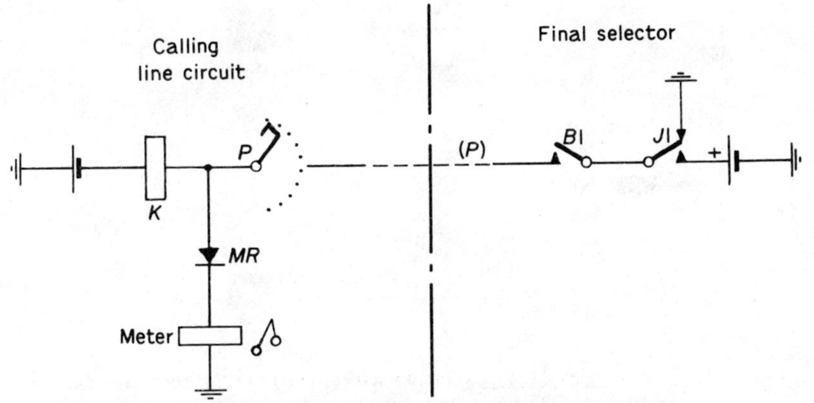

Fig. 6.17 Positive battery metering principle (6.4.7).

In the simplest type of Strowger exchange, the transmission bridge is in the final selector. The meter pulse is therefore provided by that circuit and consists of a short pulse of positive battery potential on the *P*-wire. This potential is obtained from a battery provided in the exchange for this purpose which has its negative terminal earthed instead of the positive terminal as the main negative battery has.

This is shown in Fig. 6.17 and the operation is briefly as follows:

Initially, a holding earth is provided via *B*1 and *J*1, relay *J* being unoperated. This holds *K* but has no effect on the meter.

When the called subscriber answers, *J*1 operates to connect the positive battery to the *P*-wire. This operates the meter via the rectifier *MR*1 and also holds relay *K*, current flowing from the positive battery to the negative battery at *K* as well as to the earth at the meter.

After about 250 *ms*, *J*1 releases and releases the meter which, having a reverse-acting mechanism, then steps its number wheel on. *K* remains held to earth as before.

When the caller clears, *B*1 removes the holding earth to release *K* and rectifier *MR*1 prevents the meter operating in series with *K*. (*MR*1 presents its high resistance to current in this direction.)

6.5 Selector release circuits

At the end of each call, the selectors must be restored to normal to be ready for carrying other calls. As we have already seen, the end of the call is indicated by a clearing signal from the caller in the form of a disconnexion of the calling loop. This releases

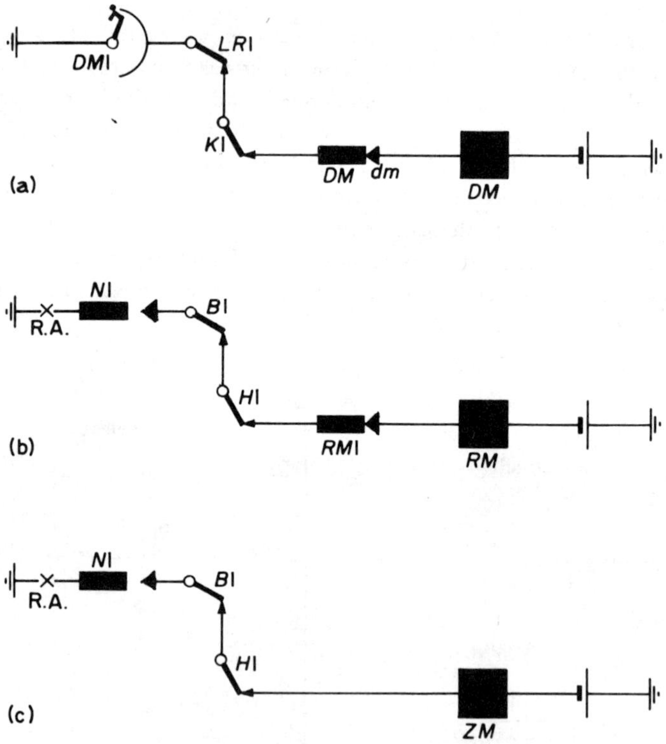

Fig. 6.18 Selector release circuits. (a) uniselector, (b) two-magnet, two-motion selector (2000-type) and (c) three-magnet, two-motion selector (4000-type). *R.A.* indicates 'release alarm relay' (6.5).

relay *A* for long enough to release relay *B*. A break contact of relay *B* can be used to connect a release circuit for its own selector (**6.3.3**), and a make contact of relay *B* disconnects the holding earth to release the *H* or *K* relay of the other selectors in the connexion (**6.4.1**). Break contacts of *H* or *K* are then used to connect the release circuits of these other selectors.

The essential feature of a release circuit is that current is connected to the appropriate magnet via a contact which will disconnect it when the mechanism has restored to normal. In the case of a uniselector, this consists of a homing arc (**5.3.2**) and, in the case of a two-motion selector, the vertical off-normal (*N*) contacts are used. If the selector has to self-drive to normal, the magnet must be connected in series with its interrupter contacts: a release magnet, as provided on pre-2000 type or 4000 type selectors, is simply held continuously energized during release. Typical release circuits are shown in Fig. 6.18. In addition to the relay contacts shown, a contact of any other relay such as *CD* must also be included if necessary in a complete circuit to ensure that release does not take place until the line wipers (− and +) have been disconnected. This prevents interference with other calls in progress via other selectors on the outlets over which the wipers must pass as they restore to normal.

Now turn to page 171 and test yourself on the questions there before reading the next part.

PART 3 TYPICAL COMPLETE CIRCUITS

6.6 Telephone exchange selector circuits

Examples of each of the three main types of selector circuit are given here. These are a subscriber's uniselector line circuit, a group selector and a final selector. Between them these illustrate all the important circuit functions necessary for the operation of a simple Strowger exchange as described in a previous chapter (Fig. 4.7).

Only a brief description of each circuit is given because a detailed discussion of all the design features is more properly dealt with at a later stage (*Telephony 'B' and 'C'*).

6.6.1 Subscribers' uniselector line circuit. This is shown in Fig. 6.19. Its main purpose is to connect the line which it serves to a free first selector whenever the subscriber wishes to make a call (**4.3.4**). It also provides for busying the final selector *P*-wire multiple when the line is engaged and a meter for recording the calls made from the line.

This circuit employs absence-of-earth testing using the circuit element given in Fig. 6.10 and positive battery *P*-wire metering as shown in Fig. 6.17. It provides for battery testing from the final selectors as shown in Fig. 6.16. The line signals are loop calling and disconnexion for clearing (**6.2**).

The operation of the circuit on a call originated from the line to which it is connected is as follows:

Subscriber calls (see also **6.4.2**).

LR operates to calling loop. Magnet *DM* operates to *LR*2 earth via home contact of *DM*4 arc. *K* does not operate (short circuit). Interrupter releases *DM* to step the wipers one step.

*DM*1 arc connects earth to the Final Selector multiple *P*-wire to mark it engaged.

Busy outlets (see also **6.4.2**).

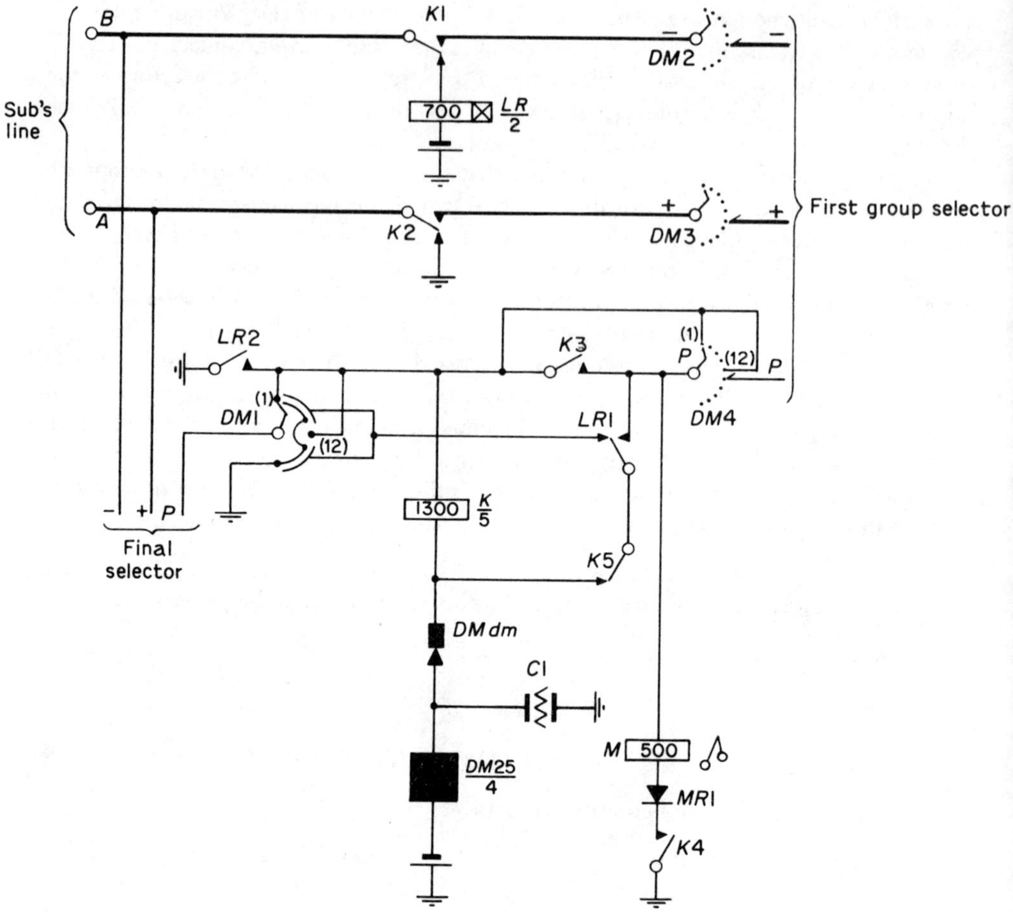

Fig. 6.19 Subscriber's uniselector line circuits (6.6.1).

P-wire earth maintains *DM* self-drive circuit and *K* short circuit. Wipers step to the next contact, *K* does not operate.

Free outlet (see also **6.4.2**).

No earth on *P*-wire.

K now operates to *LR*2 earth in series with *DM* which does not operate.

*K*1 and *K*2 disconnect *LR* and switch the calling loop through to the group selector.

*K*3 busies the outlet.

Holding (see also **6.4.1** and **6.4.2**).

LR has a front-end slug and releases slowly, allowing time for the group selector to return an earth on its *P*-wire to hold *K* via *K*3 once *LR*2 has released. *Only relay K is now operated.* The uniselector wipers are off-normal.

Metering (see also **6.4.7**).

Positive battery pulses on the *P*-wire operate *M* via *MR*1 and *K*4 and also hold *K* via *K*3.

Release (see also **6.5**).

When the calling subscriber clears, disconnexion of the calling loop causes the appropriate selector, normally the final selector, to disconnect the holding earth. This causes relay *K* to release.

*K*1 and *K*2 connect *LR* to the line but there is no loop to operate it at this time. *K*3 disconnects the *P*-wiper, and *K*4 the meter. *K*5 connects the uniselector release or homing circuit.

The uniselector self-drives on earth via *DM*1 arc, *LR*1, *K*5 and *DMdm* until the *DM*1 wiper reaches the next home position.

Capacitor *C*1 is constructed to provide its own inherent series resistance to function as a spark quench on the magnet (**1.4.2**).

Two home positions are provided so that the caller is not deprived of service by one group selector being faulty. If his first attempt at a call is connected to the selector on, say, outlet 2, the uniselector will home to contact 12 at the end of the call. If traffic on the exchange is light, his next attempt will connect him to the selector on outlet 13. Under heavy traffic conditions he is unlikely to get the same selector twice in succession anyway.

The operation of the circuit on a call from the final selector to the line is now as follows:

Final selector tests in (see also **6.4.6**).

The free condition, with both relays and the uniselector at normal is battery potential via *DM* magnet, *DMdm*, *K* relay coil (1 300 Ω), and the home contact of *DM*1 arc to the *P*-wire of the final selector multiple.

The final selector *H* relay operates to this battery and relay *K* in the line circuit operates in series with it.

*K*1 and *K*2 disconnect *LR* from the line so that it does not interfere with the operation of the final selector.

Release. The final selector releases *K* and the circuit is once more at normal.

6.6.2 Group selector (absence-of-earth testing). The example shown in Fig. 6.20, employs a 2000 type mechanism (**5.4.3**) and provides 100 outlets. This is less commonly used than the 200 outlet version, but as can be seen from Figs 6.11 and 6.12, it is simpler and will illustrate the important elements just as well.

This circuit would be connected, at the left-hand side of the diagram, either to the multipled outlets of subscriber's line circuits as Fig. 6.14 or to the multipled outlets of other group selectors in a preceding rank. These connexions are shown in Fig. 4.7.

The outlets of this circuit, at the right-hand side of the diagram would be connected either to other group selectors or to final selectors (see also Fig. 4.7).

The purpose of the circuit is to find a free selector in the appropriate group by stepping the selector wipers vertically under the control of one dialled digit from the caller and then to rotate the wipers on the selected level.

The vertical stepping circuit employs short-circuit type relays as shown in Fig. 6.5b and absence-of-earth testing is provided by a circuit element of the type shown in Fig. 6.11. The line signals are loop calling, loop-disconnect pulsing and disconnexion clearing. Dial Tone or Busy Tone may be returned to the caller when required using a signalling element as shown in Fig. 3.9.

The operation of the group selector circuit is as follows:

Seizure (see also **6.3.3**).

With the circuit disengaged, all relays are at normal and the *P*-wire is therefore disconnected at *B*3 and *CD*6. The preceding circuit tests in to this condition and extends the calling loop through on the − and + wires to operate relay *A*. *A*1 operates *B* via *R*1 and *VM*.

*B*2 operates *CD* on both windings in series via *R*2 and *H*5. *B*3 connects a holding

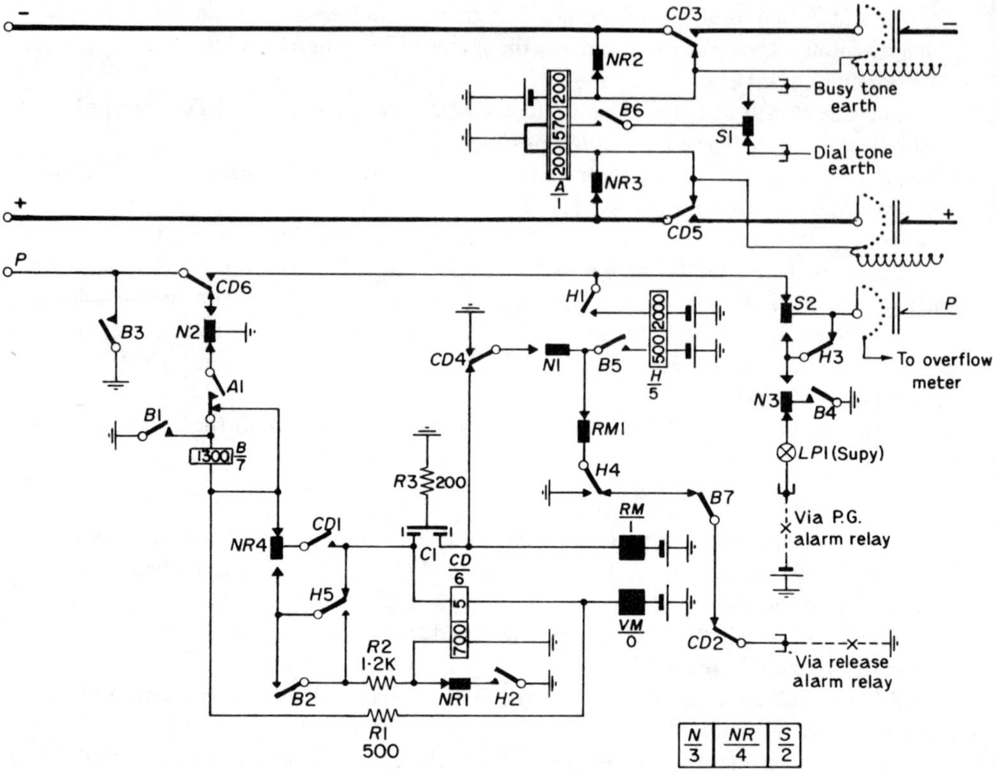

Fig. 6.20 Group selector, 100-outlet, 2000-type (6.6.2).

earth on the *P*-wire. *B*6 connects Dial Tone to the caller (only provided on first selectors).

Dialling (see also **6.3.3**).

A releases once for each break pulse, stepping the vertical magnet via *CD*1 and *NR*4 in series with *CD*.

On first vertical step, *N*1 operates *H* via *CD*4 and *B*5.

*H*2 short circuits *CD* 700 Ω winding, making it slow to release.

B and *CD* hold during pulsing.

End of dialled digit (see also **6.3.3**).

A re-operates at the end of the last pulse for long enough to release *CD*.

The release of *CD*4 contact starts the rotary hunting. Relay *B* remains operated.

Rotary hunting (see also **6.4.3**).

The self-drive circuit for *RM* is via *CD*4, *N*1, *RM*1 (interrupter) and *H*4 (operated) to earth. The selector *NR* contacts operate.

Busy outlets (see also **6.4.3**).

As the wipers step, *RM*1 opens releasing *RM*, but *H* holds on its other winding to the busying earth on the *P*-wire of the engaged outlet. The wipers self-drive to the next outlet.

Free outlet (see also **6.4.3**).

As the wipers step on to the contact, *RM*1 releases both *RM* and *H* as there is no earth on the *P*-wire contact to hold *H*.

*H*4 releasing prevents *RM* operating to step the wipers on.

Switching (see also **6.4.3**).

When *H* releases on the free outlet, *H*5 re-operates *CD* via *R*2 and *B*2. Because *NR*2 and *NR*3 are operated, *CD*3 and *CD*5 disconnect *A* and switch the calling loop to the next selector in the connexion. The operation of *CD* re-operates *H* at *CD*4.

CD now holds via *CD*1, *NR*4, and *H*5 (operated).

All four relays are now operated, but relay *A* has been disconnected and is releasing.

Holding (see also **6.4.1** and **6.4.3**).

*A*1 short circuits *B*, releasing it slowly allowing time for a holding earth to be returned on the *P*-wire from the next selector to hold *H* once *B*3 and *B*5 have released.

Relay *H* holds via *H*1 and *S*2 to the *P*-wire earth for the rest of the call and holds *CD* at *H*5. *Relays H and CD are the only relays held throughout the call.*

All outlets engaged. If there are no free outlets on the selected level, the wipers self-drive over the busy outlets until they reach the eleventh step where the selector *S* springs operate.

*S*1 connects Busy Tone.

The break contact of *S*2 disconnects *H* from the *P* wiper so that it releases. The make contact of *S*2 connects *B*4 earth to operate the overflow meter.

The release of *H* operates *CD* which re-operates *H* as on a free outlet, but *A* is held on the calling loop via the eleventh contact of the – and + banks.

Release (see also **6.5**).

The earth is disconnected from the *P*-wire to release relay *H* which in turn releases *CD* at *H*5.

*CD*6 keeps an earth on the incoming *P*-wire during release via *N*2 to prevent the circuit being seized again before it has completed its release.

The release of *CD* completes a release circuit via *CD*2, *B*7, *H*4, *RM*1, *N*1 and *CD*4 so that the rotary magnet self-drives until the wipers reach the twelfth step where the wiper carriage restores to normal.

*N*2 disconnects the guarding earth from the *P*-wire.

*C*1 and *R*3 function as a spark quench on the magnets (**1.4.2**). Dial Tone is only provided when the circuit is used as a first selector. The lamp *LP*1 and the 'P.G.' alarm are also used only on first selectors and indicate the receipt of a calling loop without subsequent dial pulses implying a faulty line condition.

6.6.3 Final selector. The example shown here, in Fig. 6.21, employs a 2000 type mechanism. For simplicity, the 100 line version is given in preference to the more commonly used 200 line final selector.

The circuit would be connected between the group selector outlets and the subscribers' lines via their line circuits as shown in Fig. 4.7.

The purpose of the circuit is to select the required subscriber's line by stepping the selector wipers vertically under the control of one dialled digit and then rotating them under the control of the next digit. These two digits are the last two digits of the called number as dialled by the caller. Having selected the required line, the selector then transmits ringing current if the line is free and provides a transmission bridge when the called party answers.

The stepping circuit employs short-circuit type relays as shown in Fig. 6.7 and battery testing of the special type referred to in **6.4.6**. The transmission bridge is of the capacitor type (Stone bridge), relays *A* and *D* providing the necessary inductance (**1.3.6**).

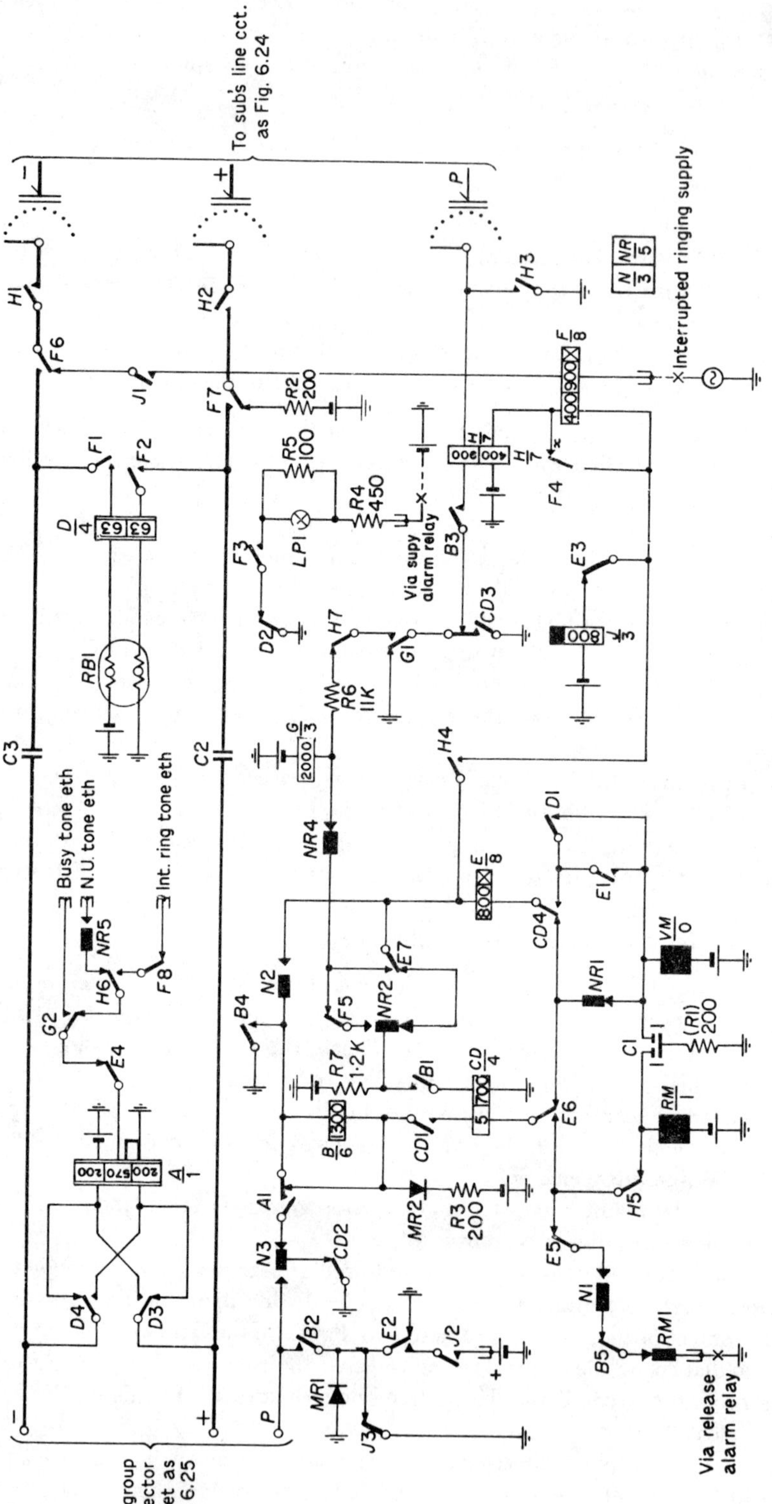

Fig. 6.21 Final selector circuit, 100-line, 2000-type (6.6.3).

The line signals are loop calling, loop-disconnect pulsing, disconnexion clearing, a.c. ringing, and loop answer.

Ring Tone, Busy Tone or Number Unobtainable (N.U.) Tone may be returned to the caller when required using a signalling element as shown in Fig. 3.9b.

The operation of the final selector circuit is as follows:

Seizure (see also **6.3.4**).

With the circuit disengaged, all relays are at normal and the *P*-wire is therefore disconnected at *B*2 and *N*3. The preceding group selector tests in to his condition and extends the calling loop through on the – and + wires to operate relay *A*.

*A*1 operates *B* via *MR*2 and *R*3.

*B*1 operates *CD* via *R*7. *B*2 connects holding earth.

1st digit (see also **6.3.4**).

*A*1 pulses the vertical magnet via *CD*1, *E*6 and *NR*1. *B* and *CD* hold. (*N*2 provides the short circuit on *CD*.)

At the end of the digit *CD* releases, *B* remains held.

*CD*4 releasing, operates *E* via *B*4, *N*2, *NR*1 in series with the vertical magnet *VM*.

*E*6 changes over the pulsing circuit from the vertical magnet to the rotary magnet.

*E*7 removes the short circuit from *CD* which now re-operates. *E*7 also operates *G* which then holds at *G*1 via *R*6 and *CD*3.

Relays A, B, CD, E and G are now operated.

2nd digit (see also **6.3.4**).

*A*1 pulses the rotary magnet.

CD releases at the end of the digit (the short circuit on *CD* is provided by *NR*2 via *E*7). The release of *CD*4 starts to release relay *E* and *CD*3 connects *G* and *H* to test the line circuit *P*-wire.

Called line busy: The *P*-wire earth holds *G* via *R*6, *H*7, *G*1, *CD*3, *B*3 in series with *H*. The high resistance of *R*6 prevents the operation of *H*.

When *E* completes its release, it connects Busy Tone at *E*4 via *G*2.

*E*7 re-operates *CD* and *CD*3 then holds *G* independently of the *P*-wire conditions.

Called line free (see also **6.4.6**).

This is indicated by a 1 300 Ω battery potential on the line circuit *P*-wire. There is thus no earth to hold *G*. Relay *G* releases, and at *G*1 completes a circuit for the operation of *H*, from earth at *G*1 via *CD*3, *B*3 to the *P*-wire battery. *H* operates.

When *E* completes its release, *E*3 operates *J*, via *B*4, *N*2 and *H*4. *J*1 connects ringing to the called line via *H*1 and *H*2 and the 'ring trip' relay *F* (**3.6.1**). *H*4 also holds *H* and *E*7 operates *CD*. *H*3 busies the line to other final selectors.

Ring Tone is connected by *E*4 via *H*6 and *F*8.

Called line spare. The *P*-wire of unused numbers is disconnected so neither *H* nor *G* are operated by the time *E* has completed its release.

*E*4 connects N.U. tone via *G*2, *H*6 and *NR*5.

Ringing (see also **3.5.1**).

(Relays *A*, *B*, *CD*, *H* and *J* are already operated.)

The called subscriber's bell rings on a.c. from the earthed supply in series with relay *F* via *J*1, *F*6 and *H*1, over the line, returning to earth at the exchange battery via *H*2, *F*7 and *R*2. The capacitor in series with the bell in the telephone instrument prevents operation of *F*.

Called subscriber answers (see also **3.5.1**).

The loop operates *F* which locks at *F*4, and disconnects the ringing at *F*6 and *F*7.

*F*5 re-operates *CD*. *F*8 disconnects Ringing Tone.

*F*1 and *F*2 connect *D* to control metering.

Metering (see also **6.4.7**).

Relay *D* operates to the called subscriber's loop. *D*1 operates *E* via *CD*4.

*E*2 connects positive battery to the *P*-wire to operate the meter.

*E*3 releases *J*.

J releases slowly, disconnecting the positive battery at *J*2 to terminate the meter pulse.

Release (see also **6.5**).

When the caller clears, the disconnexion releases *A* for long enough to release *B*.

*B*1 releases *CD*.

*B*4 and *CD*3 release the other relays.

A release circuit is completed via *RM*1, *N*1, *E*5 and *H*5 to the rotary magnet which self-drives until the wipers reach the twelfth step where the wiper carriage restores to normal.

*N*3 and *CD*2 guard the circuit against re-seizure during release.

*C*1 and *R*1 function as a spark quench on the magnets (**1.4.2**). The lamp *LP*1 and the *SUPY* alarm indicate the 'called subscriber held' condition (C.S.H.) which means that the called line is locked out of service by the caller failing to clear. *MR*1 maintains a holding earth during the transit time of the metering contacts. *MR*2 prevents *R*3 acting as a shunt which would slow the release of the magnets during pulsing. Its direction is such as to oppose the maintenance of current by the back e.m.f. of the magnets.

*RB*1 is a ballast resistor (barretter) which presents a higher resistance to large currents than it does to small currents. This enables a low resistance *D* relay to be used to give adequate current for the carbon microphone on long lines without permitting excessive current on short lines.

Contacts *D*3 and *D*4 provide an answer signal to the originating exchange on calls received over a junction using the signalling system described in **3.6.1**.

PART 4 SUMMARY

6.7.1 Pulse stepping circuits

B relay: slow release to hold during 'break' pulses but release on clearing signal. Short circuit release gives faster rise of flux, more winding space and less sparking than with slug.

CD relay: slow release to hold during 'make' pulses but release at end of pulse train (digit).

6.7.2 P-wire circuits

Absence-of-earth testing: simple circuit but unguarded on testing in and again on release.

Battery testing: more expensive but proof against double connexion.

Now turn to page 172 and test yourself on the questions there before leaving this chapter.

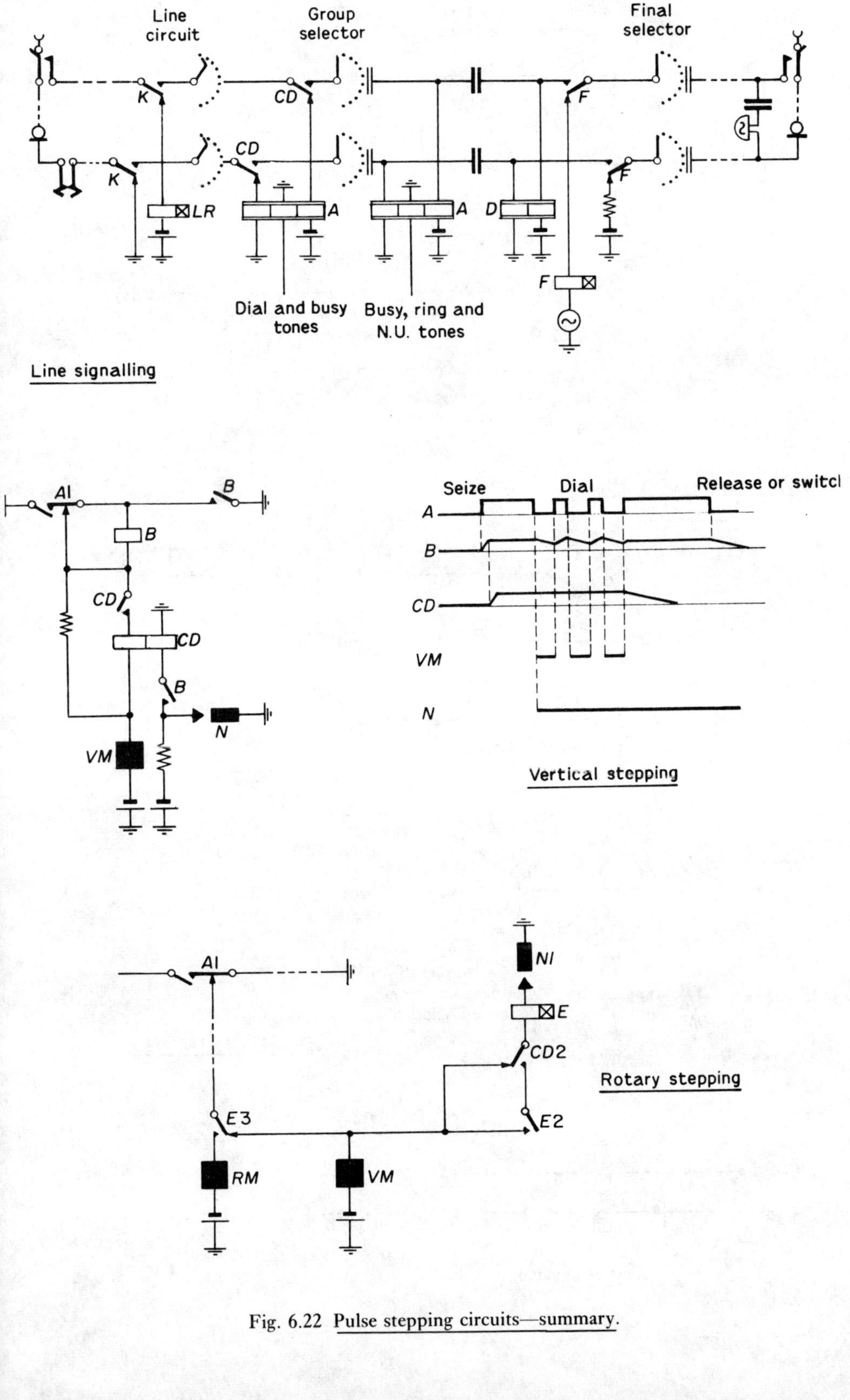

Fig. 6.22 Pulse stepping circuits—summary.

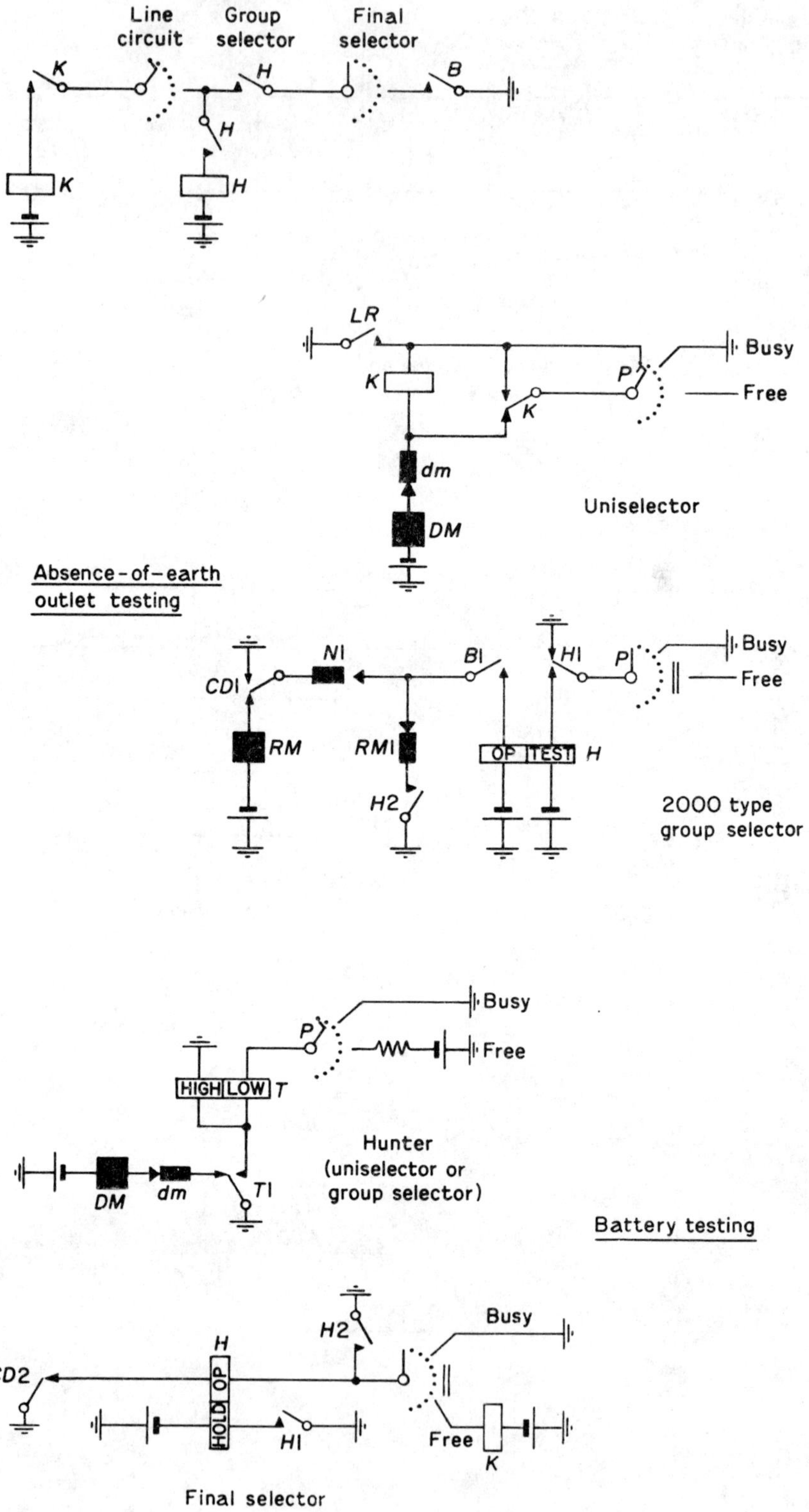

Fig. 6.23 P-wire circuits—summary.

PART 5 SELF-TEACHING NOTES

6.8.1 First read **6.1.1** to **6.2.**

If you find this hard to follow, revise the earlier work on signalling systems in Chapter 3, and also, if necessary, the work on relays and relay circuits in Chapter 2.

Then carry on with the rest of Part 2 of this chapter (**6.3** to **6.5**).

6.8.2 Having read **6.1.1** to **6.5**, try the following fifteen questions:

Q.1 What line condition is used as a calling signal from a subscriber to an automatic telephone exchange?

Q.2 What clearing signal is used?

Q.3 Why does the line relay *LR* in the subscriber's line circuit *not* have balanced windings like the pulsing relay *A* of a group selector?

Q.4 Which selector receives the last two digits of the dialled number?

Q.5 Why does the guarding relay *B* *not* release when it is disconnected (or short circuited) during each break pulse?

Q.6 Give two advantages of releasing this relay *B* by means of a short circuit rather than a disconnexion.

Q.7 The dialling control relay *CD* is pre-operated before the commencement of dialling. What is used to render this operating path ineffective so that *CD* can release when necessary later?

Q.8 When does relay *CD* release?

Q.9 When a selector is designed to be pulse-controlled for both vertical and rotary stepping, a relay *E* is provided which is arranged to operate after vertical stepping. What two functions must it perform to prepare the circuit for rotary stepping?

Q.10 What is absence-of-earth testing?

Q.11 In the absence-of-earth testing circuit used with uniselectors, relay *K* is connected in series with the magnet which self-drives over busy outlets. Why does relay *K* *not* operate on these busy outlets?

Q.12 Why is the simple outlet testing circuit used with uniselectors *not* suitable for use with two-motion selectors?

Q.13 What is the advantage of using a pre-operated testing relay in an absence-of-earth testing circuit?

Q.14 What is the advantage of battery testing compared with absence-of earth testing?

Q.15 Why can a final selector use a slower battery testing relay than is necessary on a group selector using battery testing?

Now check your answers on page 264.

6.8.3 You should have got at least ten of these fifteen questions correct. If not, read the appropriate parts of the previous chapters as suggested on the answer page (and **6.8.1** above); then have another go at this chapter.

Learn the circuit elements as suggested on the answer page.

Then read the rest of this chapter (**6.6** to **6.7.2**) and look for the circuit elements we have dealt with in the practical selector circuits.

6.8.4 The whole of Chapter 6 should have been studied a few days before attempting the following questions.

Write a complete answer to each and check your answers on page 265 before going on to the next question.

Q.16 Draw and explain the operation of a circuit element suitable for controlling the rotary hunting of a two-motion selector using the earth-testing principle. What advantages arise from the use of battery testing compared with earth testing? (C & G 1962)

Q.17 The vertical stepping circuit of a two-motion selector includes two slow release relays.
Explain why these relays need to be slow to release, illustrating your answer with a sketch of the circuit element concerned. (C & G 1963)

Q.18 Sketch and describe suitable circuit elements for the rotary hunting action of a uniselector using:
(*a*) earth testing
(*b*) battery testing
Explain the reason for any difference in the wipers required for (*a*) and (*b*).
(C & G 1964)

Q.19 Sketch a circuit element to show how a final selector is stepped in a rotary direction under the control of a subscriber's dial, and explain how the circuit functions.
Why does a group selector have a different arrangement for the control of its rotary action?

7 · TRAFFIC AND TRUNKING

For self-teaching notes, see page 203.

PART 1 INTRODUCTION

7.1.1 What is traffic? Telephone or telegraph traffic consists of calls (or messages) which can be regarded as passing or flowing through the system.

This traffic is like road traffic in some ways. If you are going to build a road, for example, you need to know how much traffic it is expected to have to carry. If there are few vehicles, or if each vehicle only wants to use the road for short periods, you would build only a narrow road.

So it is with an exchange; you need to know how many calls are made at the busiest time of the day and their average duration in order to determine how many selectors should be provided.

Traffic is therefore defined as the aggregate of calls passing over a group of circuits or trunks. The term 'trunk' means a connecting circuit between two switching stages (e.g. between 1st selectors and 2nd selectors).

7.1.2 Trunking is the provision of adequate plant to carry the traffic and the interconnexion of the selectors in such a way as to route this traffic in the most efficient and economical manner.

In addition to the general ideas discussed earlier (**4.3.4**, **4.3.7**, and **4.3.8**), this involves interconnecting the selectors to distribute the traffic uniformly and the calculation of the number of selectors to be provided to give the required grade of service (**7.3.1**).

The advantage of distributing the traffic uniformly can be illustrated by referring once again to the case of road traffic. At peak traffic periods in many towns, the main street is congested and traffic is at a standstill but the side streets are empty. There are often good reasons why the side streets should not be used, but traffic would flow more easily and quickly through the town if it could use them.

The equivalent situation in a telephone exchange would occur if all the calls in one part of the exchange were to be routed to one group of selectors while another equally suitable group of selectors in the same rank were standing idle. In practice, such a state of affairs is largely avoided in the Strowger system and some crossbar systems by a method of interconnexion known as grading.

7.1.3 In this chapter, you will learn about the nature of telephone and telegraph traffic, the way in which it varies, how it is measured and its effect on plant provision. You will also have the opportunity to try some simple traffic calculations for yourself which will enable you to gain an understanding of the principles on which practical trunking schemes such as grading are based. We shall then see how to connect small numbers of selector groups to form the best grading in given circumstances.

Finally, there are details of the practical arrangements provided in exchanges for making the connexions between the selectors in a grading.

PART 2 THEORY OF TRAFFIC AND GRADING

7.2 The nature of traffic

7.2.1 Units of measurement. The choice of units depends on the use which is to be made of the traffic measurements.

The number of operators needed in a manual exchange, for example, depends on the number of calls to be handled in a given time. The duration of the calls is not important for this purpose, only the time spent by an operator on each call. This is 18 s on average for local calls and this is known as a *unit call*, giving a *standard operator's load* of 200 unit calls per hour.

The electrical power consumption, on the other hand, is related to the total or aggregate time of all the calls. This quantity is the *total traffic* and is measured in call-hours or call-minutes, i.e. the number of calls multiplied by their average duration.

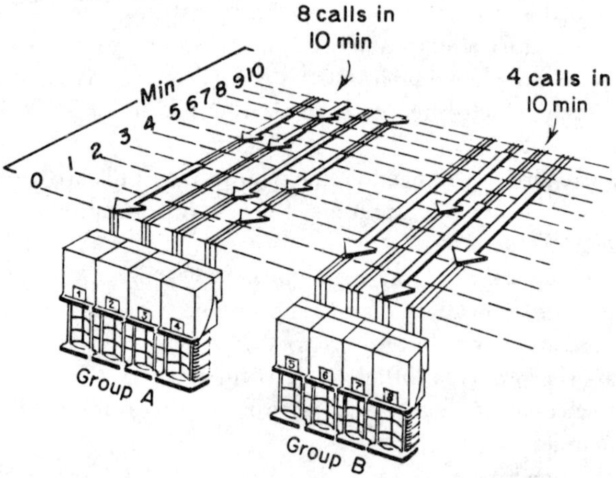

Fig. 7.1 Traffic intensity as calls multiplied by average duration. The traffic in both groups is 2 erlang. There are forty-eight calls per hour of average duration $2\frac{1}{2}$ min on group A. Group B has only twenty-four calls per hour but they have an average duration of 5 min each (7.2.1).

The most important traffic quantity, however, is the rate of flow of traffic or *traffic intensity*.

Traffic intensity is defined as the average number of calls in progress simultaneously and is therefore used in the calculation of equipment quantities such as the number of selectors to be provided. If there are N calls in progress at a time, they give rise to a total traffic of N call-minutes every minute so that traffic intensity is the same thing as the rate of flow of total traffic in unit time. The unit of measurement is therefore effectively call-minutes per minute (or call-hours per hour) and this unit is called the *erlang* after the Danish engineer and mathematician Agner Krarup Erlang. (The erlang used to be known as the *Traffic Unit* or *T.U.*) (see also **7.2.3**).

The relation between the various traffic quantities can be seen by examining Figs. 7.1 and 7.2 as follows:

In Fig. 7.1, the four selectors comprising group A carry eight cells between them in a period of 10 min. Two of these calls have a duration of 1 min, two others of 2 min, three of 3 min and the eighth one has a duration of 5 min. Adding all these together gives a total traffic of 20 call-minutes. The same answer could also be obtained by averaging the call durations and multiplying by the number of calls ($2\frac{1}{2}\times 8 = 20$).

Since this total traffic occurs in a period of 10 min, the rate of flow is $20 \div 10 = 2$ call-minutes per minute which is a traffic intensity of 2 erlang.

The traffic intensity on group B is also 2 erlang even though there are only half the number of calls as there are on group A. This is because the call duration is doubled.

These calculations of the traffic intensity can be represented by a simple formula derived as follows:

If the number of calls is C, and their average duration is t min, the total traffic $= Ct$ call-minutes. (1)

If this traffic occurs in a period of T min, the traffic intensity A is given by (1) divided by T,

$$\text{i.e.} \quad A = \frac{Ct}{T}. \tag{2}$$

(This is more easily memorized in the form $TA = Ct$.)

Since in most practical cases, $T = 60$ min (the busy hour, see **7.2.2**), this becomes $A = \dfrac{Ct}{60}$ if t is measured in minutes.

If t is expressed as a fraction of an hour, however, and T is the busy hour, then $T = 1$ h and (2) can be simplified to:

$$A = Ct. \tag{3}$$

For most purposes, this is the most useful form of the formula.

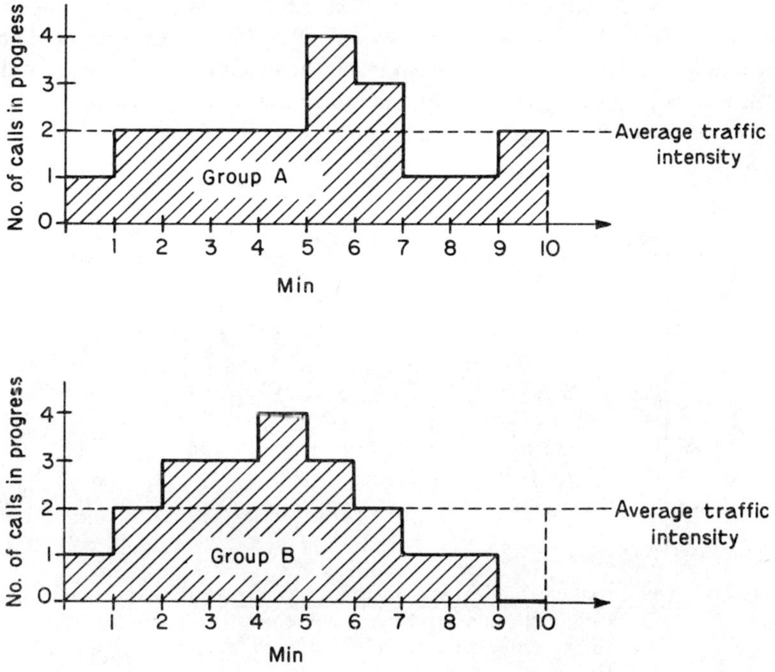

Fig. 7.2 Traffic intensity as average number of simultaneous calls. These graphs refer to the same selector groups and traffic as Fig. 7.1 (7.2.1).

Figure 7.2 illustrates traffic intensity as the number of simultaneous calls. Graphs have been drawn of the number of calls in progress at any time for the two groups of selectors shown in Fig. 7.1.

If observations of the number of selectors engaged at a time were made on group A, for example, at 1 min intervals, one selector would be found engaged during the first minute, two during each of the next 4 min and so on as shown on the graph.

The average of these observations is found by adding them all together and dividing by the number of observations. This gives an average number of simultaneous calls of 22 ÷ 11 = 2, which is equal to the traffic intensity in erlangs as calculated previously.

The same result is of course obtained if this process is repeated for group *B*.

Traffic intensity can be measured under working conditions in an exchange in this way. Counting and recording the number of switches engaged is a laborious task which would take staff away from more important work in the exchange. An automatic *traffic recorder* is therefore provided in all large exchanges. This uses uniselectors to connect a set of meters to the *P*-wires of a group of selectors at regular intervals. If a selector is engaged at the time the meter is connected, the earth on its *P*-wire operates the meter which is a ratchet operated meter of the same type as that used for recording calls at the subscribers' line circuits (**6.4.7**).

The time for which a call occupies a circuit or selector is known as the *holding time* of the call. For our present purposes this can be assumed to be equal to the total call duration. In practice, however, different circuits within the same exchange may have different holding times. The average holding time of a first selector, for example, is longer than that of a final selector because it is taken into use earlier in the call.

7.2.2 Traffic variations. The precise instant at which any given subscriber decides to make a call is determined by factors which generally affect that subscriber alone. He is unaware of the demands made by other subscribers on the system at the time and there is a wide variety of different factors affecting the different subscribers. The result is that the number of calls made varies from instant to instant in a wholly unpredictable manner. A new call is as likely to arise at one moment as at any other moment. Under these conditions the traffic is said to be *random*, or *pure chance* traffic. This is illustrated in Fig. 7.3 for a short period ($\frac{1}{4}$ h).

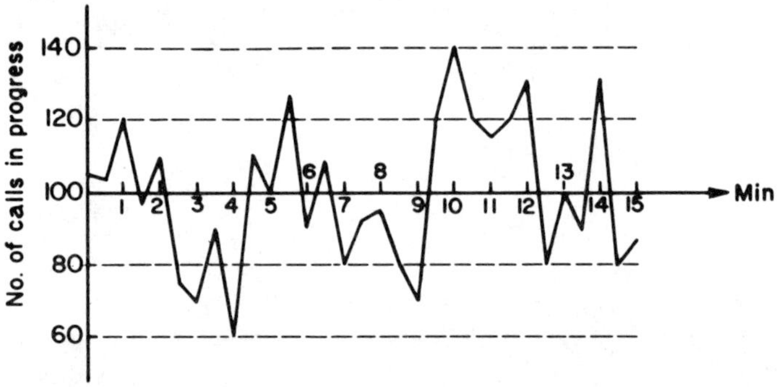

Fig. 7.3 Typical traffic variation over a period of 15 min. Pure chance traffic on a 2 000-line telephone exchange, measured at half-minute intervals. The average traffic intensity is 100 erlang (7.2.2).

Certain factors affecting the timing of calls influence a number of subscribers simultaneously. In a business area, for example, the offices all have similar working hours, during which their calls are made. Such factors are more or less constant from one day to the next over the working week, and have a greater effect on the traffic as a whole than the random variations referred to above. There is usually no traffic at all in the early hours of the morning, around 2 a.m. for example, whereas the random variations are not usually great enough to cause the traffic to fall to zero at any time during the working day.

The exact pattern of the variation of traffic over the course of a day depends upon the type of area served. A typical case of an exchange serving mainly business subscribers in an inland town is shown in Fig. 7.4. This graph represents the average traffic for successive half hour periods and it should be remembered that within each half hour period, the traffic is substantially random as in Fig. 7.3.

The principal factors affecting the incidence of calls during the day are shown in Fig. 7.4 but other local factors such as staggered working hours may alter the effects of these factors in particular exchanges.

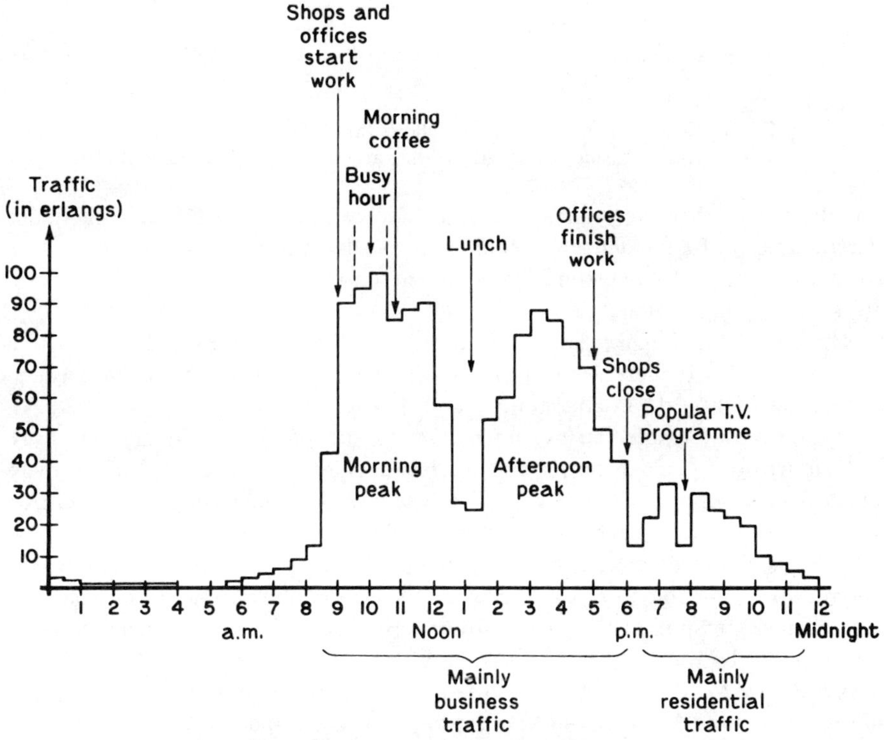

Fig. 7.4 Typical traffic variation in the course of a day. The graph represents the average traffic intensity over successive half-hour periods for a 2 000-line exchange serving a largely commercial area (7.2.2)

One very important feature can be seen, however, in Fig. 7.4 which is common to all exchanges. This is the existence of a distinct peak in the traffic at about the same time each day. The provision of sufficient plant to meet the requirements of this peak will obviously ensure more than adequate provision for the rest of the day.

For this purpose, one hour is chosen for each exchange corresponding to this traffic peak and this hour is known as the *busy hour*. The busy hour can occur at any time of the day or night depending upon the type of area served, but in most areas it is in the morning at some time within the limits of 9.30 a.m. to 11.30 a.m.

In the U.K., the busy hour is determined by measuring the traffic during the busiest $1\frac{1}{2}$ h of the day, this $1\frac{1}{2}$ h period beginning on a full hour or half-hour of the clock (e.g. 9.30 a.m., 10 a.m. etc.). The busy hour is then the busier of the two one-hour periods so measured.

As can be seen from Fig. 7.5a, the traffic does not vary much from one day to the next over the week; so unless there are special local conditions (e.g. market day), the busy hour can be based on traffic records for any weekday.

The traffic varies from one month to another due to seasonal factors and local events especially at seaside resorts where, of course, it is heaviest in the holiday seasons as illustrated in Fig. 7.5a.

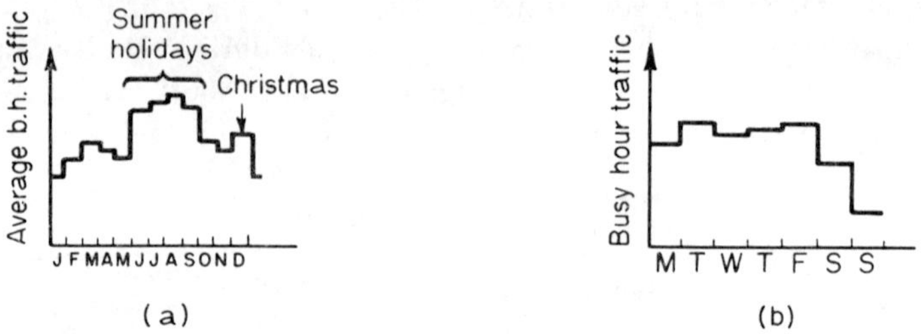

Fig. 7.5 Typical traffic variations. This shows (a) the variation during a year at a typical seaside town and (b) the variation during a week (7.2.2).

In addition to these variations, there are also differences in the type of traffic at different times. The incidence of trunk calls usually differs from that of local calls, for example, and call office traffic does not necessarily have the same busy hour as that of, say, business subscribers.

Telephone traffic, in fact, can be said to mirror all the business and social activities of the community. It increases in prosperous times, decreases in times of depression and shows sudden peaks in emergencies and disasters.

Telegraph traffic is affected by similar factors, but not always in the same way as telephone traffic. A sudden power cut in a business area, for example, may provoke a short peak of telephone calls reporting it before traffic subsides whereas telegrams and telex messages would probably cease immediately.

The main object of trunking is to provide adequate plant for present needs as economically as possible in order to provide the required service at reasonable cost to the subscriber. In spite of the various short term fluctuations, the general trend over the years is for both telephone and telegraph traffic to increase. Some provision must always be made for this growth, even though this increases present costs, in order to avoid the need to make changes at even greater expense later. Usually, selector banks and cabling are provided initially in sufficient quantity for 5 years growth at the expected rate, selectors for only 2 years and floor space for 20 years.

7.2.3 The erlang and 'probability'. We have already seen (in **7.2.1**) that the erlang is the unit of traffic intensity or flow and that *for a specified period, usually the busy hour, the traffic on a group of circuits expressed in erlangs is equal to the average number of simultaneous calls during that period.*

For a single circuit, the traffic can never be more than 1 erlang because it can carry only one call at a time. *The erlang can thus also be defined as that traffic flow which would continuously occupy one circuit.*

In a practical case, no circuit can ever be continuously occupied because it must become free for a short period after each call in order to be seized by the next call. *The traffic on one circuit is* therefore always a fraction of an erlang, this fraction being *the proportion of time for which the circuit is engaged.* This can be shown by using our earlier formula as follows:

If a single circuit carries C calls of average duration t, it will be engaged for a total time of Ct (min). If these calls occur in a period T (min), the proportion of this time for which the circuit is engaged is $\dfrac{\text{time engaged}}{\text{total period}} = \dfrac{Ct}{T}$, and this as we have seen is equal to the traffic A erlang (**7.2.1**).

For a numerical example we can refer to our previous diagram Fig. 7.1. The traffic intensity on group B as a whole is 2 erlang and all four circuits are equally loaded. The traffic flow in each circuit is therefore 0·5 erlang. Examination of the diagram will show that each circuit is in fact engaged for 0·5 of the 10 min period shown.

In the case of group A in Fig. 7.1, the circuits are not equally loaded. Circuit 1 is engaged for 0·6 of the time, circuit 2 for 0·4, circuit 3 for 0·5 and circuit 4 for 0·5 of the time. These added together give a total of 2, the traffic intensity on the group as a whole.

The important thing about this is that the chance, or *probability of finding a given circuit engaged at any time is numerically equal to the proportion of time for which the circuit is engaged and therefore to the traffic in erlangs.* In practice, it would be uneconomic to provide so many selectors that no subscriber would ever find the one he needed already engaged. It is therefore necessary to be able to calculate the probability of circuits being engaged under given conditions in order to assess whether adequate service will be given by a particular arrangement or quantity of selectors.

This idea of probability is something like betting odds except that it is convenient to express it in the form of a fraction. To take an example, when betting on dice the odds on, say, a six being thrown on a single die could be stated in the form 'six to one'. On average over a large number of throws a player would neither gain nor lose on such odds because there is one chance in six of any given number being thrown. This corresponds to a probability of $\frac{1}{6}$ or 0·16.

The probability of any given event occurring is calculated as the number of occasions on which it *does* occur divided by the number of occasions on which it *might* occur. The probability of a 'certainty' is therefore 1, and the probability of an 'impossibility' is 0. All other probabilities are fractions, i.e. they are less than 1 and greater than 0.

From this it follows that the probability p_1 of a particular selector being found to be engaged at any instant during a specified period is given by:

$$
\begin{aligned}
p_1 &= \frac{\text{length of time the selector is engaged in that period}}{\text{total duration of the period}} \\
&= \text{proportion of time for which the selector is engaged} \\
&= \text{traffic on the selector in erlangs.}
\end{aligned}
$$

The probability of a call being lost due to there not being a free selector in the required group is the same thing as the probability of all circuits in that group being engaged simultaneously. Now, if two circuits each carry 0·25 erlang, for either of these circuits the probability of its being found engaged will be 0·25. If, then, we select a number of occasions on which the first circuit is engaged, the probability of the second circuit being engaged on any one of these occasions is still 0·25. On average, therefore, we shall find the second circuit also engaged on one out of every four occasions on which the first circuit is engaged. These four occasions, however, are themselves only one quarter of the total, so that we shall only find both circuits engaged on one occasion in sixteen, a probability of 0·0625 ($= 0{\cdot}25 \times 0{\cdot}25$).

The probability that two independent things will occur simultaneously is always equal to the product of their separate probabilities in this way.

It will be appreciated from this that the relationship between the traffic intensity on a group of circuits and the probability of finding all circuits in the group simultaneously engaged is not a simple one. The mathematics required for the calculation of probabilities of this sort is beyond the scope of this book, but Fig. 7.6 illustrates the relationship graphically.

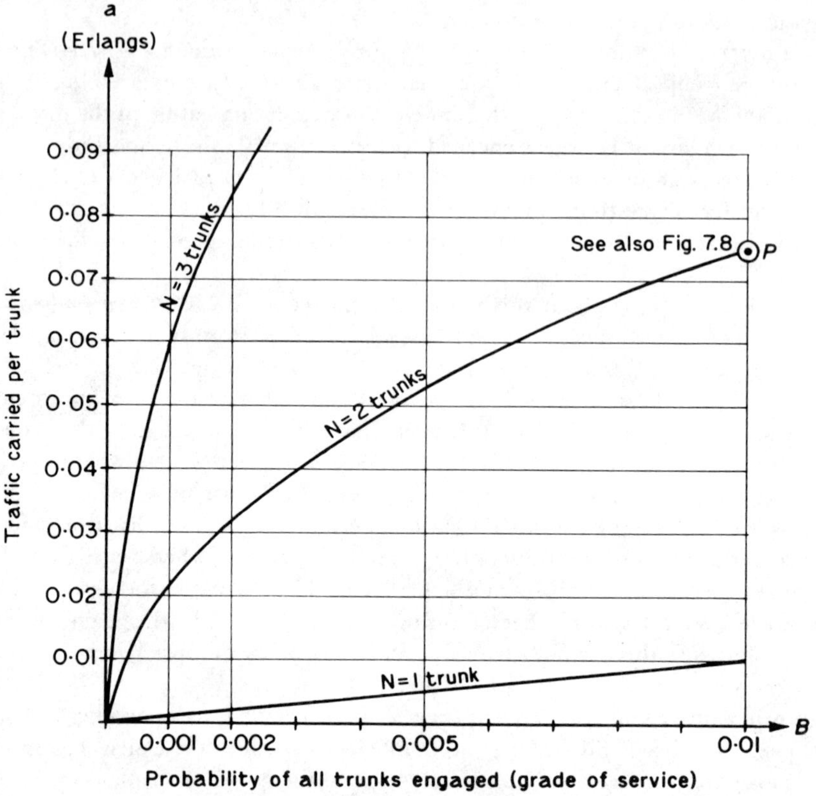

Fig. 7.6 Probability of all trunks simultaneously engaged. The graph is derived from Erlang's full availability formula:

$$B = \frac{A^N/N!}{1 + A + A^2/2! \ldots + A^N/N!}$$

where $N! =$ factorial $N = N \times (N-1) \times (N-2) \ldots \times 3 \times 2 \times 1$ and A is the traffic offered to the group (7.2.3).

7.3 Effect of traffic on plant provision

7.3.1 Grade of service. This is defined as *the proportion of calls which are allowed to fail in the busy hour owing to the limitation, for economic reasons, of the amount of switching plant.* Calls which fail in this way are referred to as *lost calls.*

The grade of service, B, for a group of circuits can therefore be calculated as:

$$B = \frac{\text{number of lost calls}}{\text{number of calls offered}}$$

The number of calls offered is, of course, equal to the sum of the number of calls lost and the number of calls carried by the group of circuits.

If it is assumed that the lost calls would have had the same average duration as the calls carried, the grade of service is also given by:

$$B = \frac{\text{traffic lost}}{\text{traffic offered}}$$

Since the number of lost calls is the same thing as the number of occasions on which a call was offered and found all circuits in the group engaged,

$$B = \frac{\text{number of occasions all circuits are engaged}}{\text{total number of possible such occasions}}$$

$$\therefore\ B = \text{probability of finding all circuits engaged.}$$

This is therefore an alternative way of defining grade of service and the relationship between this and the traffic offered has already been illustrated in Fig. 7.6.

For a single circuit, the probability of its being engaged is equal to the traffic it carries expressed in erlangs. The grade of service on a single circuit is thus also equal to the traffic in erlangs, in the busy hour.

A typical practical value for the grade of service on a group of selectors in an automatic telephone exchange is 0·005, representing 1 lost call in 200.

Example: If the busy hour traffic offered to a rank of selectors is 200 erlang, the average call duration is 3 min and 20 calls are lost in the course of the busy hour, what is the grade of service and what is the traffic carried?

First express the call duration t as a fraction of an hour:

$$t = \frac{3}{60} = \frac{1}{20} \text{ or } 0{\cdot}05 \text{ h.}$$

Using $A = Ct$, the num er of calls is

$$C = \frac{A}{t} = \frac{200}{\frac{1}{20}} = 200 \times 20 = 4\,000 \text{ calls offered.}$$

The grade of service, $B = \dfrac{\text{calls lost}}{\text{calls offered}}$.

$$\therefore\ B = \frac{20}{4\,000} = 0{\cdot}005 \ \underline{\text{(or 1 lost call in 200.)}}$$

The traffic carried = number of calls carried × t

$$= \frac{\text{calls offered} - \text{calls lost}}{20} = \frac{4\,000 - 20}{20} = \frac{3\,980}{20}$$

$$= 199 \text{ erlang traffic carried.}$$

7.3.2 Traffic efficiency of large groups. It is a very important trunking principle that *large groups are more efficient than small groups.* For a given grade of service, fewer circuits are required in a large group to carry a certain traffic than are required if the same traffic is to be shared over a number of smaller groups.

This can be seen by comparing five separate groups, each consisting of a single circuit, with a group of five circuits. If each of the single circuits carries 0·1 erlang, each will be engaged for 0·1 of the busy hour and calls offered in this time are lost. The traffic carried altogether is 0·5 erlang and the grade of service is 0·1. If the same traffic were offered to the group of five circuits, calls would be lost only when all five circuits were simultaneously engaged. Although each circuit will be engaged for 0·1 of the busy hour, the group as a whole will be fully engaged for only a fraction of this time, giving a much better grade of service than the five separate circuits (see **7.2.3**).

Provided the trunking is so arranged that any free circuit in the group is made available to any new call when the other circuits are engaged, the group needs only two circuits to carry our 0·5 erlang at a grade of service of 0·1. This type of trunking constitutes a full availability group (**7.4.1**).

This same principle is shown in more general terms in Fig. 7.7a. From this graph it can be seen that as the size of the group is increased, the traffic carrying capacity per circuit increases for a given grade of service. The larger the group, the more nearly the traffic per circuit approaches the theoretical maximum of 1 erlang. This means that there is less unoccupied time per circuit to absorb overloads. The effect of this is shown in Fig. 7.7b where it can be seen that the grade of service is more severely affected on large groups than on the smaller ones.

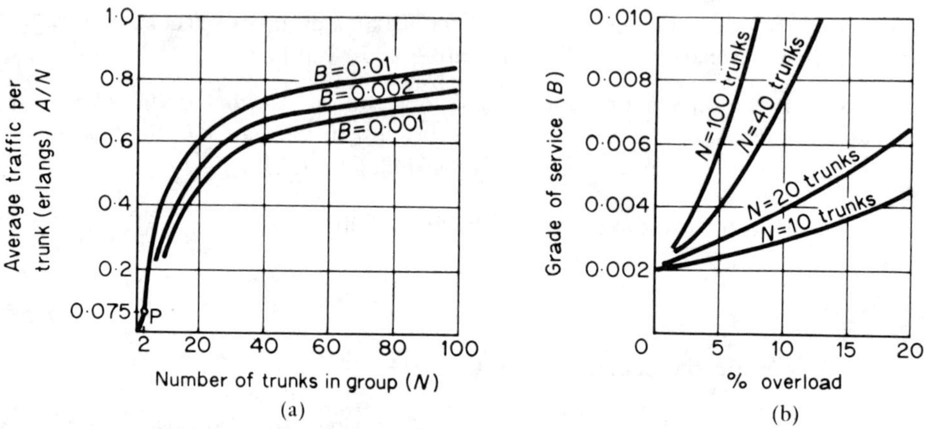

Fig. 7.7 Efficiency of large groups. Both these graphs are derived from the same formula as Fig. 7.6. Compare the values of A/N, N and B at point P here and in Fig. 7.6 (7.3.2).

7.3.3 Traffic data for new exchanges is obtained from traffic records taken on comparable existing exchanges in the same area or another similar area. These existing exchanges are unlikely to have exactly the same number of each type of line as are required on the new exchange so it is convenient to express the data on a per line basis. The busy hour traffic can thus be divided by the number of lines giving 'average traffic per line', typical values for which are as follows:

Residential	0·05 erlang per line
Coin box	0·10 erlang per line
Business	0·15 erlang per line

Not all rates of provision of apparatus are directly related to traffic intensity in the same way as selector quantities (**7.2.1**) and so it is found more convenient to work in terms of the *calling rate* (calls per line) and to calculate traffic from this using the appropriate holding time for each item (**7.2.1**).

For calculating selector quantities and peak current consumption, the *busy hour calling rate* is used. This is defined as the average number of calls originated per line in the busy hour. Typical values are:

Residential area	0·5 calls per line
Business area	1·5 calls per line

For calculating reserve power to be supplied by the battery in the event of mains power failure, the *day calling rate* is used. This is defined as the average number of calls originated per line in the course of 24 h. Typical values are:

Residential area	3 calls per line
Business area	14 calls per line

It is also useful to obtain the *day to busy hour calling rate ratio* by dividing the day calling rate by the busy hour calling rate. This enables future estimates of day calling rates to be calculated from busy hour traffic records, or busy hour traffic to be estimated for any group of subscribers from the total number of calls indicated on the subscribers' meters. Typical values for this ratio, based on the values of calling rate given above are:

Residential area	6 (6:1)
Business area	9·33 (9·33:1).

Example: A 2 000 line exchange has a day to busy hour calling rate ratio of 8. If the total number of calls originated during the day is 24 000, determine the busy hour calling rate and the average traffic intensity on the exchange in the busy hour assuming the average call duration to be 2 min.

$$\text{Day calling rate} = \frac{24\ 000}{200} = 12 \text{ calls per line per day.}$$

$$\text{Busy hour calling rate} = \frac{\text{day calling rate}}{\text{calling rate ratio}}$$

$$= \frac{12}{8} = \underline{\underline{1{\cdot}5 \text{ calls per line}}}$$

Busy hour traffic, $A = Ct$,

$$t = 2 \text{ min} = \frac{1}{30} \text{ h}, \ C = 1{\cdot}5 \times 2\ 000 = 3\ 000.$$

$$\therefore\ A = 3\ 000 \times \frac{1}{30} = \underline{\underline{100 \text{ erlang.}}}$$

Example: A 10 000 line exchange has a busy hour calling rate of 2·0 and an average call duration of 3 min. How many first selectors should be provided if the average traffic per first selector is not to exceed 0·8 erlang?

$$C = \text{number of calls in busy hour} = 10\ 000 \times 2{\cdot}0 = 20\ 000,$$

$$t = 3 \text{ min} - \frac{1}{20} \text{ h.}$$

$$\text{Traffic, } A = Ct = 20\ 000 \times \frac{1}{20} = 1\ 000 \text{ erlang.}$$

But traffic also equals traffic per selector × number of selectors,

$$\therefore \text{ number of selectors} \times 0{\cdot}8 = 1\ 000,$$

$$\text{i.e. number of first selectors} = \frac{1\ 000}{0{\cdot}8} = \underline{\underline{1\ 250 \text{ selectors.}}}$$

Now turn to page 203 and try some examples on 'traffic' for yourself before going on to grading.

7.4 Principles of grading

7.4.1 Full availability and limited availability. The number of trunks available to a selector on a particular route is called its *availability*. For a two-motion selector, this would be the number of outlets per level, a 100 outlet selector having an availability of 10 and a 200 outlet selector having an availability of 20 (**4.3.3**).

On any route (i.e. level) where the availability of a selector is less than the number trunks provided, the selector is said to have *limited availability*. Under this condition, *some of the trunks are not available* to the selector.

On those routes where the selector availability is either equal to, or greater than, the number of trunks provided, a selector is said to have *full availability*. Under this condition, *all of the trunks are available* to the selector.

Example: The trunking of first selectors at a certain exchange is as shown in Fig. 7.8. State which first selector levels have full availability and which have limited availability:

(*a*) if the first selectors are 100 outlet, and

(*b*) if they are 200 outlet selectors.

In case (*a*), all levels having 11 or more trunks are limited availability levels, i.e. *levels* 1, 2, 3, 4 *and* 0 have limited availability and *levels* 8 *and* 9 have full availability.

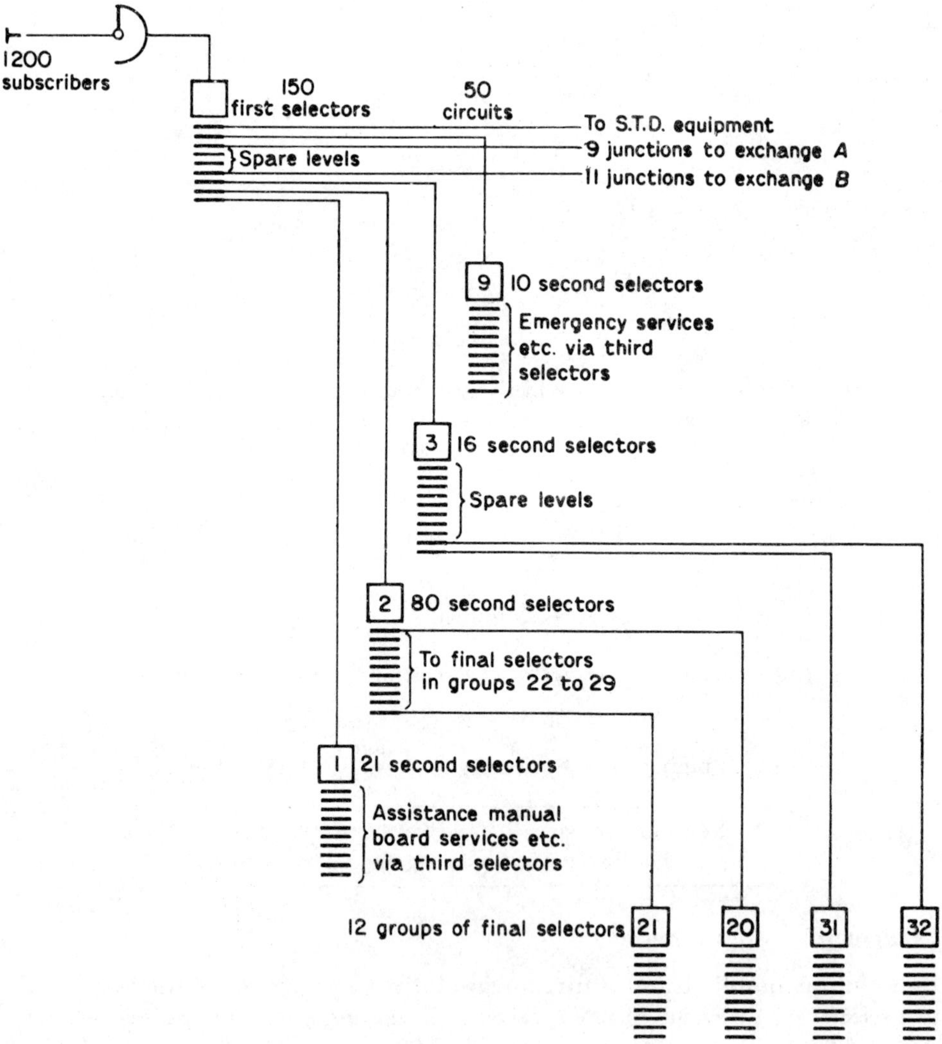

Fig. 7.8 Full and limited availability (7.4.1).

In case (*b*), all levels having 21 or more trunks are limited availability levels, i.e. *levels* 1, 2, *and* 0 have limited availability and *levels* 3, 4, 8 *and* 9 have full availability. (The spare levels on the first and second selectors provide for future growth—**7.2.2.**)

7.4.2 Interconnecting is a method of connecting together two or more limited availability level multiples so that the sets of trunks available from the different multiples are partially common to one another.

Selectors are divided into a number of groups on those levels on which they have limited availability, and these groups are connected as separate multiples. This gives an increased number of multiple outlets over which the available trunks can then be distributed. In Fig. 7.8, for example, level 3 of the first selectors is to be connected to 16 second selectors. If 100 outlet selectors are used, this means dividing the first selectors into at least two groups, giving 10 outlets on level 3 from each group, a total of 20 outlets to serve the 16 trunks.

One way of connecting these 16 trunks is as two full availability groups as shown in Fig. 7.9. These two smaller groups are very much less efficient than a single group of 16 trunks because a call can be lost on one group even though there may be a free trunk at that time on the other group (**7.3.2** and Fig. 7.7).

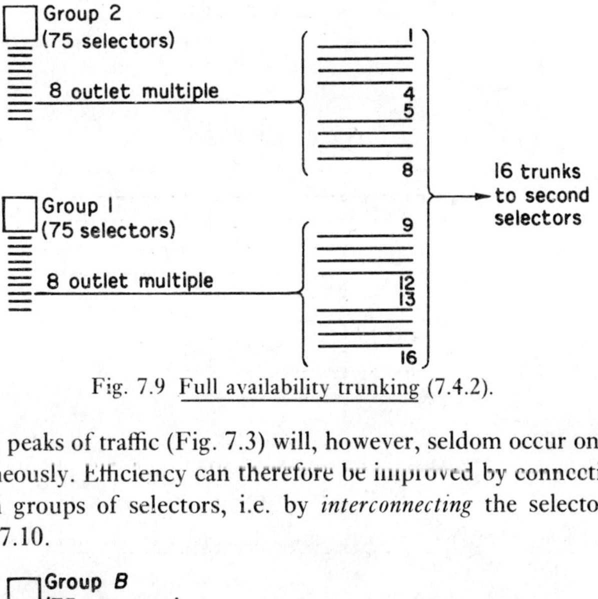

Fig. 7.9 Full availability trunking (7.4.2).

The random peaks of traffic (Fig. 7.3) will, however, seldom occur on more than one group simultaneously. Efficiency can therefore be improved by connecting some of the trunks to both groups of selectors, i.e. by *interconnecting* the selector multiples, as shown in Fig. 7.10.

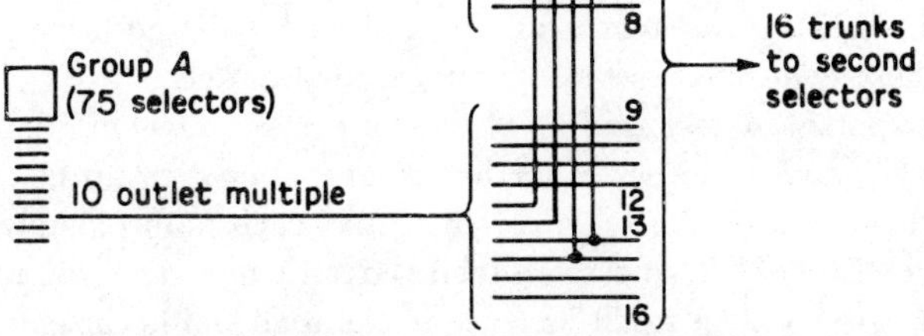

Fig. 7.10 Interconnecting. Compare this with Fig. 7.9 (7.4.2).

The method of interconnecting which is found to be the most effective for the subscribers' uniselectors and the group selectors in Strowger type exchanges is known as *grading*.

7.4.3 Grading is the method of connecting limited availability level multiples together so that a group of selectors is given access to individual trunks on the early choices, but on later choices, shares access to trunks with other groups. (The term 'grading' is also used to describe an arrangement of trunks connected to the banks of a selector by the method of grading.) The same interconnected groups as Fig. 7.10 are shown in the form of a grading in Fig. 7.11.

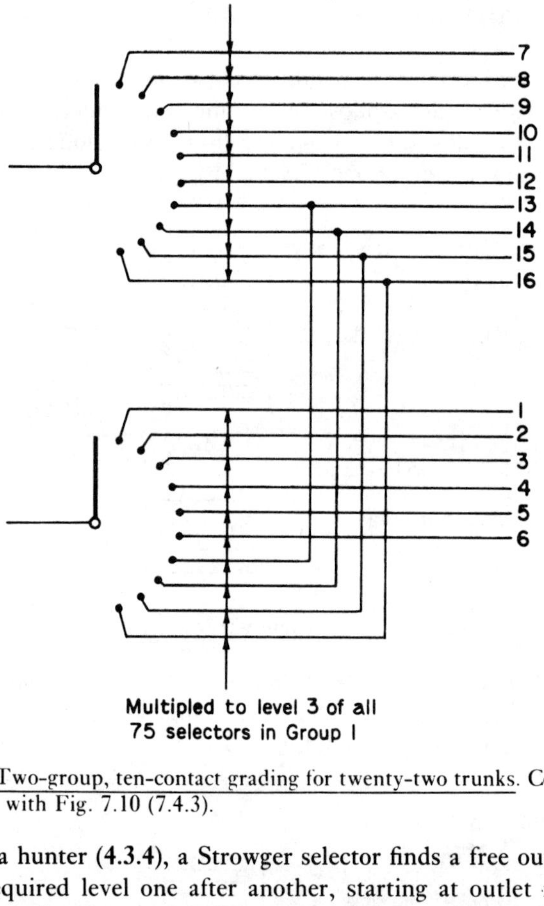

Fig. 7.11 Two-group, ten-contact grading for twenty-two trunks. Compare this with Fig. 7.10 (7.4.3).

When used as a hunter (4.3.4), a Strowger selector finds a free outlet by testing the outlets on the required level one after another, starting at outlet number 1 (6.6.2). Calls are therefore offered to a particular outlet only when all the preceding outlets are engaged, the later choice outlets in a multiple being offered less traffic than the earlier choices (see Fig. 7.12). If any trunks are to be shared with another multiple, it should obviously be those connected to these lightly loaded outlets.

The two-group, ten-contact grading of sixteen trunks shown in Fig. 7.11 has a traffic capacity roughly equal to either a single full availability group of fifteen trunks or two separate full availability groups of ten trunks each (approximately 6 erlang at a grade of service of 0·002). Using more than sixteen trunks to form a two-group, ten-contact grading gives only a small increase in the total traffic capacity. The efficiency in terms of average traffic per trunk is then actually reduced because there is a smaller

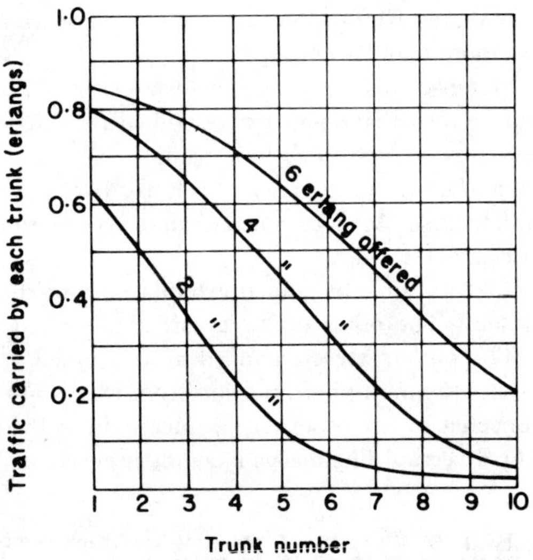

Fig. 7.12 Distribution of traffic over full availability group for 2, 4 and 6 erlang of traffic offered (7.4.3).

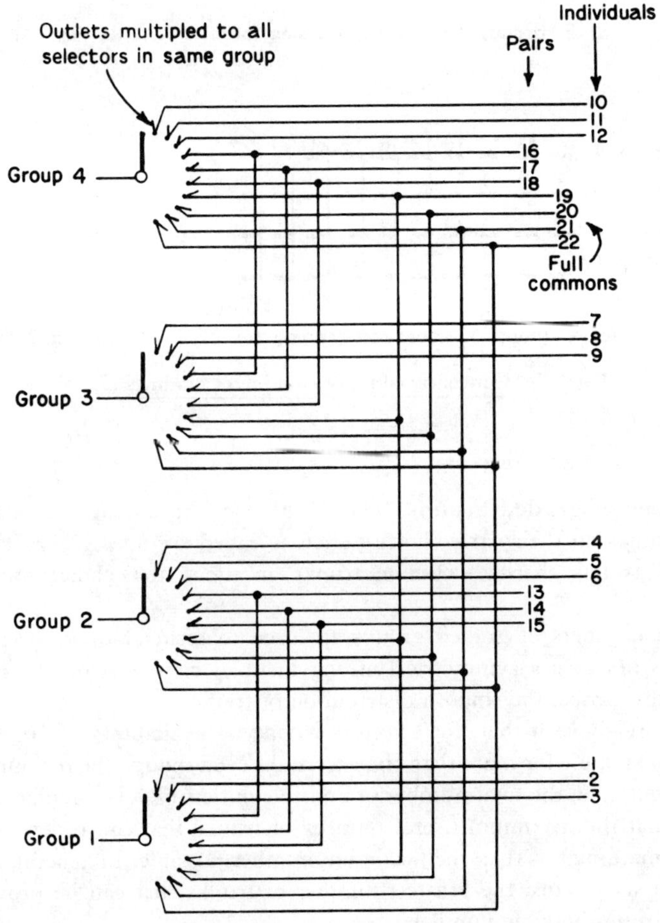

Fig. 7.13 A four-group, ten-contact grading for twenty-two trunks (7.4.3).

number of common or shared trunks. For larger numbers of trunks, the selectors are therefore divided into more than two groups.

A typical four-group, ten-contact grading is shown in Fig. 7.13. This is for twenty-two trunks and it will be seen that some are individual trunks, some are common to two groups and others are common to all four groups.

The drawing of such gradings in the form of selector banks as we have done so far is unnecessarily cumbersome. A more convenient convention is therefore usually adopted as shown in Fig. 7.14.

The outlets of the first multiple or graded group are represented in the form of a horizontal row of dashes. The other multiples are represented by similar rows of dashes above these. The outlets are commoned as required by vertical lines. Each dash therefore represents the multiple connexion from one group of selectors.

Each dash not connected to any other dashes means an individual trunk and each set of dashes joined by a vertical line means a common or shared trunk.

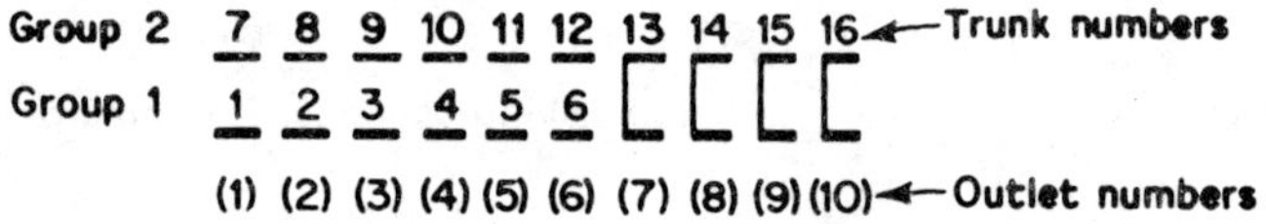

(a) 2-group, 10-contact grading for 16 trunks as Fig. 7.11

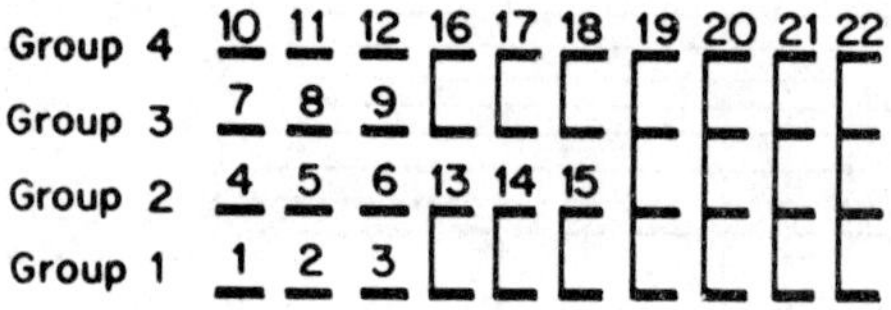

(b) 4-group, 10-contact grading for 22 trunks as Fig. 7.13

Fig. 7.14 Conventional representation of gradings (7.4.3).

7.5 The design of gradings

7.5.1 Number of graded groups. The set of selectors having access to the same individual trunks on their early choice outlets is called a *graded group*. Two or more such graded groups sharing common trunks on their later choice outlets form a grading.

Only even numbers of graded groups are used to form gradings because an odd number does not give a symmetrical arrangement of commons on the intermediate outlets and this upsets the smooth distribution of traffic.

We have already seen that, for selectors having an availability of 10, a two-group grading is not suitable for more than sixteen trunks even though the maximum possible would be twenty, i.e. the total number of multiple outlets (7.4.3). Similarly, it is found in practice that the maximum useful number of trunks that can be provided by any grading is something less than the maximum number possible. In general, for gradings of more than six groups, the greatest number of trunks that can be provided is one half the maximum possible number.

Thus, if the number of trunks is n, the availability a and the number of groups g:

$$n = \frac{g \times a}{2} \tag{4}$$

The minimum number of groups that must be used to provide a grading for n trunks is therefore:

$$g = \frac{2n}{a} \tag{5}$$

The actual number used in practice would be the next even number higher than the value of g calculated from equation (5).

The maximum number of trunks for which gradings of two, four or six groups have been found to be suitable are given in the following table:

g	max. value of n			
	$a=10$	$a=20$	$a=24$	
2	16	32	38	60% more than equation (5)
4	25	50	60	25% more than equation (5)
6	32	64	76	7% more than equation (5)

Example: How many graded groups should be used to provide a grading for seventy-four trunks,
(*a*) from a level of 200 outlet selectors,
(*b*) from a level of 100 outlet selectors, and
(*c*) from twenty-four uniselectors?

For (*a*), the availability is 20, therefore,

$$g = \frac{2n}{a} = \frac{2 \times 74}{20} = 7{\cdot}4.$$

8 groups should therefore be used.

For (*b*), the availability is 10, therefore,

$$g = \frac{2 \times 74}{10} = 14{\cdot}8.$$

16 groups should therefore be used.

For (*c*), the availability is 24, for which the table above allows up to seventy-six trunks to be provided by a six group grading.
6 groups should therefore be used.

7.5.2 'Best' gradings. For a given number of trunks and availability, there is only one possible two-group grading. For gradings of more than two groups, there is, however, more than one possible grading, and not all these possibilities have the same traffic capacity. In these cases, it is necessary to choose the grading with the greatest traffic capacity (i.e. the best grading).

That several alternative gradings may be possible can be seen from the four-group grading of Fig. 7.14. If trunks numbered 3 and 6 are joined to form a pair and 9 and 12 to form another and commons 19 and 20 are replaced by pairs, another grading is formed. This is shown in Fig. 7.15 where it can be seen that this is still a four-group, ten-contact grading for twenty-two trunks.

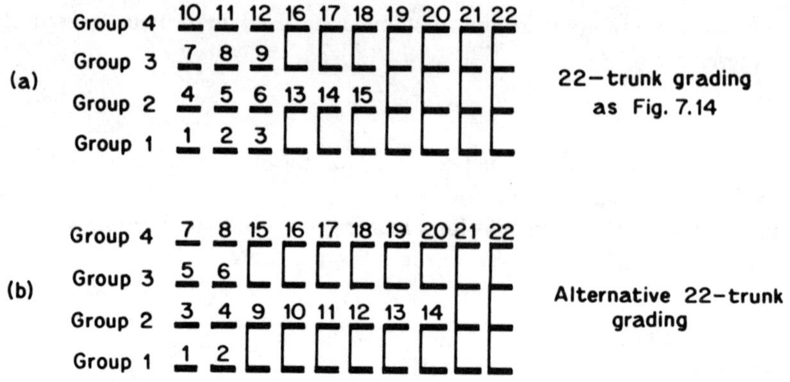

Fig. 7.15 Different gradings for twenty-two trunks (7.5.2).

It is found that the best grading is obtained when there is a smooth progression from individuals to partial commons and from partial commons to full commons. This is to be expected because it means that outlets which are only a few contacts apart, and are therefore offered only slightly different values of traffic (Fig. 7.12) are not given access to very different sizes of common.

To determine the best grading, we must work out all the possible alternative gradings and then select the one with the smoothest progression of commons. For this purpose a measure of this 'smoothness' is required, and this is provided by a quantity known as the 'sum of successive differences'.

The *sum of successive differences* for a grading is obtained by adding together the differences between the number of outlets per graded group connected to each possible type of common taken in order. For a four-group grading, for example, this means subtracting the number of individuals from the number of pairs, and subtracting the number of pairs from the number of full commons. These two differences are then added together, disregarding the signs. *The 'best' (smoothest) grading is the one with the smallest sum of successive differences.*

Example: Determine the sum of successive differences for each of the gradings shown in Fig. 7.15 and hence choose the better of the two gradings.

For the grading of Fig. 7.15a,

No. of individuals $(a) = 3,$
No. of pairs $(b) = 3.$
First difference, therefore $= 0.$

No. of pairs $(b) = 3,$
No. of commons $(c) = 4.$
Second difference, therefore $= 1.$

Sum of differences (S) $= 0 + 1 = 1.$

A convenient way of setting this out is in the form of a table as follows:

	a	b	c	S
Grading⟶	3	3	4	
Differences⟶	0	+ 1		= 1

The corresponding table for the grading of Fig. 7.17b is as follows:

	a	b	c	S
Grading ⟶	2	6	2	
Differences ⟶	4	+	4	= 8

The grading of Fig. 7.15a, having a sum of successive differences of 1 is better than the grading of Fig. 7.15b which has a sum of successive differences of 8.

7.5.3 Four-group gradings (algebraic method). The various possible four-group gradings for a given number of trunks and availability can be worked out by a process of trial and error. This is not a very satisfactory method, however, because it is difficult to be sure that *all* the possible gradings have in fact been found.

The same process can be performed more systematically by making use of simple algebra as follows:

Outlets can be connected as individuals, pairs or full commons:

Let a = number of outlets per group connected as individuals.
Let b = number of outlets per group connected as pairs.
Let c = number of outlets per group connected as full commons.

The total number of outlets per group is then:

$$a + b + c = \text{availability}. \qquad (6)$$

For each outlet connected as individuals, 4 trunks are required for the grading, (a total of $4a$ trunks). For each outlet connected as pairs, 2 trunks are required for the grading, (a total of $2b$ trunks). For each outlet connected as a full common, 1 trunk is required for the grading, (a total of c trunks).

The sum of these is the total number of trunks,

$$\text{i.e. } 4a + 2b + c = \text{number of trunks}. \qquad (7)$$

Subtracting (6) from (7), gives:

$$3a + b = (\text{number of trunks}) - (\text{availability}) \qquad (8)$$

From (8), it is possible to determine the value of b for each possible value of a. From (6), it is then possible to find the value of c for each value of a. Each such set of values for a, b, and c gives one possible grading.

The sum of successive differences for each of these gradings can be calculated from the values of a, b, and c:

$S = (a - b$, or $b - a) + (b - c$, or $c - b)$ taking only positive values for the differences (7.5.2).

Example: Determine all possible four-group, ten-contact gradings for twenty-two trunks. Which of these is the most efficient?

From (7), $4a + 2b + c = 22$.
From (6), $a + b + c = 10$,
Subtract, $3a + b = 12$,
(This last equation is equivalent to (8) above.)

First, let $a = 1$,
then $b = 12 - 3a = 12 - 3 = 9$,
and $c = 10 - (a + b) = 10 - 10 = 0$.

These values give us one possible grading, the sum of successive differences being $(9 - 1) + (9 - 0) = 17$.

Next, let $a = 2$, then 3 and so on. This can be done in the form of a similar table to the last example as follows, starting with the grading for $a = 1$ which we have just derived:

	a	$b = 12 - 3a$	$c = 10 - (a + b)$	S
Grading (i) ⟶	1	9	0	
Differences ⟶	8	+	9	= 17
Grading (ii) ⟶	2	6	2	
Differences ⟶	4	+	4	= 8
Grading (iii) ⟶	3	3	4	
Differences ⟶	0	+	1	= 1
Grading (iv) ⟶	4	0	6	
Differences ⟶	4	+	6	= 10

Higher values of a give negative values for b, so there are no possible gradings having more than four contacts per group as individuals.

The best grading is grading (iii) which has a sum of successive differences of 1. This is the grading shown in Fig. 7.15a.

7.5.4 Six-group gradings (algebraic method). The same basic method as that described above for four-group gradings can be used for six-group gradings.

There are four factors of six, so that there are four ways in which the outlets can be connected, namely as individuals, pairs, threes or full commons. For these we use the letters a, b, c and d respectively, giving a similar equation to (6) in **7.5.3**:

$$\underline{a + b + c + d = \text{availability.}} \tag{9}$$

The number of trunks required for each is 6, 3, 2 and 1 respectively, giving a similar equation to (7) in **7.5.3**:

$$6a + 3b + 2c + d = \text{number of trunks.} \tag{10}$$

Subtracting (9) from (10) gives:

$$\underline{5a + 2b + c = \text{no. of trunks} - \text{availability.}} \tag{11}$$

For every value of a, there are now a set of values of b and each of these must be tried separately. For each pair of values of a and b, however, there will only be one pair of values of c and d which can be calculated from the two equations.

Example: Determine the best six-group grading for thirty-one trunks from selectors with an availability of 10.

From (10), $6a + 3b + 2c + d = 31$.
From (9), $\underline{a + b + c + d = 10}$, i.e. $d = 10 - (a + b + c)$. (12)
Subtract $\underline{5a + 2b + c \quad = 21}$, i.e. $c = 21 - (5a + 2b)$. (13)
(This last equation is equivalent to (11) above.)

From $c = 21 - (5a + 2b)$, the highest possible value for a is seen to be 4; higher values give negative values for c.

When $a = 4$, the only possible value of b is 0,
so that $c = 21 - (5a + 2b) = 21\ (20 + 0) = 1$,
and, $d = 10 - (a + b + c) = 10 - (4 + 0 + 1) = 5$.

This gives one grading: $a = 4$, $b = 0$, $c = 1$, $d = 5$. (i)

Next try $a = 3$, the highest possible value for b is then $b = 3$. Gradings can therefore be derived by calculating values for c and d corresponding to $b = 3$, $b = 2$ and $b = 1$.

Gradings for $a = 2$ and $a = 1$ can then be derived for various values of b similarly and the sum of successive differences calculated for each grading as follows:

	Individuals		Pairs		Threes		Full commons	Sum of successive differences
	a		b		c From (13)		d From (12)	S
Grading (i) →	4		0		1		5	
		4		1		4		9
Grading (ii) →	3		3		0		4	
Differences →		0		3		4		7
Grading (iii) →	3		2		2		3	
Differences →		1		0		1		2 (Best grading)
Grading (iv) →	3		1		4		2	
Differences →		2		3		2		7
Grading (v) →	3		0		6		1	
Differences →		3		6		5		14
Grading (vi) →	2		5		1		2	
Differences →		3		4		1		8
Grading (vii) →	2		4		3		1	
Differences →		2		1		2		5
Grading (viii) →	2		3		5		0	
Differences →		1		2		5		8
Grading (ix) →	1		8		0		1	
Differences →		7		8		1		16
Grading (x) →	1		7		2		0	
Differences →		6		5		1		13

The best of these gradings is (iii), having a sum of successive differences of 2. This grading is illustrated in Fig. 7.16.

7.5.5 Larger gradings (transposition method). The algebraic method so far described is the only system for reliably determining all possible gradings for a given number of trunks, groups and availability. With more than six graded groups, the possible number of gradings can be very large especially with availabilities of 20 or 24, and many of these gradings would not use all the available types of common for

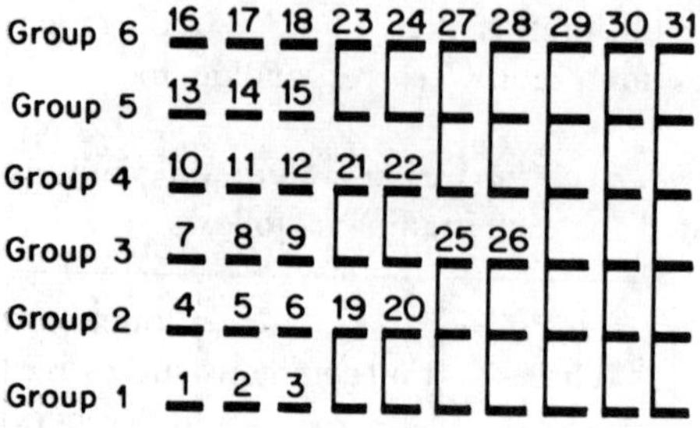

Fig. 7.16 Six-group, ten-contact grading for thirty-one trunks (7.5.4).

the number of groups in use. Of the smoother gradings, the difference in efficiency between near identical gradings is often quite small. In these circumstances, a satisfactory grading can be designed without working out all the possible arrangements.

The method consists of finding a smooth grading for a convenient number of trunks and then changing it to provide the number of trunks actually required while making as little difference to the sum of successive differences as possible.

This method is known as the 'transposition' method, a transposition being the alteration of the type of common on one outlet at a time. Each transposition increases the number of one type of common by one and reduces the number of one other type of common by one at the same time.

This was the method used in **7.5.2** to derive the four-group grading of Fig. 7.15b from the four-group grading of Fig. 7.15a by one transposition from individuals to pairs on outlet 3, a second transposition on outlet 7 from a full common to pairs and a similar transposition on outlet 8. In this particular case, the object was to keep the number of trunks the same without regard to the sum of successive differences.

In the event of two gradings for the same number of trunks, groups and availability having the same sum of successive differences, the one having the greater number of the larger commons gives the most interconnexion between the multiples resulting in the greater traffic carrying capacity.

If none of the possible arrangements uses all the different types of common, it is best that there should always be at least one full common because it would otherwise consist of two or more smaller gradings giving a lower efficiency than one large grading.

Now turn to page 203 and try questions 11–18 there before continuing.

PART 3 TYPICAL PRACTICAL GRADING ARRANGEMENTS

7.6 Grading terminals

The selector banks are not a very convenient point at which to provide the complex interconnexion of multiples required to form a grading. The banks are therefore connected directly to form multiples, and these multiples are connected to suitable grading terminals.

The twenty outlets from a level of 200 outlet selectors, for example, are connected to twenty sets of contacts arranged in a horizontal row. Each set of terminals consists of three terminals, one each for the −, + and *P*-wires. The rows of terminals for the various multiples are provided one above the other so that commons can be formed by connecting the terminals together as required in vertical columns.

The terminals are provided on the selector racks where possible. This system is known as *decentralized trunk distribution*. Otherwise a separate frame called a *trunk distribution frame* (T.D.F.) is used. This method is known as *centralized trunk distribution*.

7.6.1 Trunk distribution frame (T.D.F.). This is a framework similar in construction to a selector rack. The size and capacity of a typical T.D.F. is shown in Fig. 7.17.

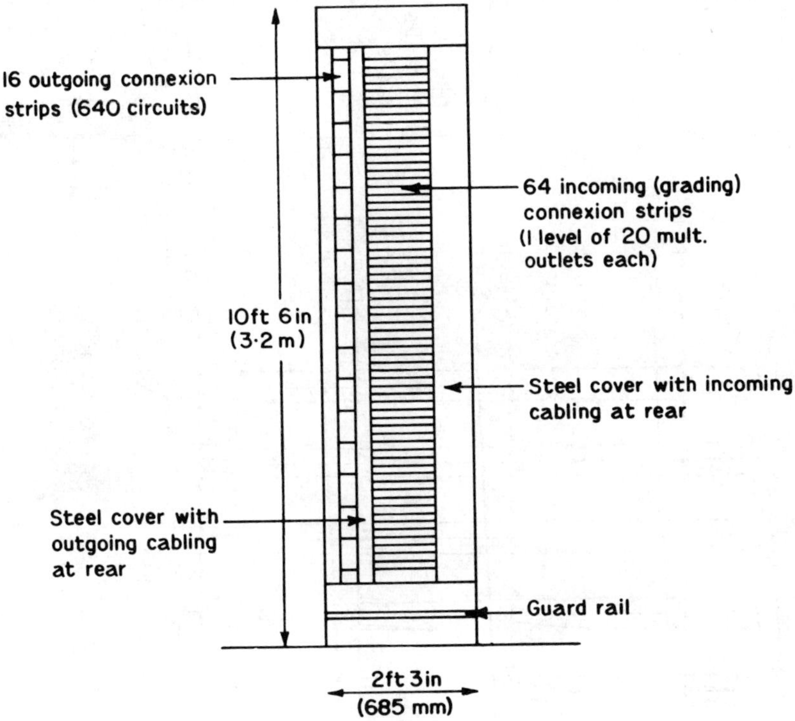

Fig. 7.17 Trunk distribution frame, front view (7.6.1).

For selectors to be graded at a T.D.F., the selector banks are usually factory wired to form a multiple over one or two shelves of selectors. The multiple outlets are wired to multiple terminals on the selector racks. When the racks are installed in the exchange, the multiple terminals are commoned as necessary to form graded groups of the required size. Cables are then provided from each of these groups to the grading terminals on the T.D.F.

The commoning of outlets from the various groups to form a grading is carried out by means of bare tinned copper wires (B.T.C straps) joining the grading terminals in vertical columns.

The trunks from the grading commons are connected by means of jumpers to the outgoing terminals on the T.D.F. These jumpers consist of a set of three wires connected to the top set of terminals in a common and passed through holes in the fanning strip and held in place at the rear of the rack by jumper rings. They are not tied in position and do not form a cable and can therefore be easily rearranged at any time during the life of the exchange.

The connexion of cables, jumpers and grading wires is shown in Fig. 7.18 where it can be seen that the terminals are specially shaped to provide a neat and orderly arrangement when connected in the order described above. The commoning and jumpering arrangements necessary to provide the four-group grading previously illustrated in Figs 7.15 and 7.16b are also shown in Fig. 7.18 for a T.D.F.

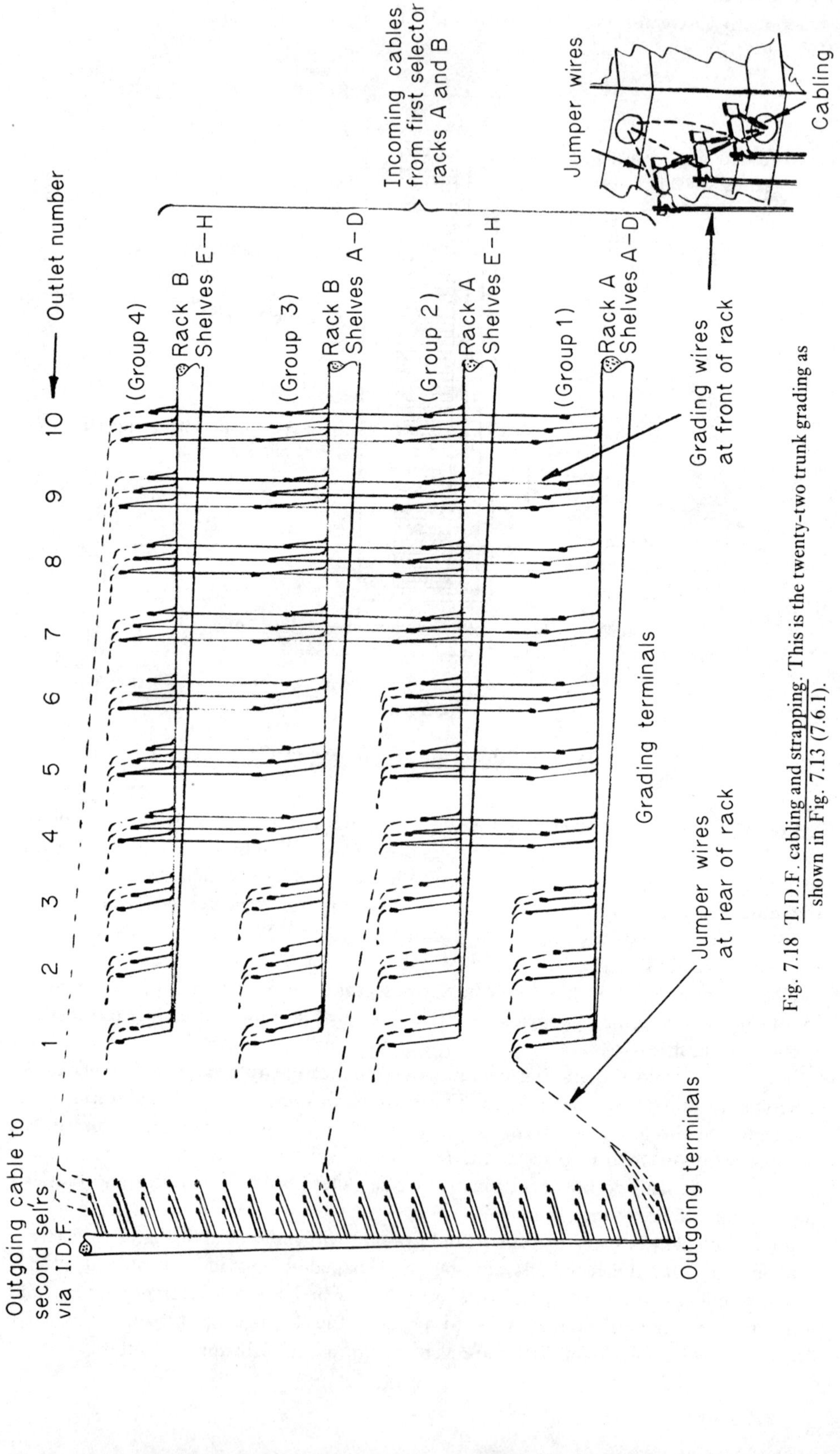

Fig. 7.18 T.D.F. cabling and strapping. This is the twenty-two trunk grading as shown in Fig. 7.13 (7.6.1).

7.6.2 Grading on uniselector rack. The general layout of a subscriber's uniselector rack with grading facilities is shown in Fig. 7.19. The grading terminals are provided on trunk distribution connexion strips on the front of the rack together with outgoing terminals similar to those on a T.D.F.

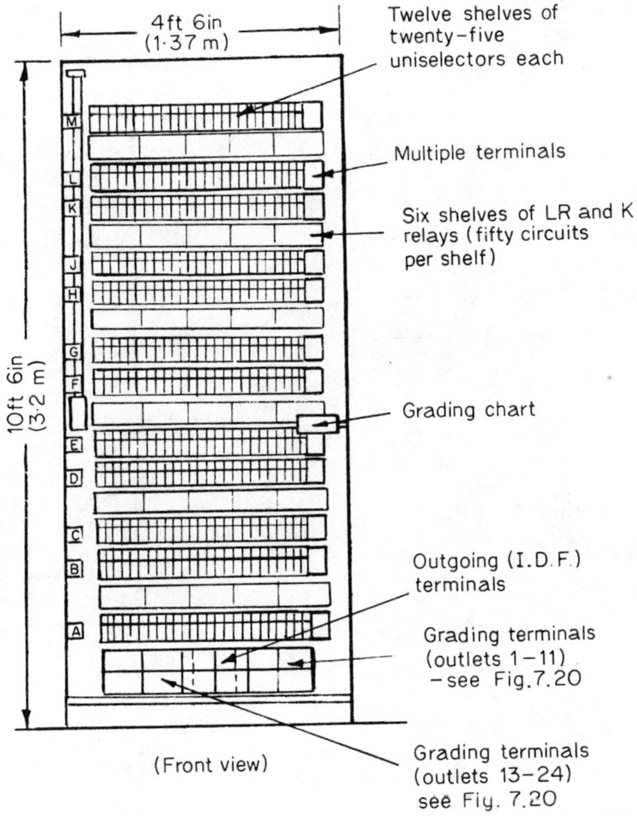

Fig. 7.19 Uniselector rack with grading facilities (7.6.2).

The arrangement of terminals on the connexion strips is shown in Fig. 7.20. Each shelf on the rack has its own row of multiple terminals on these strips so that commons can be provided by means of vertical bare tinned copper straps as in the case of the T.D.F. (**7.6.1**). The terminals, however, have three notches to take the permanent wiring from the multiple, the jumper and the strap instead of two notches and a forked end like the T.D.F. terminal.

The top row of terminals labelled *TIE* is cabled to the bottom row of terminals on the next rack (also labelled *TIE*) when the rack is installed in the exchange.

A common can thus be extended over more than one rack by extending the vertical straps to include the *TIE* terminals.

7.6.3 Grading on group selector racks. The front of a typical group selector rack is shown in Fig. 5.22. In this it can be seen that the front of the rack is completely occupied by selectors when the rack is fully equipped. The grading terminals are provided at the rear of the rack and one row of connexion strips can be seen behind the third shelf from the bottom.

These connexion strips are basically similar to those provided on uniselector racks except that they have to provide for ten separate gradings or sets of gradings, one for

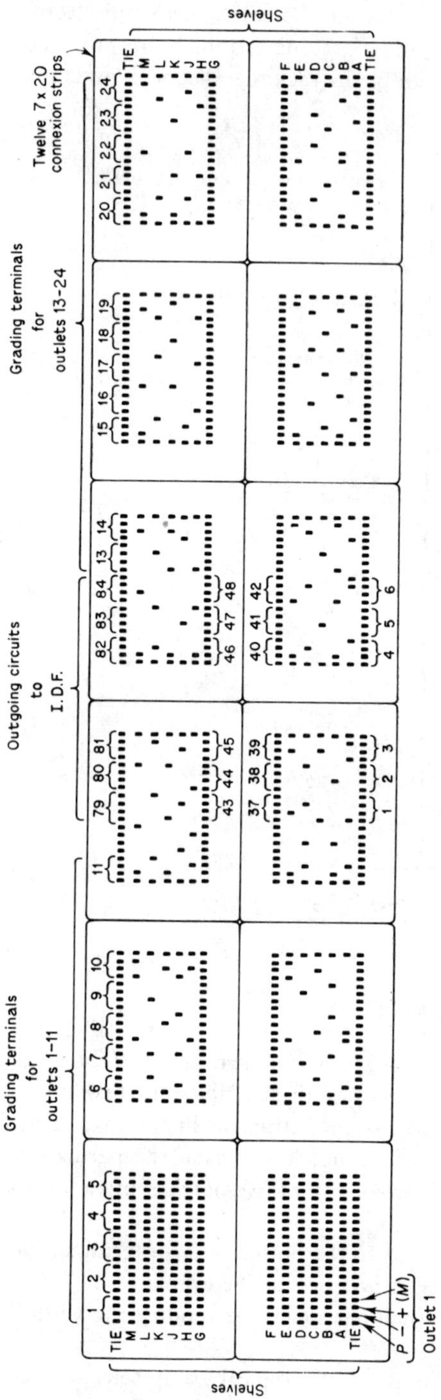

Fig. 7.20 Uniselector rack grading terminals (7.6.2).

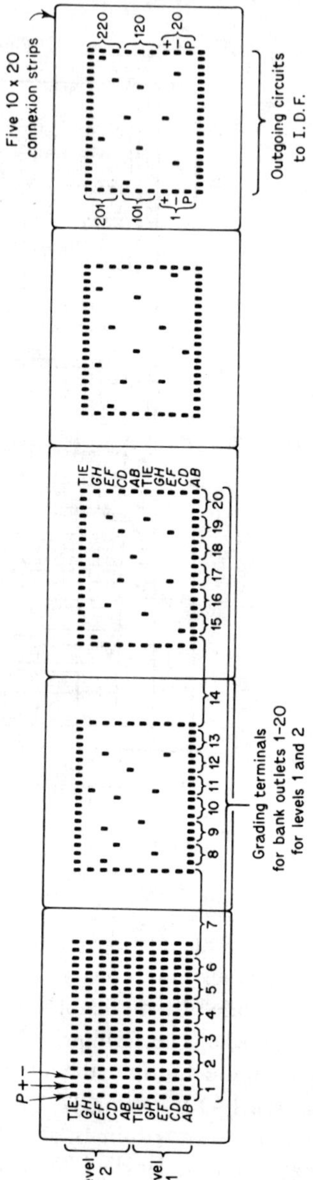

Fig. 7.21 Group selector rack grading terminals (7.6.3).

each level. For this purpose, five sets of connexion strips are provided, each serving two levels. The selectors are multipled over two shelves and each pair of shelves is connected to one horizontal row of terminals for each level.

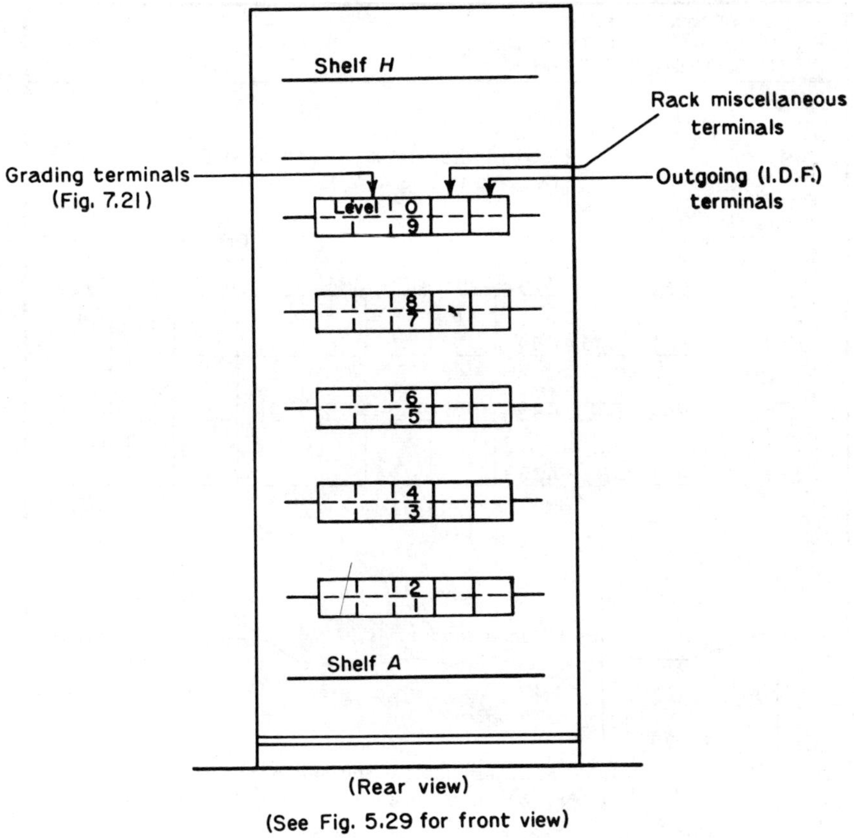

Fig. 7.22 Group selector rack with grading facilities, 200-outlet, 2000-type (7.6.3).

Every selector is therefore connected on the appropriate outlets to every set of connexion strips. Also every connexion strip is connected to the appropriate outlets of every selector on the rack, not just to the selectors on the shelf on which the connexion strip happens to be mounted.

The position of the connexion strips as seen from the rear of a 200 outlet 2000 type group selector rack is shown in Fig. 7.22, and the terminal arrangement in Fig. 7.21.

7.7 Grading charts

In order to investigate faults and service irregularities, it is often necessary to trace calls either forwards or backwards from one switching stage to another. It is easy to see on which outlet the wipers of a Strowger switch are positioned but a knowledge of the grading arrangements is necessary to find out to which circuit this outlet is connected. Charts showing these arrangements are therefore provided in glass fronted frames on the rack to which they refer. These frames are hinged so that they can be swung clear to permit access to any apparatus on that part of the rack. The position of such a chart on a group selector rack can be seen in Fig. 5.22.

Part of a typical chart for one group selector level is shown in Fig. 7.23. This is the same grading as that of Fig. 7.15a but it shows the actual selector numbers in the exchange.

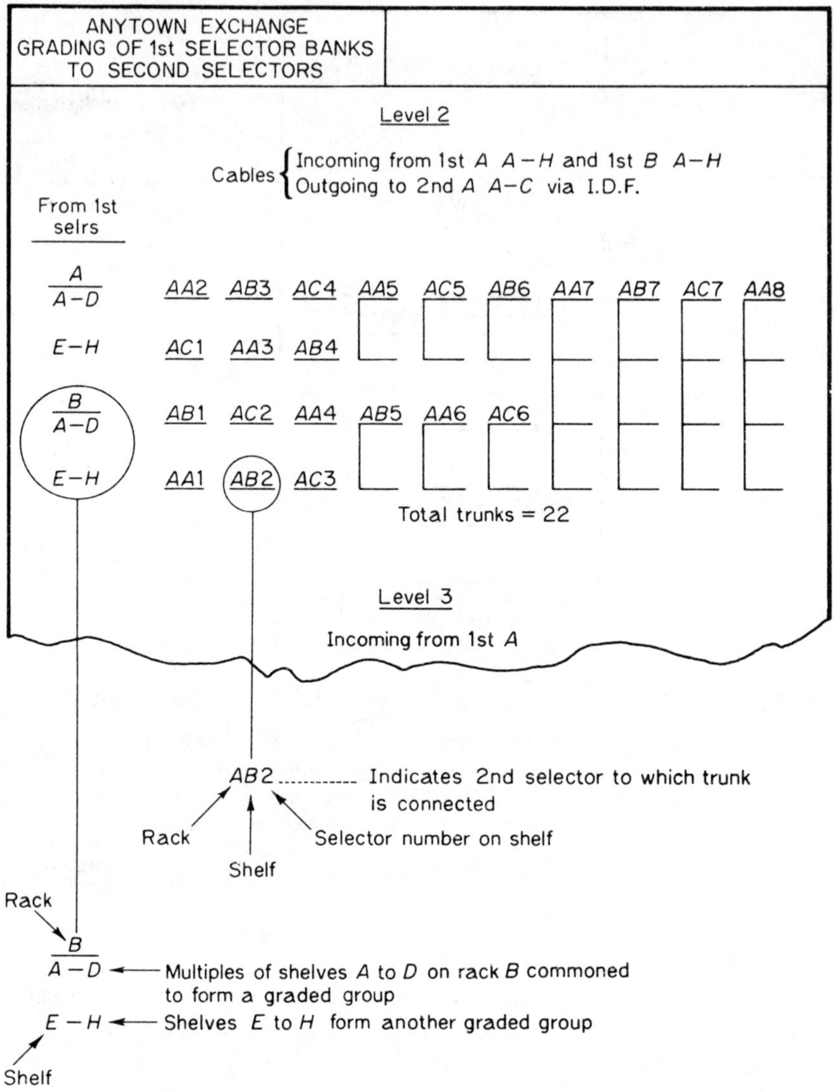

Fig. 7.23 Typical grading chart. Notice that every graded group has access to selectors on as many different shelves as possible and that selectors on the same shelf are not tested consecutively on any one group. This avoids any one group being seriously affected by one second selector shelf fuse blowing (7.7).

PART 4 SUMMARY

7.8.1 Traffic

ERLANG: unit of traffic intensity.

The traffic flow or intensity in erlangs for a specified period, usually the busy hour, is numerically equal to the average number of simultaneous calls in that period. (1 erlang = a traffic flow equivalent to one circuit continuously occupied.)

For busy hour, erlangs = calls × duration in hours ($A = Ct$).

(In general, $TA = Ct$).

Calling rate: number of calls per line in the stated period.

$$\text{Grade of service, } B = \frac{\text{number of calls lost}}{\text{number of calls offered}}$$

$$= \text{probability of all circuits engaged.}$$

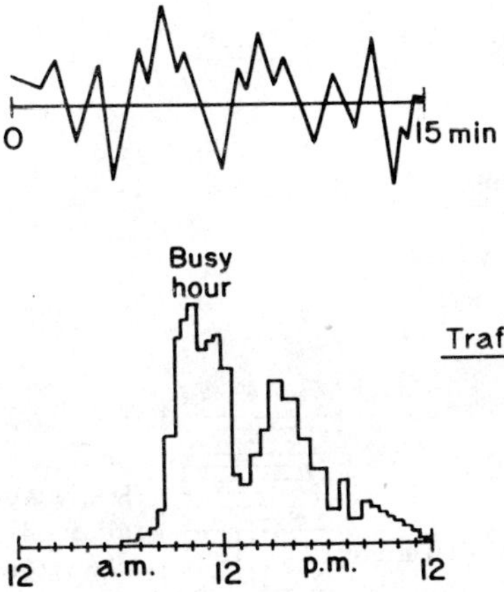

Traffic variations

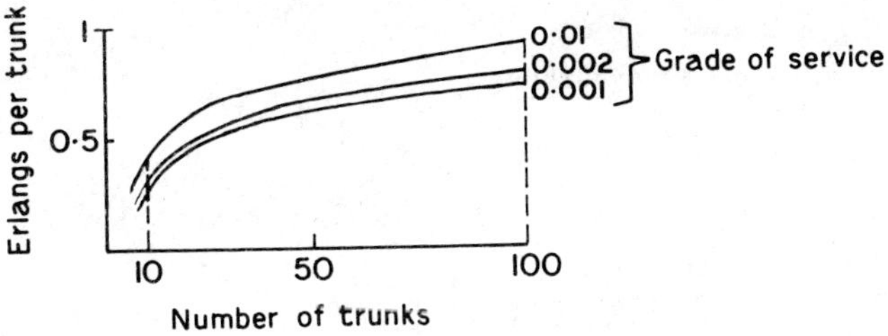

Efficiency of large groups

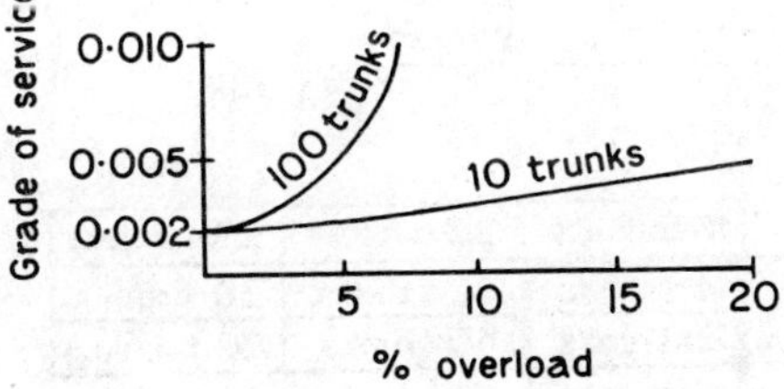

7.8.2 Grading

Limited availability: too few outlets.
Full availability: sufficient outlets.

$$\text{Number of graded groups} = \frac{2 \times \text{number of trunks}}{\text{availability}} \left(g = \frac{2n}{a}\right)$$

Best grading has smallest sum of successive differences, (e.g. for Fig. 7.25, four-group grading, sum of successive differences = (3 − 3) + (4 − 3) = 1).

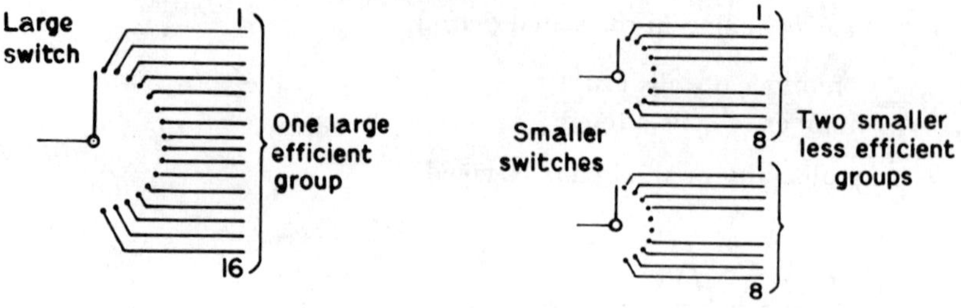

Full availability

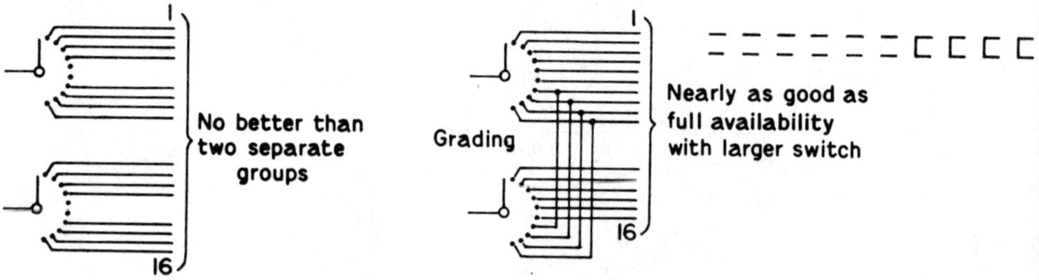

Limited availability

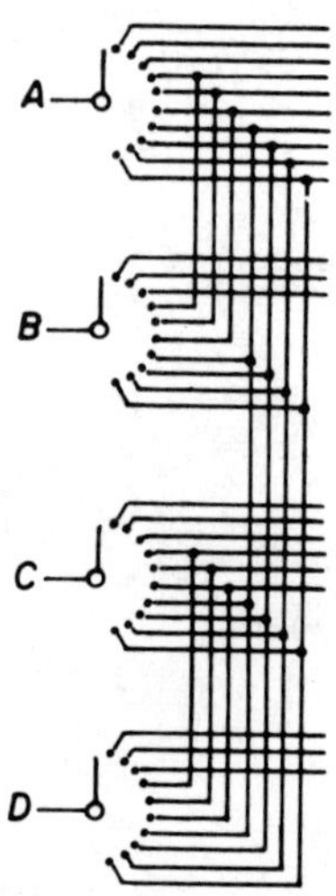

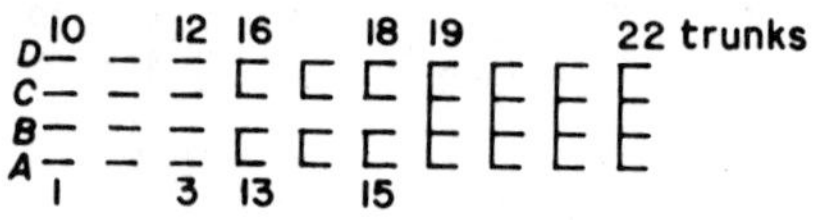

Four-group grading

	10 outlets	20 outlets	24 outlets
2 group	16 trunks	32 trunks	38 trunks
4 group	25 trunks	50 trunks	60 trunks
6 group	32 trunks	64 trunks	76 trunks

Fig. 7.25 Grading—summary.

Now turn to the questions on page 204 and test yourself before leaving this chapter.

PART 5 SELF-TEACHING NOTES

7.9.1 First read **7.1.1** to **7.3.3** inclusive. Study the worked examples carefully and make sure you see how the answers are arrived at.

7.9.2 Having read **7.1.1** to **7.3.3**, try the following questions (Q.1 to Q.10).

Q.1 Define the erlang.

Q.2 A group of twenty selectors was found to have ten selectors engaged at 10 a.m., fifteen at 10.10 a.m., sixteen at 10.20 a.m. and eleven at 10.30 a.m. Calculate the average traffic intensity during this time.

Q.3 During the busy hour, a group of selectors carries 100 calls of average duration 3 min. Calculate the average traffic intensity.

Q.4 What causes the reduction in traffic on a telephone exchange at about midday?

Q.5 For how many minutes during the busy hour is a single selector engaged if it carries an average traffic of 0·75 erlang?

Q.6 Calculate the average traffic intensity in the busy hour originated on a 2000 line exchange having a busy hour calling rate of 1·5 and an average call duration of 2 min.

Q.7 What is the probability of finding the selector of Q.5 engaged at any given time during the busy hour?

Q.8 If a group of selectors provides a grade of service of 0·005, how many calls are lost if 400 calls are offered to the group?

Q.9 If ten calls are lost out of ninety calls offered to a group of selectors, what is the traffic carried assuming an average call duration of 3 min?

Q.10 For a given grade of service, a large group of trunks will carry more traffic than the same number of trunks provided as a number of separate, smaller groups. What is the *dis*advantage of using the larger group?

Now check your answers on page 266.

7.9.3 You should have answered most of the questions Q.1 to Q.10 correctly. If you scored more than five out of ten, read the next part of the chapter (**7.4.1** to **7.5.5**) which deals with grading.

If your score was less than six, read the first part again and try Q.1 to Q.10 once more.

7.9.4 Having read up to and including **7.5.5**, try the next eight short questions:

Q.11 What is meant by 'limited availability'?

Q.12 A certain exchange has first selectors of the 100-outlet type which serve ten outgoing junctions on level 3, five outgoing junctions on level 4, and thirty-six trunks to second selectors on levels 2 and 5.

State which levels would require grading.

Q.13 Draw a two-group, ten-contact, sixteen-trunk grading (using the conventional representation of a grading).

Q.14 How many graded groups would you use to provide a grading for nineteen trunks from a selector level multiple with availability 10?

Q.15 Which of the following four-group gradings is the most efficient?

(*a*) 1 individual, 9 pairs, no full commons,

(*b*) 2 individuals, 6 pairs, 2 full commons, or
(*c*) 3 individuals, 3 pairs, 4 full commons.

Q.16 If a four-group grading has three individuals and four pairs, how many full commons must it have if the availability is 10?

Q.17 How many trunks are provided by a six-group grading having three individuals, four pairs, two threes, and one full common?

Q.18 If a four-group grading of twenty-three trunks from a selector level with availability 10 has three individual choices, how many pairs must it have?

Now check your answers on page 266.

7.9.5 If you answered fewer than five out of these eight questions (Q.11 to Q.18) correctly, read **7.4.1** to **7.5.5** again, making sure that you follow every stage of the worked examples. Then try Q.11 to Q.18 again.

With five or more right, read the rest of the chapter (**7.6** to **7.8.2**).

7.9.6 A few days after reading the whole of Chapter 7 try the following questions, writing the complete answer to each in no more than half an hour.

Check the answer to each on page 267–8 before answering the next one.

Q.19 Define the terms:
(*a*) erlang,
(*b*) busy hour, and
(*c*) grade of service.

During the busy hour a group of trunks is offered 100 calls having an average holding time of three minutes, and one of the calls fails to find a disengaged trunk. Calculate the traffic *offered* to the group and the traffic *carried* by the group. (C & G 1965)

Q.20 Give a sketch showing the appearance on a grading frame of a 4-group, 20 trunk grading from a selector level with an availability of 10. The first and last outlets of all groups should be drawn in detail to show the type of tag used and the connexion of all wires and jumpers. Intermediate outlets need only be drawn in sufficient detail to show the formation of the grading. (C & G 1962)

Q.21 Sketch a graph to show how local telephone traffic varies during a typical 24-hour period. Explain the importance of the busy hour in determining the amount of switching equipment required at an automatic telephone exchange. At an exchange having 2 000 subscribers and a busy hour calling rate of 0·5, the average holding time is 3 minutes. What is the total originated traffic during the busy hour? If these calls are routed via a switching stage giving a grade of service of 0·002, how many calls are lost during the busy hour? (C & G 1961)

Q.22 Design the best four-group grading for connecting one level of 100 outlet group selectors to a group of 18 junctions.

What re-arrangements would you propose if it became necessary to increase the size of the group to 20 junctions? (C & G 1960)

8 · DISTRIBUTION FRAMES

For self-teaching notes, see page 219.

PART 1 INTRODUCTION

8.1.1 A distribution frame is a cross-connecting point provided within an exchange to give flexibility to the trunking.

This flexibility is provided by means of semi-permanent wires called *jumpers* which can easily be changed at any time during the life of the exchange so that the exchange can be adapted to meet the changing needs of the community it serves.

One type of frame, called the *Main Distribution Frame* (*M.D.F.*), is provided for cross-connecting lines to the exchange equipment. This allows for changes in the number and location of subscribers and changes in the allocation of junction circuits. It also permits spare pairs in the cables to be terminated without having to provide line relays for them which would lie idle.

Other types of frame called *Intermediate Distribution Frames* (*I.D.F.*) and *Trunk Distribution Frames* (*T.D.F.*) (see Chapter 7), are provided for cross-connecting the apparatus within the exchange. These allow for connecting circuits into different sized groups to accommodate changes in the distribution of traffic through the exchange arising from growth of the exchange or changes in subscribers' telephone habits.

8.1.2 In this chapter, you will learn about the basic cabling arrangements for automatic exchanges, and the functions of the M.D.F. and I.D.F. We shall then see how the necessary facilities are provided on these frames, and finally, how the frames are constructed in practice.

Only the M.D.F. and I.D.F. will be dealt with in this chapter. The T.D.F. is described in Chapter 7.

PART 2 FACILITIES PROVIDED BY DISTRIBUTION FRAMES

8.2 Exchange cabling

The two wires for each subscriber's line and each junction are connected to terminals on the *line side* of the main distribution frame (M.D.F.). The order in which they appear on this side of the frame is determined solely by the position of each circuit in the external cables. The lines on the line side of the frame are therefore roughly in what might be called 'geographical' order which bears no special relation to the numerical or directory order of the subscribers.

The first purpose of the M.D.F. is therefore to convert the arrangement of the lines into numerical order. Jumpers are provided from each line to a second set of terminals arranged in numerical order on the *exchange side* of the M.D.F.

The order in which the lines and junctions are connected to the line circuits and relay sets in the exchange is arranged to distribute the traffic evenly over the first selector groups or common control equipment (e.g. registers).

The Intermediate Distribution Frame (I.D.F.) enables this order to be changed when necessary without disturbing the permanent cabling.

One side of the I.D.F. is cabled directly from the exchange side of the M.D.F. with the circuits still in numerical order. This is called the *multiple side*, because the outgoing multiple is connected here, which in a manual or a Strowger exchange has to be in directory number order for the called number to be selected. In the case of a crossbar or electronic exchange this numerical arrangement is still used for the purpose of making connexions to the subscribers' meters.

Jumpers are provided from the multiple side of the I.D.F. to terminals on the other side of the frame, which is directly cabled to the incoming line circuits on the exchange. This is called the *local side* of the I.D.F. and is arranged in equipment number order. In crossbar and electronic exchanges, the exchange side of the M.D.F. may be arranged in this equipment number order for all connexions except the meters. The I.D.F. connexions are not then required, there being only one point of connexion for each line, not two as in the case of a Strowger exchange.

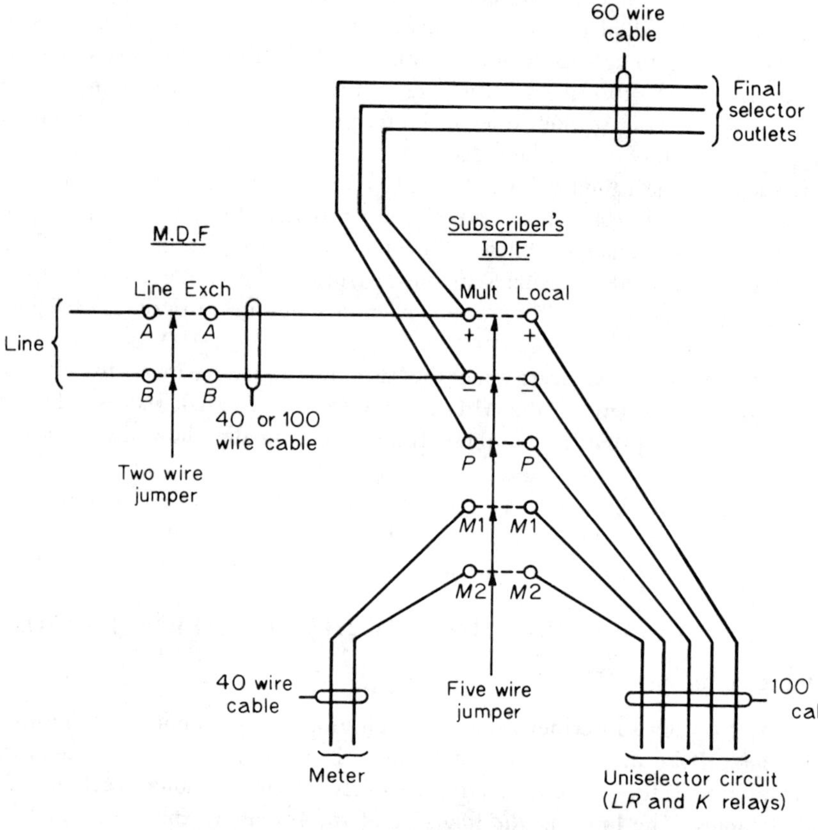

Fig. 8.1 Subscriber's line cabling in a Strowger exchange (8.2).

In those cases where grading is provided by means of Trunk Distribution Frames (T.D.F.), a sufficiently large number of outgoing circuits is usually available on each frame for the T.D.F. jumpering to give all the flexibility required between successive ranks of selectors.

With decentralized grading facilities provided by means of connexion strips on the selector racks, the number of outgoing circuits available for cross-connexion to any given trunk in a grading is very much smaller than on a T.D.F. Additional flexibility is therefore provided at an I.D.F. For maximum economy of cabling, each distribution frame should be located as near as possible to the racks to which it is to be connected.

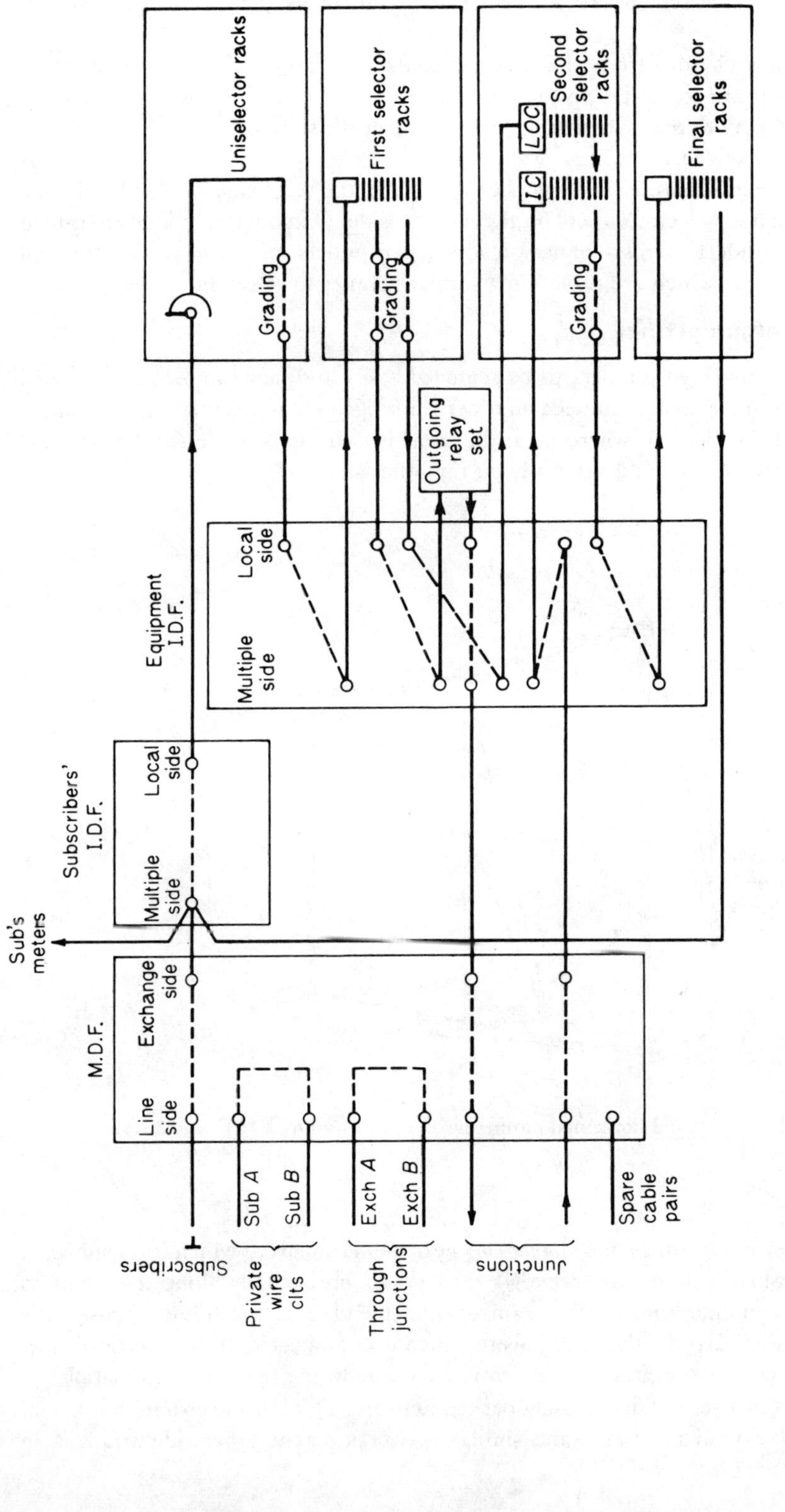

Fig. 8.2 Cabling of a Strowger exchange (without T.D.F.). Compare this with Figs 4.7, 7.8 and 8.1 (8.2).

This is usually achieved by having two or more separate I.D.F.'s. The one which is cabled to the M.D.F. for subscriber's lines is then known as the *subscriber's I.D.F.* and those which are used for connexions between selectors are known as *equipment I.D.F.'s*. The cabling scheme for the principal circuits of a typical exchange is shown in Fig. 8.2.

The outgoing circuits from the grading connexion strips on the selector racks (**7.6.2** and **7.6.3**) are connected to the *local* side of the I.D.F. and cross-connected by three-wire jumpers to the incoming or 'wiper' side of the selectors in the next rank on the *multiple* side of the I.D.F.

Incoming junctions are cabled via the M.D.F. to the *local* side of the I.D.F. for cross-connexion to selector outlets on the *multiple* side. Outgoing junctions are cabled to the *multiple* side for cross-connexion to selector outlets on the *local* side through outgoing relay sets which are cabled in a similar manner to selectors.

8.3 Method of jumpering

To allow old, unwanted jumpers to be removed easily and new jumpers to be added without disturbing existing jumpers, an orderly arrangement is essential. This arrangement is shown in Fig. 8.3, where it can be seen that all jumpers are run horizontally on one side of the frame and vertically on the other side.

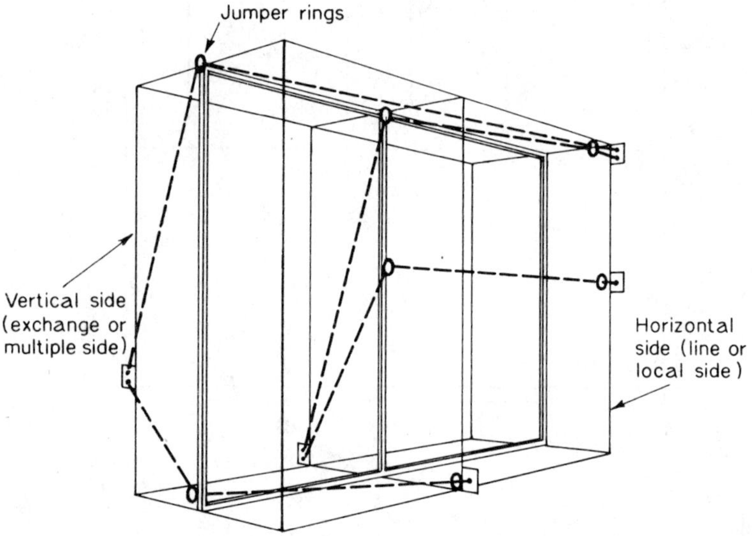

Fig. 8.3 Vertical and horizontal jumpering as used on both M.D.F. and I.D.F. (8.3).

Each jumper passes through a *jumper ring* near the terminals to which it is connected on the 'horizontal' side of the frame. It then passes horizontally along the frame to another larger jumper ring on the main upright member of the frame opposite the part of the frame carrying the terminals to which it is connected on the 'vertical' side. The jumper then passes either up or down the frame to the required termination.

This results in a set of fairly loosely packed networks of wires in separate horizontal planes on one side of the frame and similar networks on the other side arranged in separate vertical planes. The wires only bunch together at the jumper rings where the horizontal and vertical planes intersect at the centre of the frame and to a lesser extent at the smaller rings where the jumpers are fanned out from the horizontal plane to the terminals which are arranged on vertical connexion strips.

In the U.K. it is usual for the jumpers to run *horizontally* on the M.D.F. *line* (and I.D.F. *local*) side, and *vertically* on the M.D.F. *exchange* (and I.D.F. *multiple*) side. In many other countries the vertical jumpers are on the line side and the horizontal jumpers are on the exchange side.

8.4 Protection and testing

Access must be provided at the exchange for testing all junctions and subscribers' lines and the exchange apparatus associated with them. Protective apparatus must also be provided to guard against damage to the exchange equipment due to the effects of lightning or faulty power lines on the external line plant. It is convenient to provide both this testing access and protective apparatus at the M.D.F. for each line. Important junction circuits are usually also connected to a separate test jack frame (T.J.F.) to provide a more convenient access point for the more frequent routine testing necessary on such routes.

Lightning consists of short duration electrical discharges involving currents of thousands of amperes and potentials of millions of volts. The use of lightning conductors on the poles on overhead lines, and the fact that line insulators or cable insulation on the line cannot withstand such high voltages, considerably reduces the effects at the exchange. On lines constructed mainly as open wire overhead lines, the potential at the exchange end of the line due to lightning may still be in excess of 1 000 V, however, so to protect both the exchange equipment and the maintenance personnel, *protectors* are provided on the exchange side of the M.D.F. for lines of this type. In the U.K., protectors are provided only for lines having more than three spans of overhead wire in their construction.

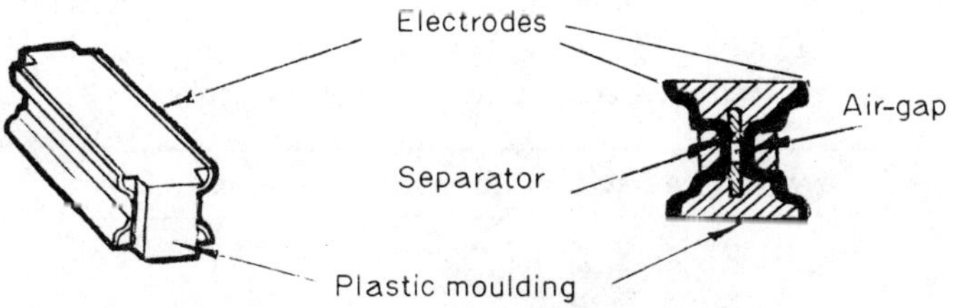

Fig. 8.4 M.D.F. protector. The air-gap breaks down at a potential of 750 V ± 150 V to earth the line when connected as shown in Fig. 8.6 (8.4).

A typical protector is shown in Fig. 8.4 and incorporates an air-gap which breaks down under a potential of 750 V (± 150 V). A protector of this type is connected from each wire of the line to earth.

Modern power distribution systems use line potentials of many thousands of volts and, although every precaution is taken, accidental contacts between such power lines and telephone lines do sometimes occur. If an earth fault occurs on an a.c. power line, the circuit is unbalanced and the resultant external magnetic field can induce potentials along a nearby telephone line even without any physical contact between the two.

So far as high voltage conditions due to power lines are concerned, the protectors will function as for a lightning discharge but the prolonged flow of current through the protector will permanently damage it. The protector normally fully recovers from a short duration lightning discharge.

Lower voltage conditions cause currents to flow which could damage the exchange equipment by overheating it and may constitute a fire risk. Fuses are therefore connected in each wire of every line on the line side of the M.D.F. A typical fuse is shown in Fig. 8.5.

This overheating of exchange apparatus can be caused either by a large current flowing for a short time, or by a smaller current flowing for a longer time. The effect of the heavier current is easily guarded against by using a fuse wire of the appropriate rating. This method cannot be used as protection in the case of small current flowing for a longer time because it would require the use of very thin fuse wire. This would break easily, and as it aged, it would become liable to blow on normal working values of line current. The slow rise in temperature produced in the higher rated fuse wire by smaller fault currents is therefore used to melt a small bead of solder so that the spring (Fig. 8.5) pulls away from the fuse wire, disconnecting the line. This type of fuse is known as a *delayed-action fuse* and replaces the older method of achieving the same result by means of an ordinary fuse on the line side and a separate 'heat coil' for low current faults on the exchange side.

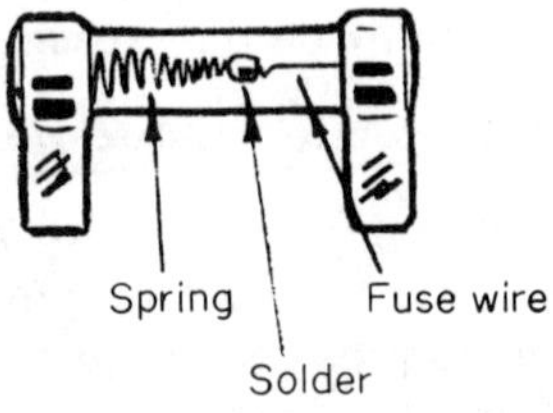

Fig. 8.5 M.D.F. line fuse. This is a delayed action fuse rated at 0.2 A. A current of 0.35 A will melt the solder in 5 s to 300 s. Large currents melt the fuse wire (8.4).

The circuit connexions of the protective apparatus on the M.D.F. are shown in Fig. 8.8. The test jack contacts shown on the exchange side of the frame are arranged to open when a test plug is inserted. This separates the line connexions from the exchange apparatus connexions so that each may be tested independently.

The test jacks also provide an access point at which interception circuits can be connected when required for such purposes as checking on reported service difficulties or advising callers of a changed number. N.U. tone can be provided on calls incoming to a line under repair by circuits connected in at the test jacks.

Now test yourself on questions 1–9 on page 219.

PART 3

CONSTRUCTION AND APPARATUS OF TYPICAL FRAMES

8.5 Main distribution frame

Many existing exchanges have the older type of M.D.F. which consists of a rigid iron framework with its base embedded in concrete to floor level. Vertical members at $6\frac{3}{4}$ in intervals along the frame carry transverse arms to which are attached the fuse mountings on the line side, and the heat coil, protector and test jack mountings on the exchange side. Each vertical provides apparatus in this way for 220 circuits on the line side and 200 on the exchange side.

Modern practice in the U.K. is to use a *rack type M.D.F.* This requires no special concrete base and can be installed in suite with selector or relay set racks.

The mounting used for fuses of the type shown in Fig. 8.5 is made up of twenty

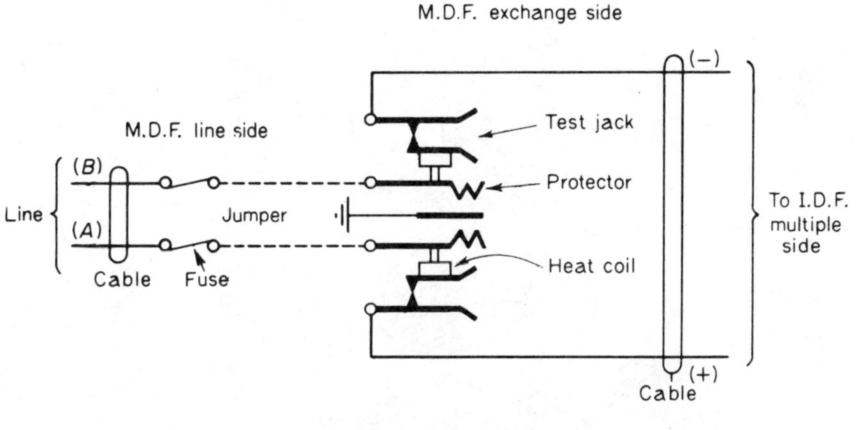

(a) With heat coils and protectors

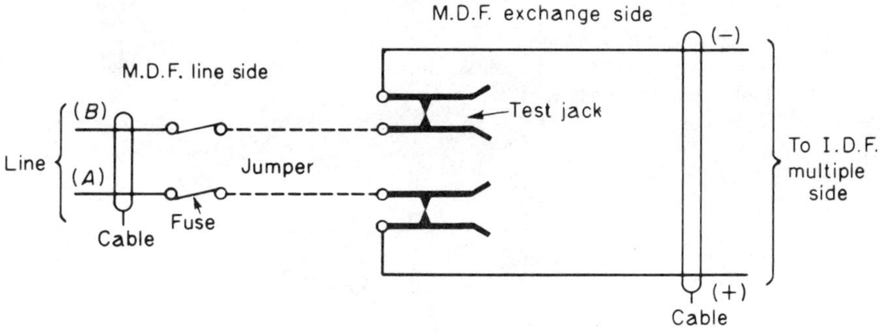

(b) With test jack only

Fig. 8.6 Typical M.D.F. circuits. Dummy components can be used in (a) on those circuits connected to this type of mounting which do not require heat coils and/or protectors (8.5).

layers, each consisting of a plastics moulding fitted with eight pairs of nickel-silver contact springs. Each layer thus accommodates four fuses and serves two circuits, giving a total capacity of forty circuits as shown in Fig. 8.7.

Where protectors are required (8.4), heat coil, protector and test jack mountings accommodating twenty circuits per unit are fitted on the exchange side. The more usual provision, however, is a unit of comparable size but accommodating fifty circuits and having test jack facilities only (i.e. without protectors or heat coil mountings). This type of test jack unit is made up of twenty-five layers, each consisting of a plastics moulding fitted with four pairs of nickel-silver contact springs. Each layer thus provides test and interception facilities for two circuits giving a total capacity of fifty circuits for the whole unit. This is illustrated in Fig. 8.8.

The rack type of M.D.F. has capacity for ten fuse mounting units per vertical on the line side and eight test jack units per vertical on the exchange side, giving a capacity of 400 circuits per vertical on each side. Each rack unit has four verticals, the exchange side apparatus being fixed directly to vertical members of the framework

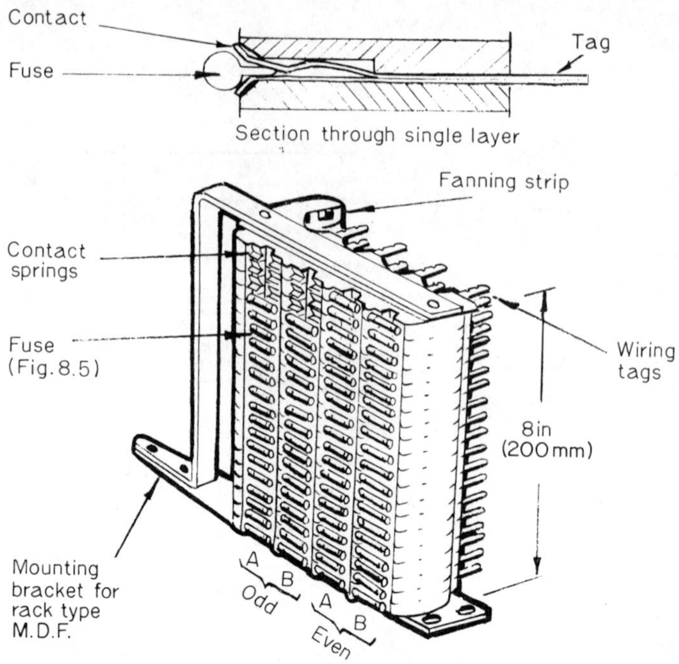

Fig. 8.7 M.D.F. fuse mounting unit. This provides for eighty fuses, serving forty circuits. The cables are terminated on the short tags and the jumpers on the long tags (8.5).

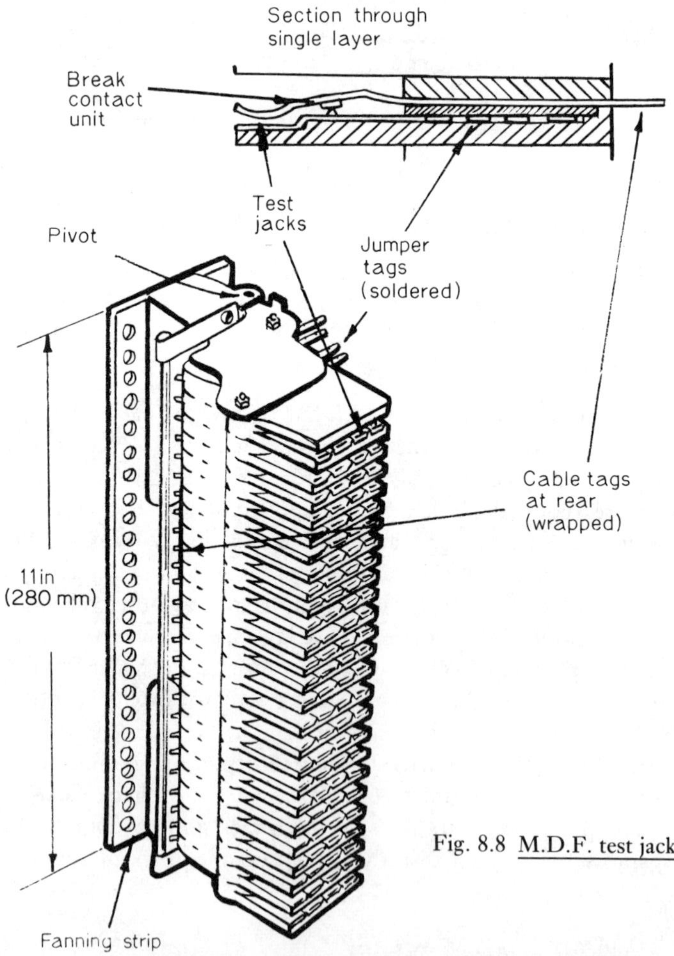

Fig. 8.8 M.D.F. test jack unit (8.5).

and the line side fuse units being mounted on horizontal shelves. This allows space for vertical jumpering between the verticals on the exchange side and space for horizontal jumpering along the shelves on the line side. An end view of the frame showing this general construction is given in Fig. 8.9.

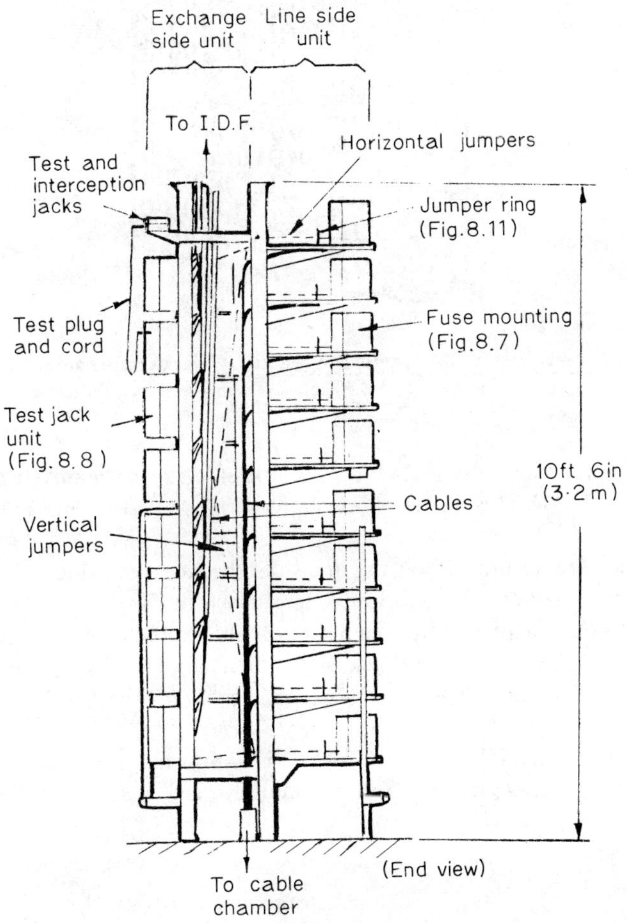

Fig. 8.9 Rack-type M.D.F. (8.5).

A frame of any required capacity can be provided by simply fixing a number of rack units end-to-end. This is illustrated in Fig. 8.10. The ironwork for the line and exchange sides are made as separate units which are normally bolted together at top and bottom to form a complete rack unit. The line side can, however, be provided on its own if necessary to accommodate extra external cables. The need for this arises because there are often a number of spare pairs in cables which require terminating on the M.D.F. for future use but do not need to be jumpered to the exchange side initially. There are also some junction cables between adjacent exchanges which are used as sections of junction circuits between distant exchanges. These are cross-connected at the intermediate exchanges on the line side only with no connexion to the exchange side (see Fig. 8.2).

External cables enter the exchange at the cable chamber in the basement of the building. The M.D.F. is normally positioned as near as possible directly above the cable chamber, the external cables being brought up through the floor under the frame.

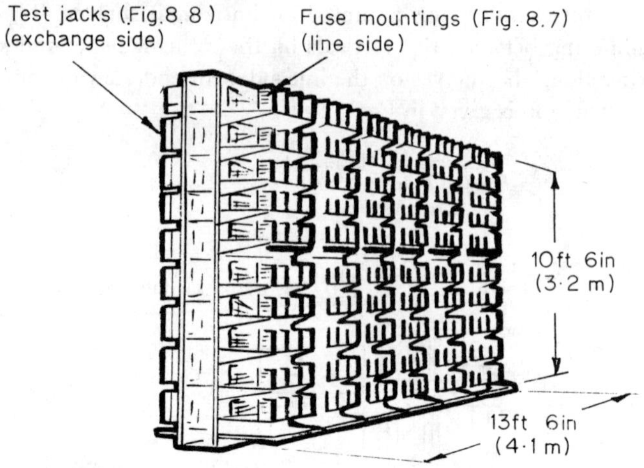

Fig. 8.10 Rack-type M.D.F., unit construction. The example shown is a 9 600 line frame of six racks each 2 ft 3 in (0·69 m) wide. Compare this with Fig. 8.9 (8.5).

The ironwork below the bottom shelf on the line side can be removed to allow cables to be laid along the floor from the point of entry to the vertical on which they are to be terminated.

Although they are terminated on the line side, the external cables come in vertically and therefore have to be on the 'vertical' side of the framework, which is of course, the exchange side. This is achieved by attaching the cable to the rear of the line side framework as shown in Fig. 8.1. As the rack unit is made in two parts, it is convenient to erect the line side unit and terminate the external cable first and add the exchange side unit afterwards.

The cabling to other racks or to the I.D.F. is supported overhead in the apparatus rooms and therefore these internal cables come into the M.D.F. from the top on the exchange side.

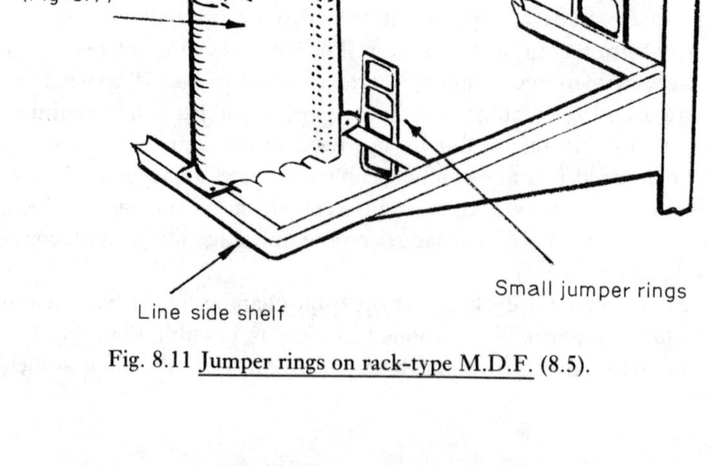

Fig. 8.11 Jumper rings on rack-type M.D.F. (8.5).

Jumper ring units of moulded nylon are fitted on the line side shelf at the rear of the fuse mountings and at the rear of the line side framework as shown in Fig. 8.11. Jumper rings are not provided on the exchange side at the test jack or protector mounting units.

Soldered connexions are used for cabling and jumpers at the fuse units, but the test jack units have terminals designed for wrapped joints. Both the fuse units and the test jack units are pivotted on the frame so that they can be swung round into a convenient position for wiring.

8.6 Intermediate distribution frame

The I.D.F. provides for cross-connexion of internal exchange cabling only and therefore no protective apparatus such as fuses is required. Both the multiple side and the local side are fitted with simple connexion strips. These are not of the flat type used for such purposes as grading on racks, but are specially arranged for cabling connexions on one side and jumpers on the other side. Double-ended terminals are used for this purpose and the block stands out at right angles to the line of the distribution frame so that both sides are accessible. A fanning strip is provided at the rear of the block to keep the wiring tidy.

I.D.F. connexion strips are built up in layers according to the number of wires to be connected in each circuit and may accommodate either twenty or twenty-five circuits. The terminals can be of the type used for soldered wiring or of the type used for wrapped connexions, as shown in Fig. 8.12.

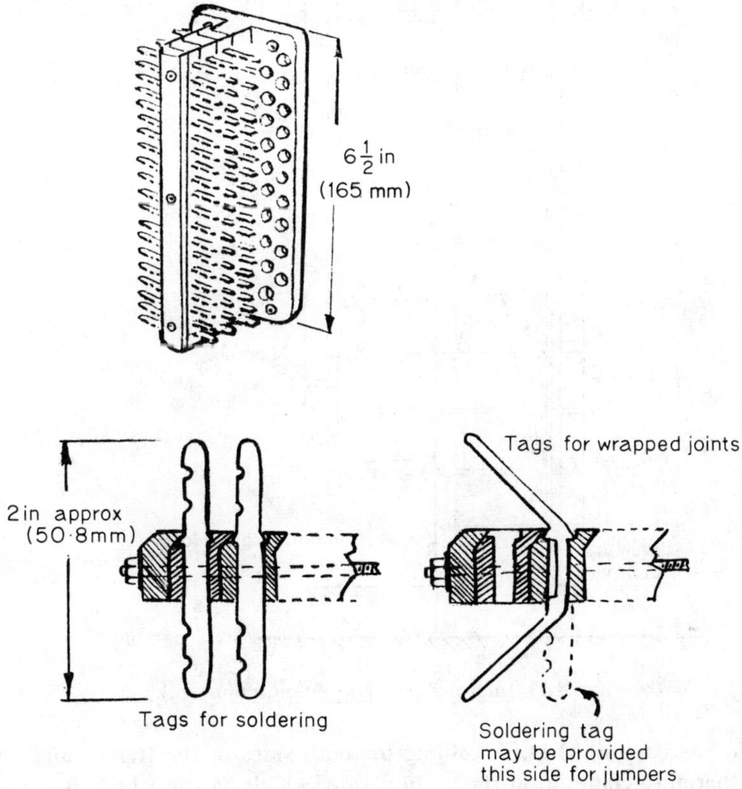

Fig. 8.12 I.D.F. connexion strips. The tags for wrapped joints are angled forward to allow easy access for the wrapping tool (8.6)

The frame is constructed in a similar manner to the older type of M.D.F. referred to in **8.5**. Vertical members at $6\frac{3}{4}$ in (160 mm) intervals along the length of the frame carry up to ten transverse members each, to the ends of which are attached the connexion strips. The verticals are joined at the bottom by longitudinal members which form a base which is bolted to the floor.

The connexion strips on the local side are attached by brackets at their centres to the ends of transverse members of the frame, leaving a clearance between the ends of the strips on the same vertical to allow for running the horizontal jumpers. On the multiple side, the strips are fixed at their ends between the transverse members, the jumper ring being vertical on this side.

The general construction is shown in Fig. 8.13.

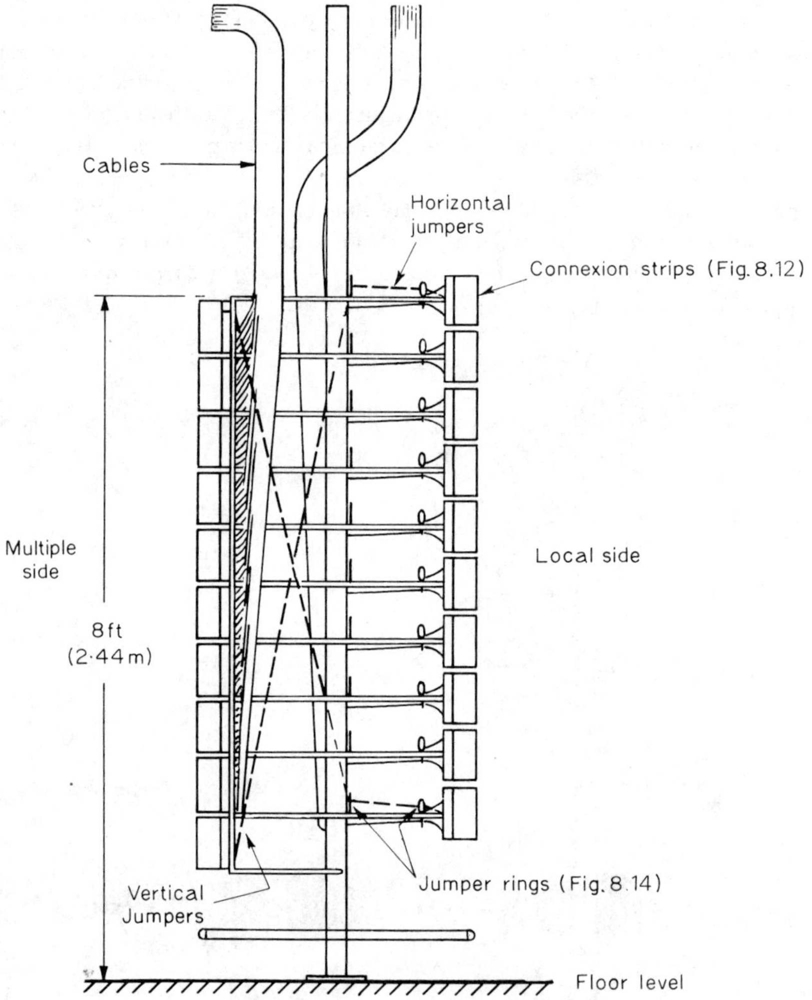

Fig. 8.13 Intermediate distribution frame (8.6).

As in the case of the M.D.F., cabling for both sides of the frame runs vertically, and must therefore come in on the vertical side, which on the I.D.F. is the multiple side. As all cabling is internal cable, no cables are fed through the floor, as they are on the M.D.F.

The large jumper rings are fitted between the verticals, adding strength to the structure. Horizontal stiffening bars are provided along the length of the frame behind the connexion strips and the small jumper rings are fitted to these stiffening bars as shown in Fig. 8.14.

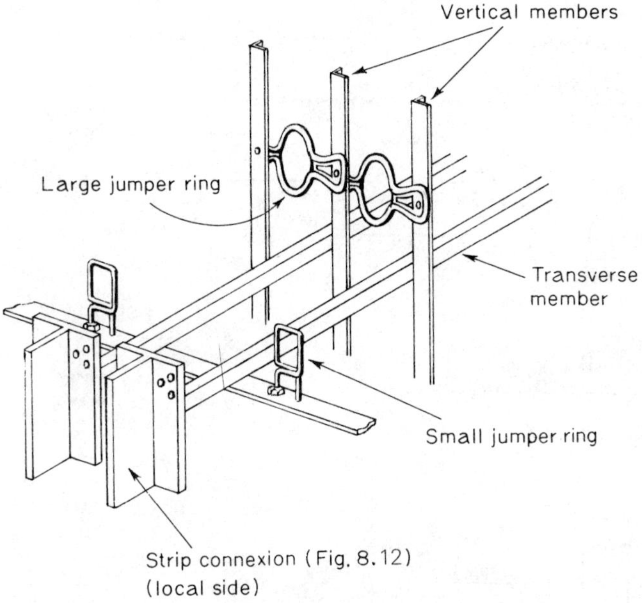

Fig. 8.14 Jumper rings on I.D.F. (8.6).

8.7 Jumpers

Jumper wire is supplied in the form of a twisted flex of up to seven wires according to the number of conductors required in the jumpers (see Fig. 8.1). Each wire in the jumper is of $9\frac{1}{4}$ lb gauge tinned copper, insulated with PVC plastic and covered with a lacquered cotton braiding colour coded so that each conductor can be identified to avoid reversals.

When running a jumper between two circuits on opposite sides of a frame, the jumper wire is accommodated on a specially designed drum called a 'vertical swift'. This is placed beside the frame on the 'horizontal' side of the frame opposite the large jumper ring through which the jumper is to pass. The end of the wire is then passed through the large jumper ring from the 'horizontal' side and then up or down the vertical jumper field where it is then terminated (i.e. connected to the required terminals). The wire on the swift is then taken along the horizontal jumper field to the required terminal on this side, cut and threaded through the small jumper ring and fanning strip and terminated.

PART 4 SUMMARY

8.8 Distribution frames

M.D.F.: Line side —fuses, cable termination order.
Horizontal jumpers (HORIZON: LINE).
Exchange side—test jacks, subs' number order.
(Also heat coil and protectors if required.)
Vertical jumpers ('CHANGE: UP and DOWN').

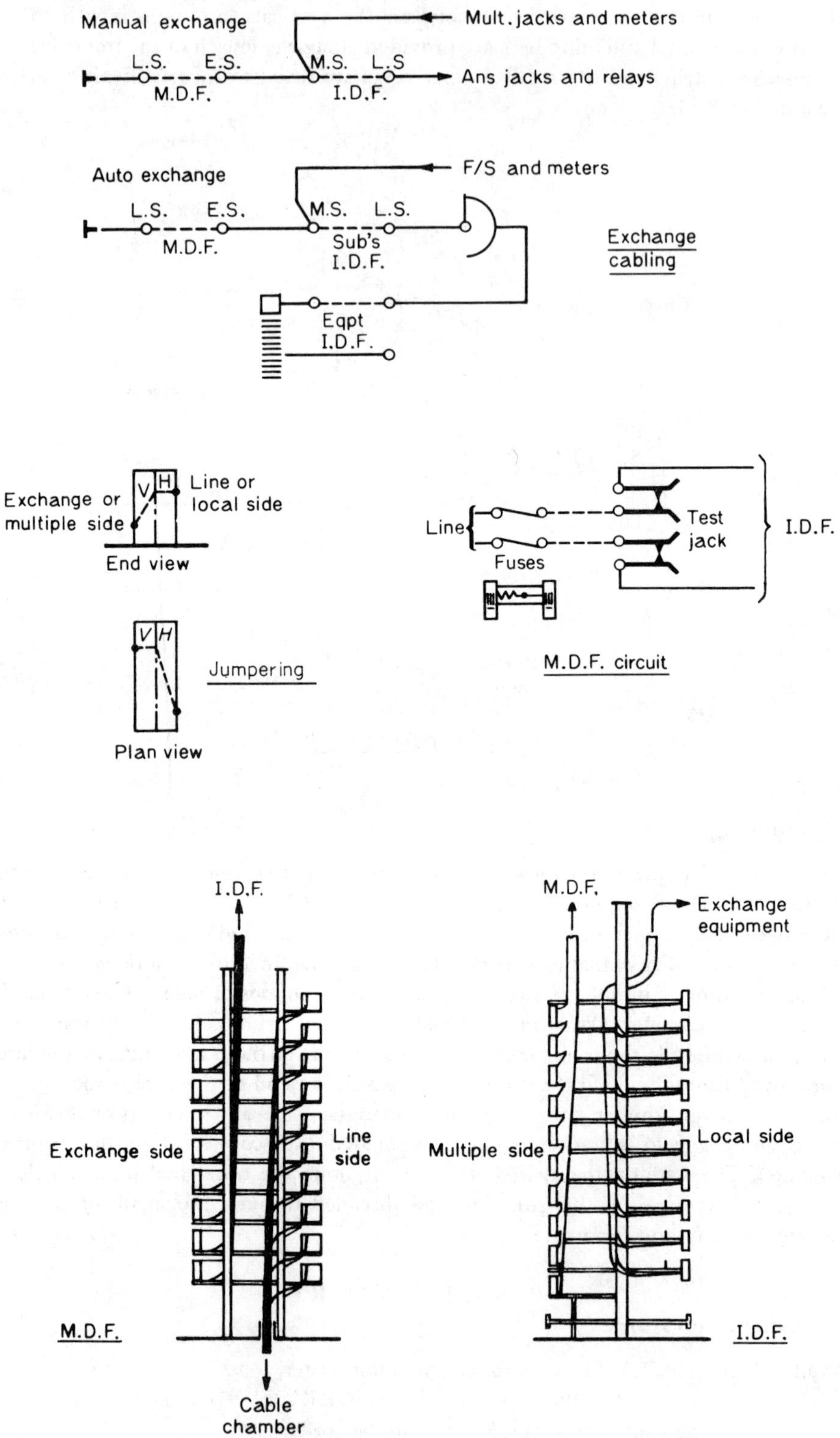

Fig. 8.15 Main and intermediate distribution frames—summary.

I.D.F.: Multiple side —subs' multiple, M.D.F. cabling (number) order.
Vertical jumpers (same as M.D.F. exchange side to which it is cabled).
Local side —calling equipment, loading order.
Horizontal jumpers (*L*ocal as for *L*ine on M.D.F.).

Now test yourself on the questions below before leaving this chapter.

PART 5 SELF-TEACHING NOTES

8.9.1 First read **8.1.1** to **8.4** inclusive.

8.9.2 After reading **8.1.1** to **8.4**, try the following nine questions:

Q.1 What names are given to the two sides of the intermediate distribution frame (I.D.F.)?

Q.2 One side of the M.D.F. is directly cabled to one side of the I.D.F. Which sides of the two frames are these?

Q.3 Why are the meters and the multiple (jacks or final selectors) connected to the I.D.F. rather than to the exchange side of the M.D.F. which is also in subscribers' number order?

Q.4 At a certain exchange, one group of first selectors is found to be carrying much more traffic than other similar groups in the same exchange.
What can be done to distribute the load more evenly?

Q.5 To which side of the equipment I.D.F. in a Strowger exchange are the banks of group selectors cabled?

Q.6 On which side of the M.D.F. are the jumpers horizontal?

Q.7 Where are the line fuses mounted?

Q.8 On what types of line are protectors required?

Q.9 Where in the exchange are the protectors mounted?

Now check your answers on page 268.

You should have answered most of these nine questions correctly.

If you scored less than six out of nine, read **8.1.1** to **8.4** again and try the questions once more.

With at least six of Q.1 to Q.9 right, read the rest of the chapter (**8.5** to **8.8**).

8.9.3 The whole of Chapter 8 should have been studied before attempting the following questions.

A few days should preferably have passed after reading **8.1.1** to **8.7** before trying these questions.

Q.10 Show schematically the cabling and jumpering arrangements at an automatic exchange for subscribers' lines on:
(*a*) the main distribution frame (M.D.F.) and
(*b*) the intermediate distribution frame (I.D.F.).
What devices are used on the M.D.F. to prevent damage to the exchange equipment?

What possible sources of damage do they guard against? (C & G 1962)

Now check your answer on page 269.

Q.11 What are the functions of the main distribution frame (M.D.F.), and the subscribers' intermediate distribution frame (I.D.F.) at an automatic exchange? Describe the fittings provided to carry the jumpers on the M.D.F. and the means by which an orderly system of jumpers is achieved.
Why does an I.D.F. jumper have more wires than an M.D.F. jumper? (C & G 1965)

Now check your answers on page 269.

9 · EXCHANGE POWER PLANT

For self-teaching notes, see page 247.

PART 1 INTRODUCTION

9.1.1 Power requirements. Telephones and teleprinters run on electricity and so do their exchanges.

Most of the apparatus used requires direct current (d.c.) at relatively low voltages (e.g. $\pm$ 80 V for teleprinters, and 24 V for manual telephone exchanges). The public electricity supply or 'mains', however, usually provides alternating current (a.c.) at a much higher voltage (e.g. 240 V) and one of the main functions of the power plant at an exchange is to convert this to a d.c. supply at the appropriate voltage.

Automatic exchanges use 50 V d.c. Some variation in this voltage is unavoidable in practice, so the selectors and relays are designed to operate within the range 46 V to 52 V. Storage batteries are used to maintain the correct voltage even if the mains supply fails completely.

This 50 V supply is negative with respect to earth. A subsidiary supply is often provided in addition to this, mainly for operating subscribers' meters. This is at 50 V also, but is positive with respect to earth.

Teleprinters with double-current signalling (3.3.6), need one supply at 80 V positive and one supply at 80 V negative. These are maintained within the range 75 V to 92 V with respect to earth, and with the positive and negative values differing by not more than 4 V.

Additional low voltage d.c. supplies are provided for the electronic logic used in the control circuits of electronic exchanges and for some specialized functions in electromechanical exchanges. These are typically at 5 or 12 volts positive and negative, provided by d.c./d.c. converters near the point of use, driven from the -50 V exchange supply.

In addition to these d.c. supplies, some a.c. supplies are also required. These include low voltage supplies at the same frequency as the mains (e.g. 6 V, 50 Hz for switchboard lamps), low frequency signalling currents (e.g. 75 V, $16\frac{2}{3}$ Hz ringing), and audio frequency tones (e.g. 400 Hz at about 1 V).

9.1.2 In this chapter, you will learn about the methods employed to generate these various supplies, the use of batteries and the control of the supply voltage. We shall then see how these ideas are combined in practical exchange power plants.

This is the last chapter of this book. It is by no means the least important, however, because the other apparatus we have discussed will not operate without adequate power supplies. The limitations and characteristics of practical power plants are also an important factor in the design and maintenance of the rest of the exchange.

PART 2 PRINCIPLES OF EXCHANGE POWER SUPPLIES

9.2 Sources of energy and power

9.2.1 Units. *Power* is defined as the rate of doing work or of consuming energy.

The SI unit of *energy* or *work* is the *joule*, the energy required to accelerate a mass of 1 kilogramme at 1 metre per second per second through a distance of 1 metre (1 newton acting over 1 metre = 1 Nm).

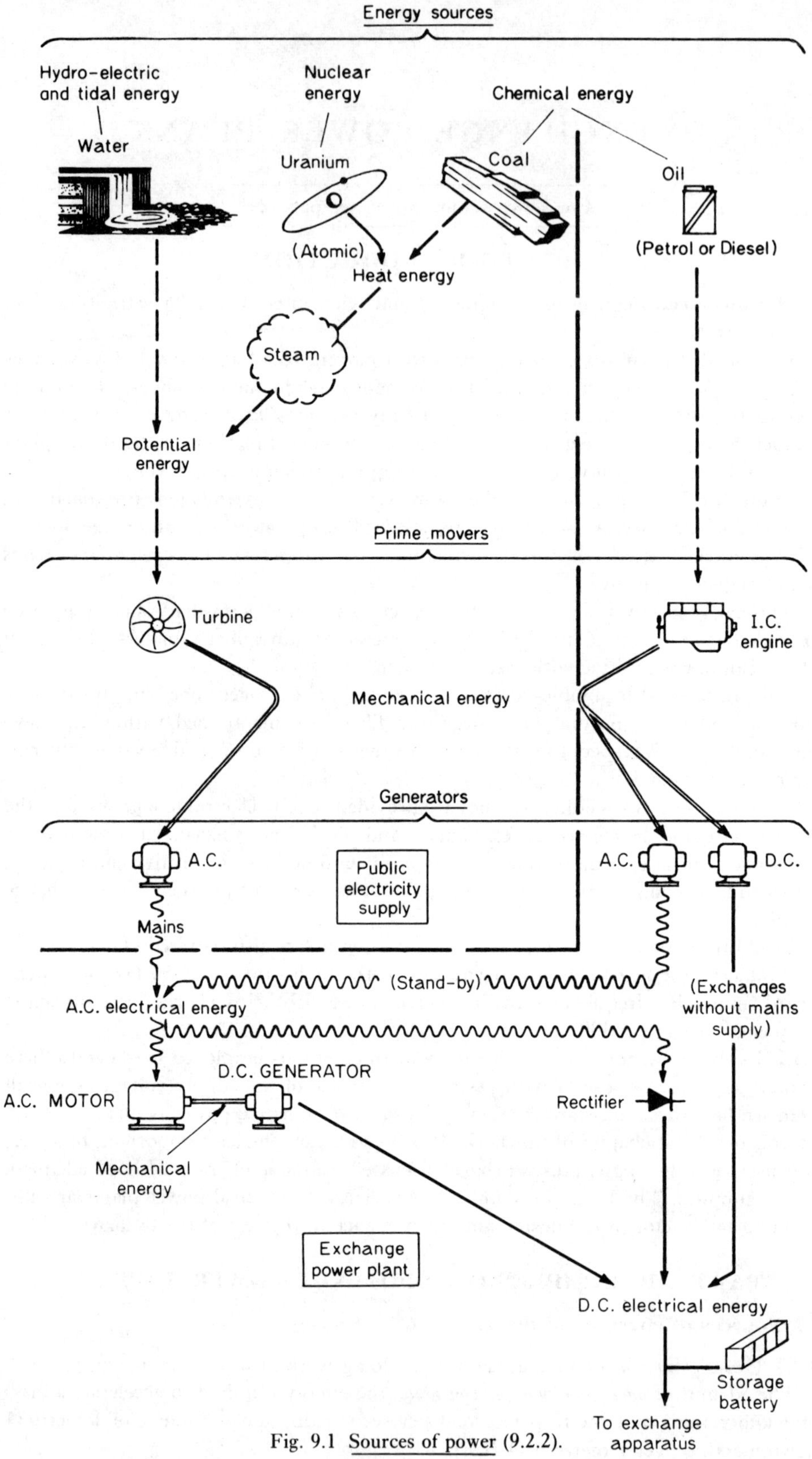

Fig. 9.1 Sources of power (9.2.2).

The SI unit of *power* is the *watt*, and this is equivalent to 1 joule per second = 1 Nm/s.

The power dissipated in an electric circuit is equal to the product of current times voltage (i.e. watts = volts × amps.). The higher the voltage, therefore, the smaller is the current for a given value of power.

9.2.2 Prime movers. Energy takes many forms, and given the necessary apparatus, we can convert it freely from one form to another. The principal methods of converting the readily available forms of energy into electrical energy suitable for an exchange are shown diagrammatically in Fig. 9.1.

The generators used to produce the electrical energy are rotating machines which need an engine of some description to drive them. These engines are known as *prime movers*. It is important to remember that energy is not something we can create, nor for that matter can we destroy it. This is the principle known as the 'conservation of energy'. The generation of power in electrical form is only a process of energy conversion.

The generation of electricity is more economical in large quantities. For this reason, the usual source of power for exchanges is the public supply or mains. In the event of failure of this supply for a longer period than the capacity of the exchange storage batteries can cover, or in the few cases where no public supply is available, a prime mover must be provided at the exchange. This usually takes the form of a petrol engine or a diesel engine.

9.2.3 Mains supplies and a.c. machines. The mains supply is usually a.c. so that power can be transmitted at high voltages over the main lines from the power station and then converted by means of transformers (**1.3.4**) to a safer low voltage for local distribution to the consumers. High voltage transmission reduces the current required for a given power (**9.2.1**), thus reducing the losses (I^2R) in the line.

Machines for generating alternating current are known as alternators. The principle of operation of such machines is shown in Fig. 9.2.

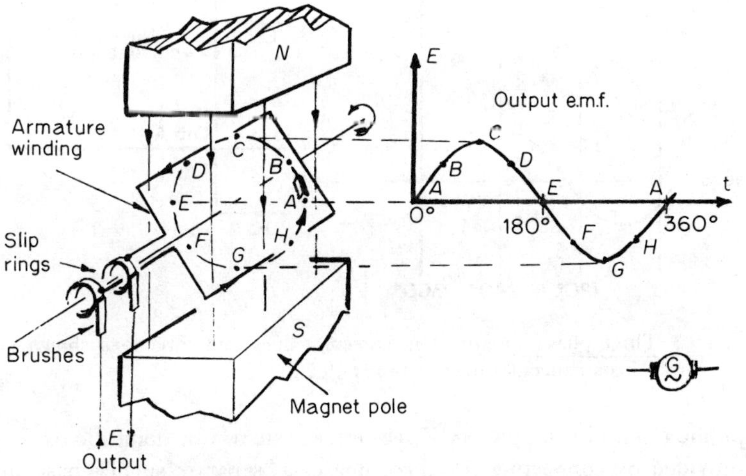

Fig. 9.2 Principle of alternator (single-phase) (9.2.3).

The magnet poles shown in Fig. 9.2 may either be part of a permanent magnet or an electromagnet. If a permanent magnet is used, the machine is known as a *magneto* generator. If electromagnets are used, their windings are known as the *field windings* of the machine.

At one time, only the smaller alternators such as the hand-generator of the magneto system used permanent magnet fields but with the development of improved magnetic materials, permanent magnets are also used in modern ringing machines and tone generators.

The output from this simple alternator is a *single-phase* supply. It consists of an e.m.f. across two output wires which varies with time as shown in Fig. 9.2. The current flowing varies in a similar manner.

If three coils are used in the armature winding instead of one, and these three coils are set at 120° to one another, a *three-phase* supply is obtained. This consists of three e.m.f.s varying in the same way as the single-phase a.c. supply but displaced from one another by 120° (one third of a cycle) so that their algebraic sum is always zero. It can be seen from Fig. 9.3 that if the three phases use a common return, the resultant current in the return conductor *O* is zero in a balanced circuit. Three conductors can therefore carry the same power as would require six conductors using the same value of current on separate single phase supplies. This saves not only the copper which would be used for the additional conductors but also the power which would be dissipated in the resistance of these conductors.

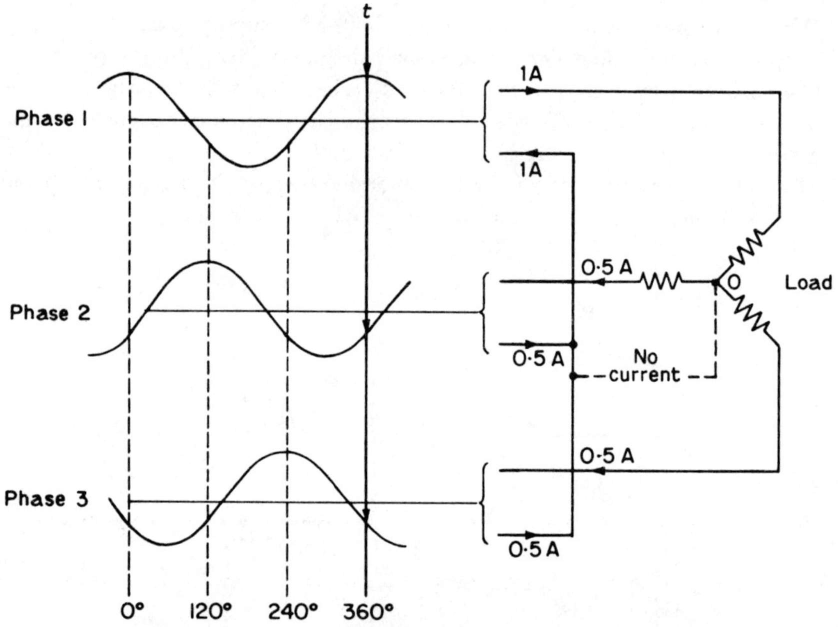

Fig. 9.3 Three-phase supply. The current values and directions shown are instantaneous values at time *t* (9.2.3).

The public electricity supply is a three-phase system. For domestic use, a balanced load is provided by connecting the three phases as separate single-phase supplies to equal groups of consumers.

Large users, such as factories or exchanges, receive the three-phase supply and must provide a balanced load themselves. This is achieved by using motors etc. designed for three-phase operation for the heavier applications and distributing any smaller single-phase loads evenly over the three phases.

One way of producing the d.c. supply required for an exchange is by means of a motor-generator set. This consists of a d.c. generator driven by an a.c. (mains) motor. This is the biggest single load put on the mains supply by the exchange and must therefore use a three-phase motor. This is of the type known as an *induction motor*, the principle of which is illustrated in Fig. 9.4. A rotating field is produced by connecting the three phases to the field windings of consecutive poles. This field induces alternating currents in the conductors of the armature winding at the frequency at which the field alternates past each conductor. A force is therefore exerted on these current carrying conductors at right angles to the field, and this force causes the rotor to rotate.

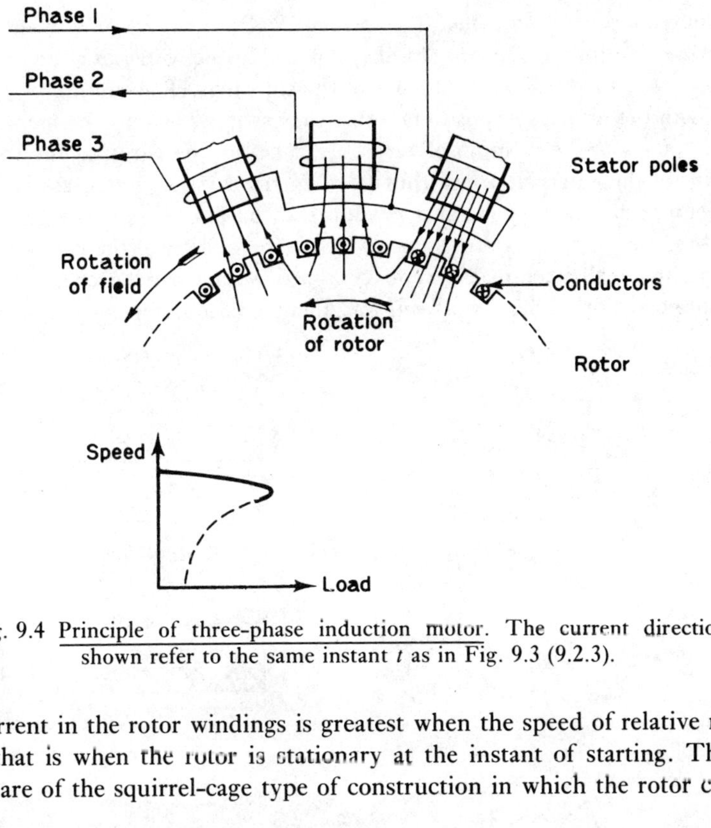

Fig. 9.4 Principle of three-phase induction motor. The current directions shown refer to the same instant t as in Fig. 9.3 (9.2.3).

The current in the rotor windings is greatest when the speed of relative rotation is greatest, that is when the rotor is stationary at the instant of starting. The smaller machines are of the squirrel-cage type of construction in which the rotor conductors

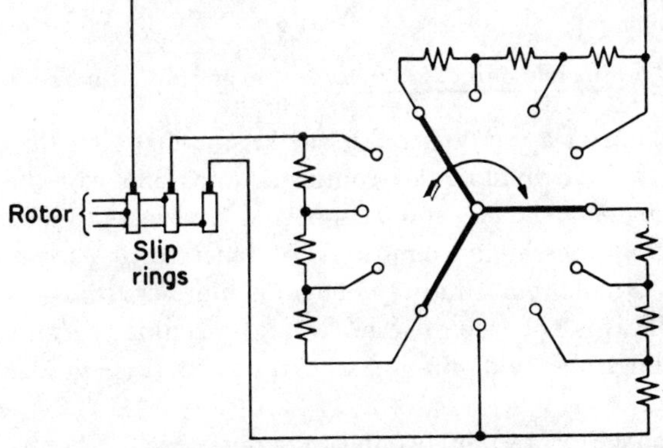

Fig. 9.5 Induction motor starter (9.2.3).

are solid copper bars short circuited at the ends. These can usually withstand the heavy starting current. If necessary, however, the rotor current can be limited by starting with reduced supply voltage on the field winding.

The larger induction motors have three-phase, wire-wound rotors, the windings being brought out on to three slip rings. The slip rings are connected by brushes to a starting resistance as shown in Fig. 9.5 which is cut out in steps as the rotor gains speed. The windings are finally short circuited at full speed.

The speed/load characteristic for an induction motor is shown in Fig. 9.4. As the load on the motor is increased, its speed falls slightly. If the load increases too far, the motor stalls and will not re-start until the load is removed.

9.2.4 Direct current machines. The principle of a *d.c. generator* is shown in Fig. 9.6. It can be seen that it is essentially the same as the a.c. generator shown previously in Fig. 9.2, but has a *commutator* in place of the slip rings. The commutator shown has segments with the brushes so positioned that they switch from one segment to the next as the coil rotates to the position of zero output before the e.m.f. in the coil reverses. The current in the external circuit thus flows in the same direction at all times.

In this simple form, the generator produces an output current which flows in one direction but rises to a maximum value twice per revolution of the coil and falls again to zero. In practical machines, several coils are wound on the armature and their outputs connected in series by a multi-segment commutator to give a steady d.c. supply.

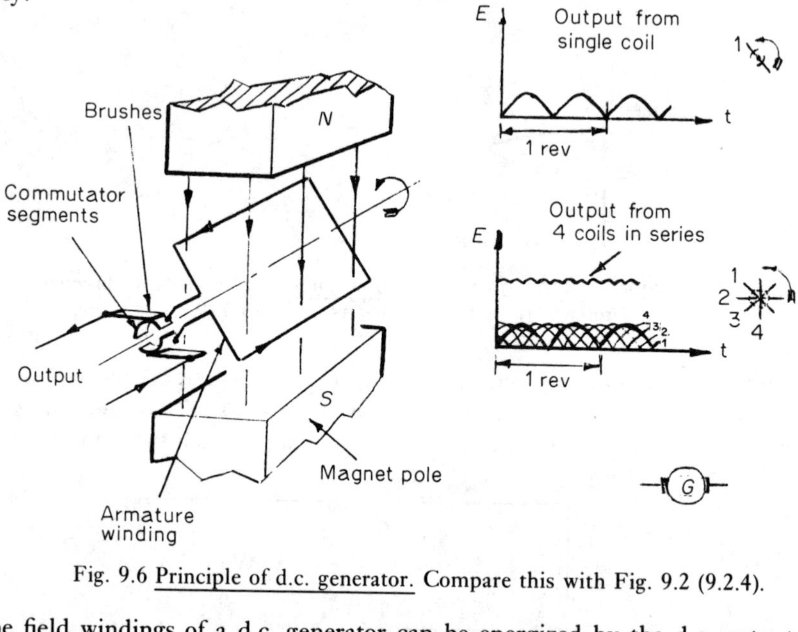

Fig. 9.6 Principle of d.c. generator. Compare this with Fig. 9.2 (9.2.4).

The field windings of a d.c. generator can be energized by the d.c. output of the machine itself. If the field windings are connected in parallel with the load, the machine is described as being *shunt-wound*, and if connected in series with the load, it is described as *series-wound*. Only the shunt-wound type of generator is used for exchange power plant because its terminal p.d. (output voltage) is more or less constant under varying load conditions. This is not true of series-wound machines because the load (output) current flows through the field winding so that any increase in load current increases the generated e.m.f.

The output voltage of a shunt-wound generator can be controlled by means of variable resistance in series with the field winding. Increasing this resistance reduces the current in the field winding, thereby reducing the strength of the magnetic field which in turn reduces the generated e.m.f. (see Fig. 9.7)

If an external source of direct current is connected to the terminals of a d.c. generator, the current flowing in the armature winding will react with the magnetic field. The resultant force on the armature winding conductors will cause the armature to rotate. This is the principle of the d.c. motor which is therefore similar in construction to a d.c. generator.

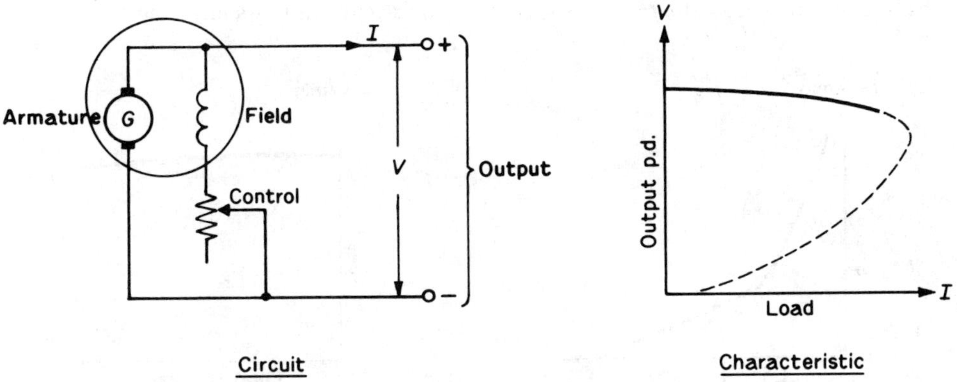

Fig. 9.7 Shunt-wound d.c. generator (9.2.4).

When the motor is running, an e.m.f. is induced in the armature windings in the same way as in a d.c. generator. This e.m.f. opposes the supply and limits the armature current. When starting, the armature is initially at rest and this back e.m.f. is absent. The armature current is therefore large when starting and decreases as the motor gains speed.

If the motor is series-wound, the large starting current flows in both the armature and the field winding, giving a large starting torque. For this reason series-wound motors are used where frequent stopping and starting under load is required as, for example, in teleprinters designed for private wire operation. A disadvantage of the series connected motor, however, is that the decrease of field current with increasing speed causes a rapid reduction of torque. Any increase in load requires increased torque which can only be provided by the motor slowing down. Series motors are not really suitable for application where a constant speed is required under widely varying load conditions.

Shunt-wound motors do not give such a high starting torque as series-wound motors, but the field current is independent of the motor speed as it is connected directly across the supply. The torque is therefore much less dependent on speed than in the series-wound motor. A more nearly constant speed is therefore maintained under varying load conditions. Shunt-wound d.c. motors are used for such purposes as driving ringing machines on the exchange 50 V d.c. (battery) supply and for teleprinters designed primarily for applications where frequent starting under load is not required. Before the general introduction of a.c. mains, shunt-wound d.c. motors were also used for driving generators from d.c. mains supplies.

The direction of rotation of a d.c. motor is determined by the direction of the armature current and by the polarity of the field. It can therefore be reversed by reversing the connexions of either the armature or of the field winding. The direction of rotation is unchanged if both the field and the armature connexions are reversed.

A d.c. motor will therefore operate on a single-phase a.c. supply. Single-phase commutator motors similar in construction to d.c. machines are used for low power applications such as driving ringing machines.

As we have seen, when starting a motor, there is initially no back e.m.f. to limit the current in the armature winding. To prevent damage to the machine through excessive armature current it is usual to connect a starting resistance in series with the armature and cut it out in steps as the motor gains speed. The circuit of a typical motor starter is shown in Fig. 9.9. This includes an electromagnet *NV* which holds the starting handle operated against a restoring spring so long as current flows in the field winding.

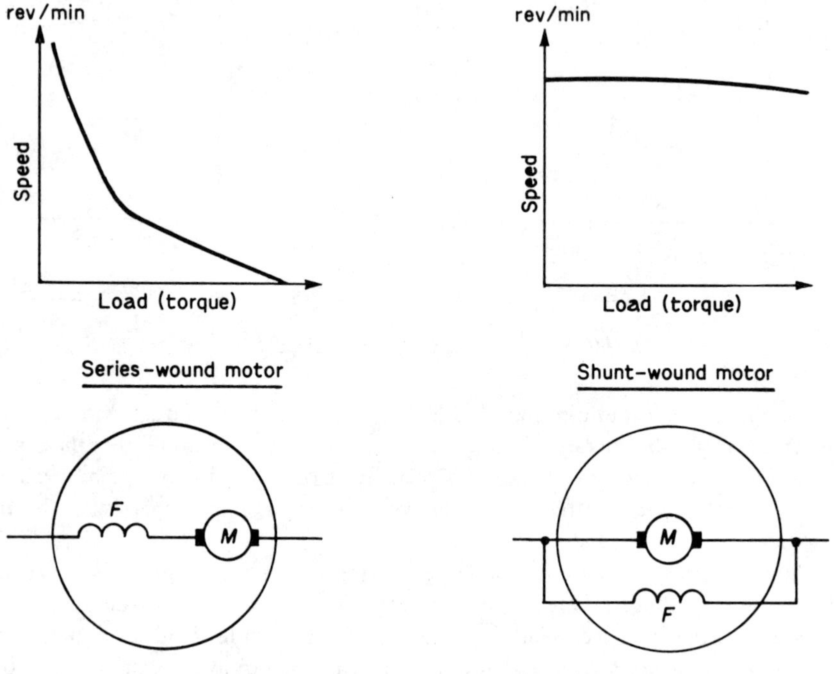

Fig. 9.8 D.C. motor characteristics (9.2.4).

Should the supply fail, this magnet is de-energized, allowing the starter to restore to normal so that all the resistance is in circuit when re-starting. In addition to this, a low resistance 'overload' relay *OL* is connected in series with the armature winding. This is adjusted to operate at a value of current just below that at which the motor would stall in the event of an excessively large mechanical load being applied. A contact of this relay de-energizes the *NV* magnet, again ensuring that the full value of resistance is in circuit when re-starting the motor.

9.2.5 Mains rectifiers. It can be seen in Fig. 9.1 that an alternative to the use of a motor-generator for producing the required d.c. supply from an a.c. mains source is the use of rectifiers. Metal rectifiers have a fairly low resistance and are widely used for relay circuits and for measuring instruments. Their relatively low current carrying capacity and low working voltage, however, make them unsuitable for power supply purposes.

Thermionic rectifiers (diode valves) can withstand higher reverse voltages than metal rectifiers and are used as power rectifiers on some small installations.

Large installations use *semiconductor rectifiers*, generally of the silicon type. These employ a junction between an n-type semiconductor (e.g. silicon with phosphorus or boron) and a p-type semiconductor (e.g. silicon with aluminium). The junction has the property that current flows more easily from p-type to n-type (electrons from n to p) than in the opposite direction.

The phosphorus in the n-type *semiconductor* has more electrons than are required to complete the silicon crystal structure. Electrical conduction in this n-type material is by the movement of these free electrons as 'carriers'.

The aluminium in the p-type, however, has too few electrons. This leaves a series of gaps in the crystal structure called 'holes'. Conduction in p-type material consists

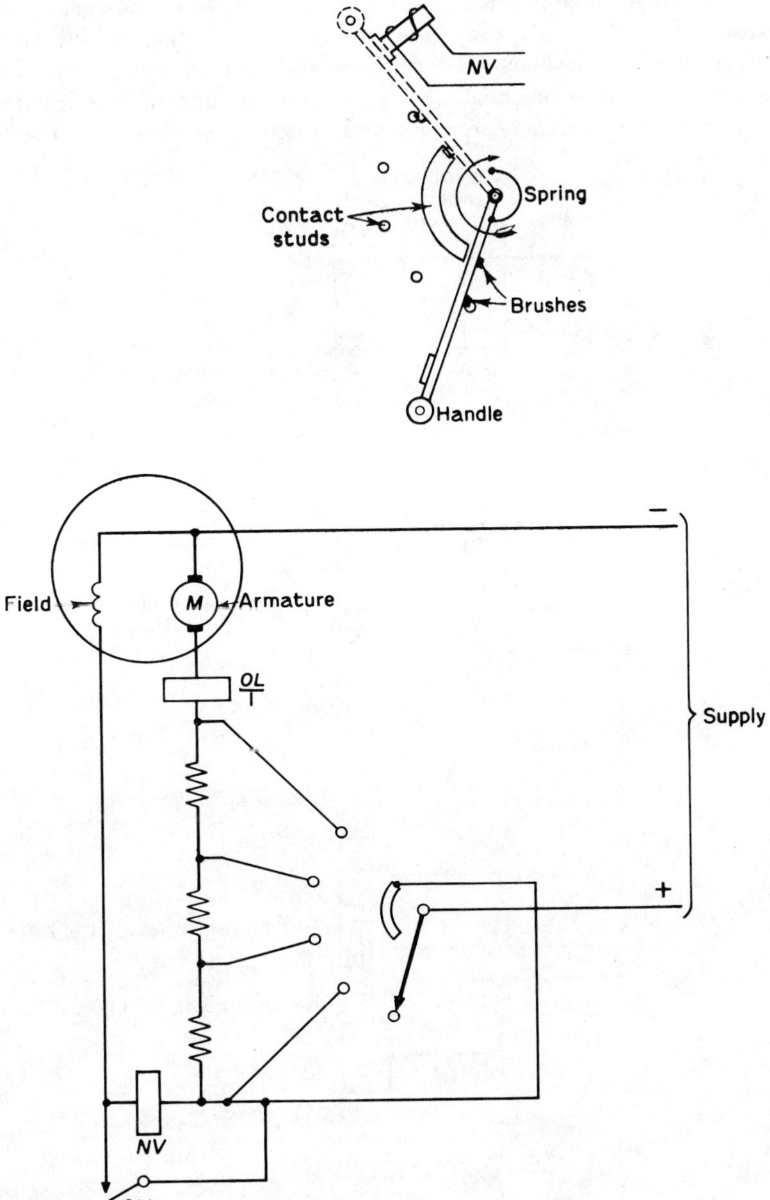

Fig. 9.9 Starter for shunt-wound d.c. motor (9.2.4).

of the movement of these 'holes' as carriers of an effective positive charge. In physical terms, this occurs by one electron moving into a hole and therefore leaving a hole in the crystal at the point it last occupied. Another electron then moves into this 'new' hole, leaving a hole further on. The hole thus moves through the crystal in the opposite direction to the movement of each of these electrons.

If we apply a potential to the rectifier such that the end composed of the p-type material becomes positive with respect to the end composed of the n-type material, electrons are attracted across the junction by the positive potential. Holes are attracted across the junction in the other direction by the negative potential and conduction takes place.

If the p-type material is made negative with respect to the n-type, however, the free electrons are attracted away from the junction by the positive potential in the n-type material. The holes in the p-type material are also attracted away from the junction by the negative potential. This forms a non-conducting barrier in the region of the junction in which there are neither free electrons nor holes (see Fig. 9.10).

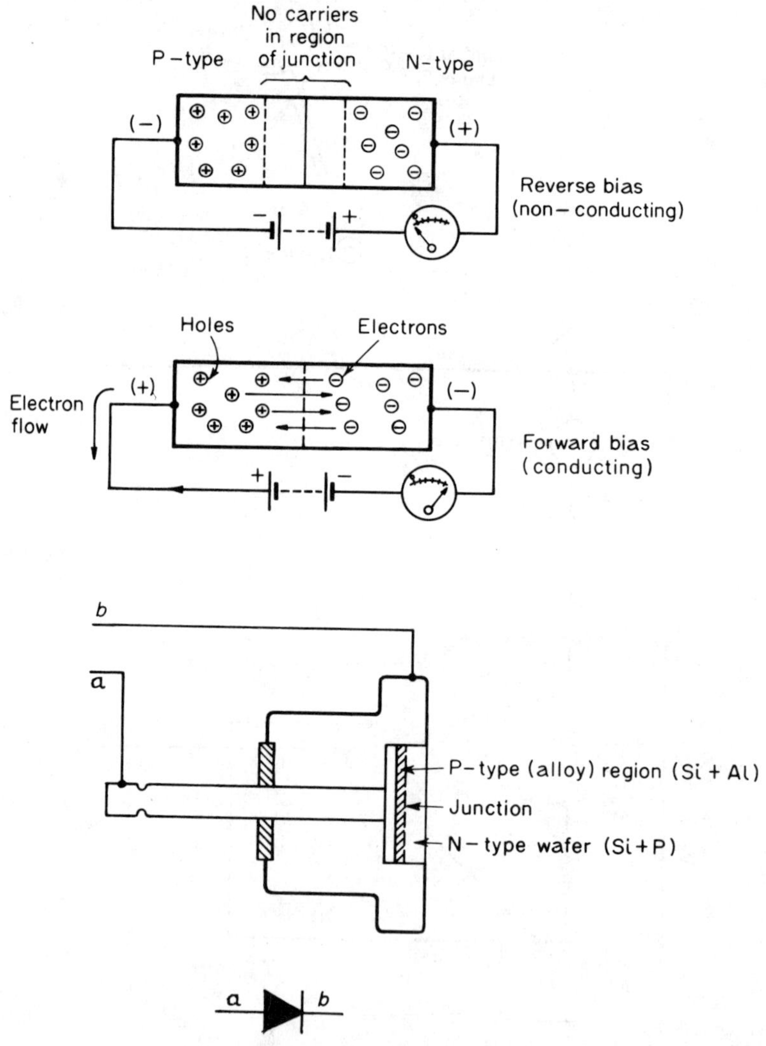

Fig. 9.10 Principle of silicon rectifier (9.2.5).

The simplest way of using a rectifier to obtain a direct current from an a.c. source is to use one rectifier in series with the output (load) as shown in Fig. 9.11a. This provides half-wave rectification, an output being provided on alternate half-cycles only so that the a.c. supply is unevenly loaded and only half the available power is used.

Practical power rectifiers are connected to give full-wave rectification, as shown in Fig. 9.11b for a single-phase supply. This uses four rectifiers connected as a bridge circuit. Current then flows via *MR*1 and *MR*2 on one half-cycle *A* and via *MR*3 and *MR*4 on the next half-cycle *B*, and the whole of the transformer winding is used each time.

The circuit of a three-phase rectifier is shown in Fig. 9.11c. The primary windings of the transformer are connected in 'star' formation like the field windings of the motor

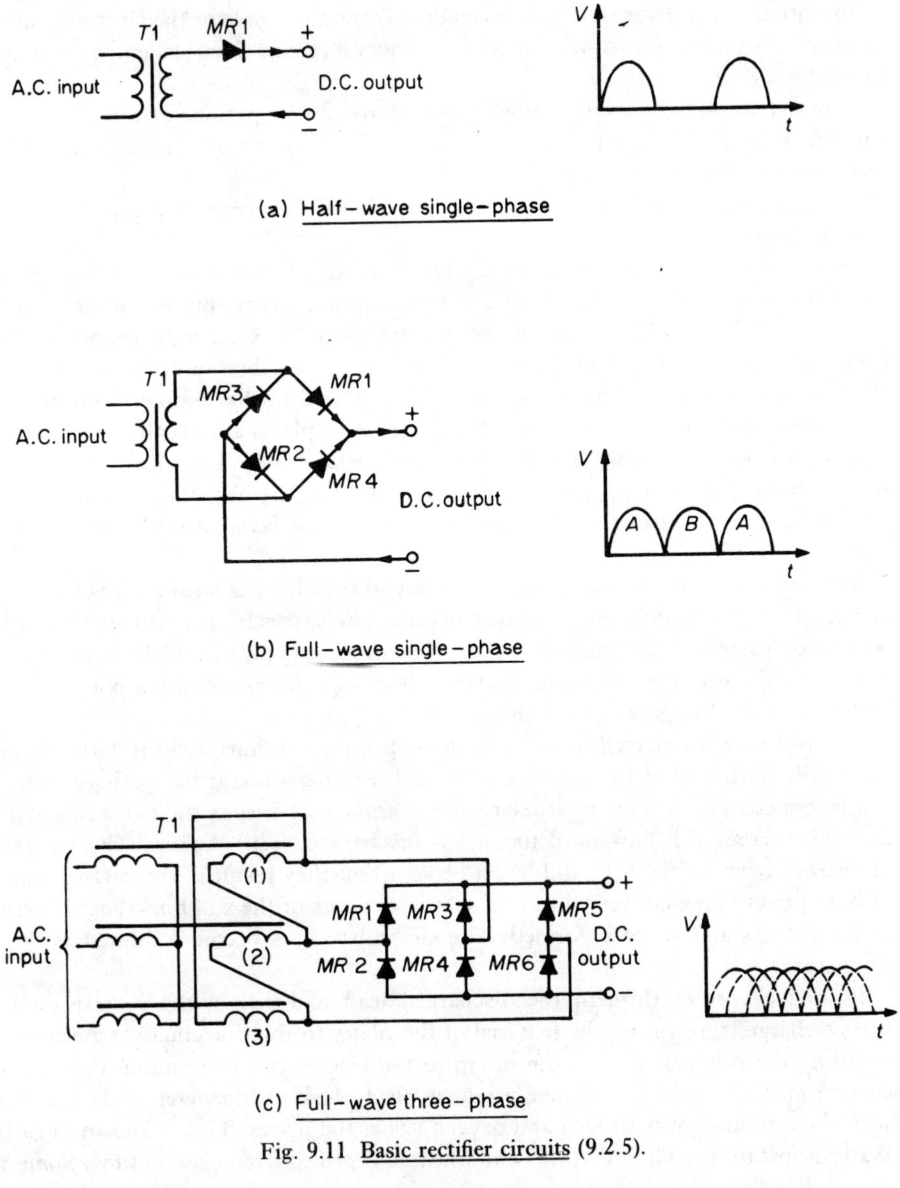

Fig. 9.11 Basic rectifier circuits (9.2.5).

shown in Fig. 9.4. The secondary windings are connected in 'mesh' formation. The three phases of the supply are transformed to the appropriate voltage and appear across the three sections of *T*1 labelled 1, 2 and 3.

Examination of the circuit in Fig. 9.11c will show that each one of these three sections of *T*1 is connected to four of the six rectifiers. In each case, the rectifiers form a bridge circuit giving full-wave rectification.

9.3 Storage batteries

9.3.1 Secondary cells. An exchange battery consists of a number of secondary cells connected in series to give the required voltage. The battery is provided in order to ensure continuity of supply to the exchange under all conditions.

In general, a cell produces electrical energy by changes occurring in the chemical composition of the material of the electrodes. When the whole of the electrode material has been converted in this way, the cell produces no more electrical energy and is said to be *discharged*.

A *secondary cell*, (or accumulator) is one in which the passage of a reverse current through it in a discharged state will re-charge it by restoring the electrode material to its original composition.

A *primary cell* is one in which this is not possible and it cannot therefore be used as a storage battery.

In its simplest form, one secondary cell consists of two lead plates immersed in sulphuric acid. This can be charged by passing a current through it to form lead peroxide (PbO_2) on the surface of the positive plate. A number of charge-discharge cycles are necessary to activate the plates fully and even then only a small fraction of the total quantity of lead in the plates is actually used in this simple secondary cell.

In practical lead-acid secondary cells, the positive plates are usually cast from pure lead with deep grooves to increase the active surface area. They are then subjected to a forming process equivalent to charging, discharging and finally re-charging an accumulator. This results in the formation of a porous layer of lead peroxide on the surface of the plates.

The negative plates are of a lead antimony alloy which is stronger than pure lead and can be formed into a box or grid structure. The spaces in this grid are filled with a lead oxide paste and the plate undergoes a forming process essentially similar to that used for the positive plates. This converts the lead oxide paste into a porous layer of spongy lead in the spaces of the grid.

The plates are immersed in an electrolyte of dilute sulphuric acid to form the complete cell. With lead at the negative plate and lead peroxide at the positive plate, the cell is in a charged condition. If a circuit is completed between the two plates outside the cell, current will flow until the cell is discharged. The chemical changes taking place are shown in Fig. 9.12 and it can be seen that they result in the surface material of both plates being converted to lead sulphate. Some of the sulphuric acid is used up in the process and water is formed. The electrolyte thus becomes diluted as the cell discharges.

Passing a current through the discharged lead-acid cell in the reverse direction will re-charge it, restoring the material of the plates to their original composition and restoring the original acid concentration in the electrolyte. Continuing this charging current after charging is complete produces electrolysis of the water in the electrolyte with the evolution of hydrogen and oxygen gas at the plates. This is known as *gassing*. Water is lost to the cell in this way and must be replaced from time to time. Some acid

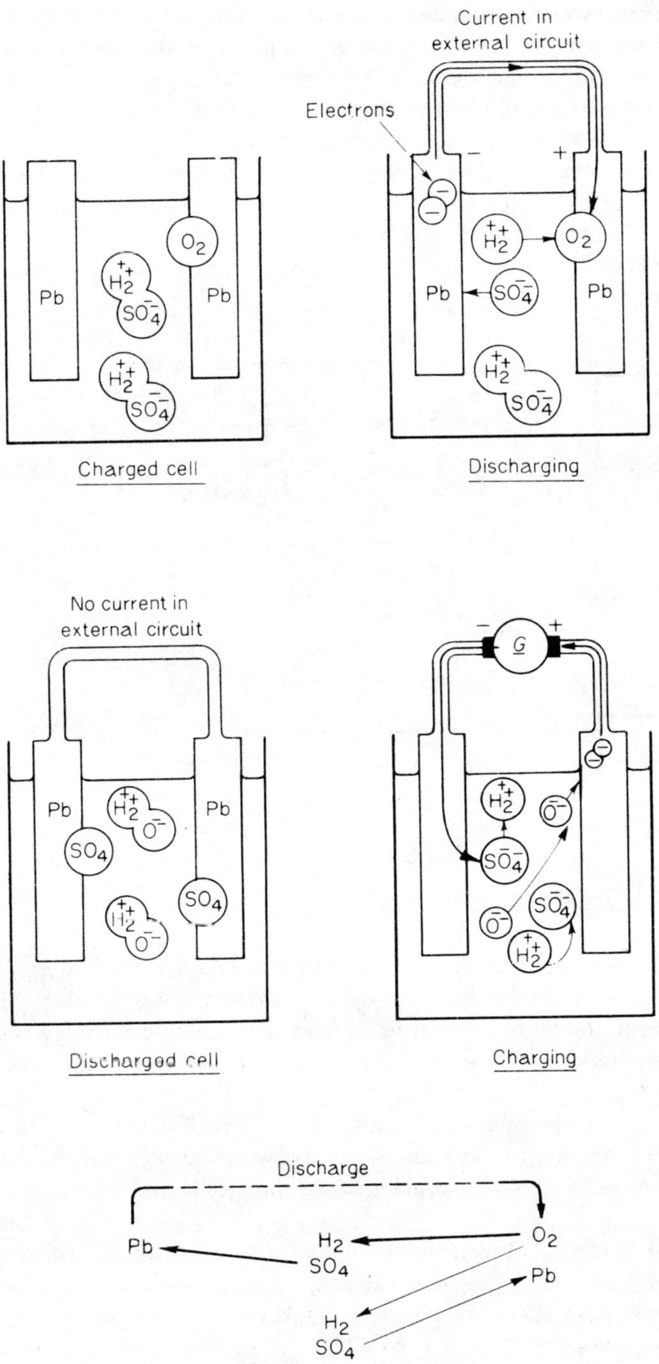

Fig. 9.12 Principle of secondary cell. The chemical symbols in these diagrams are Pb = lead, H = hydrogen, O = oxygen, SO_4 = sulphate ion, H_2SO_4 = sulphuric acid, H_2O = water, PbO_2 = lead peroxide, $PbSO_4$ = lead sulphate (9.3.1).

is also lost by splashing caused by the gassing and some of the active material on the plates may also be loosened.

9.3.2 Charge-discharge and float operation. There are two basic ways in which a storage battery may be used to supply an exchange as shown in Fig. 9.14.

In the *charge-discharge* system, each of two batteries is charged from the supply while the other is being discharged to provide current to the exchange. This system is only used where no public supply is available, a rare circumstance nowadays.

In the *float system*, the main exchange load is normally supplied directly by the rectifier or motor generator and the battery is kept permanently charged in parallel with the exchange load. The battery only supplies power to the exchange if the supply fails or if the exchange load exceeds the power available from the rectifier or generator for a short time.

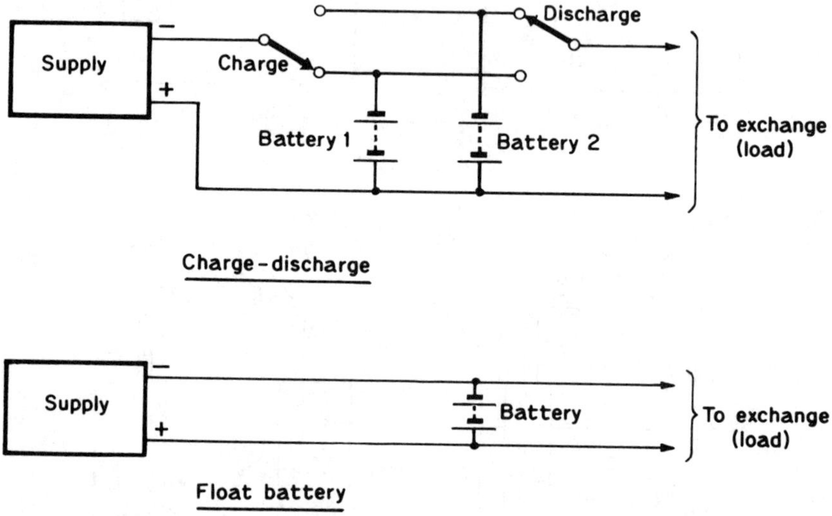

Fig. 9.13 Charge-discharge and float systems. Compare this with Fig. 9.20 (9.3.1).

The advantages of the float system arise from the fact that the battery does not have to undergo successive charge-discharge cycles. This eliminates the energy loss involved in each stage of the conversion from electrical to chemical energy and back into electrical energy and the wear and tear on the plates each time their composition is changed (**9.5.3**).

Floated battery systems need automatic voltage control to allow for variations in the load. One effect of this is to eliminate overcharging the battery with the loss of electrolyte that the resultant gassing would involve, thus reducing the maintenance effort required.

Because the battery is always fully charged, the whole of the battery capacity is available at all times to supply the exchange load in the event of a mains failure. Smaller batteries can therefore be used to provide a given reserve than with the charge-discharge system where the possibility of neither battery being fully charged at the start of a power failure must be allowed for.

Some practical systems use a floated battery which is nevertheless partially discharged and re-charged during the course of the day as the exchange load varies. These are known as 'partial float systems' (**9.4.2**).

9.4 Power switchboard

9.4.1 General. The power switchboard accommodates the control apparatus for maintaining a supply to the exchange. This includes switching facilities, measuring instruments and voltage regulating apparatus.

The switchboard itself consists of a panel of a suitable insulating material. Slate, hardwood (e.g. ebony), or other locally obtainable materials (e.g. marble in Greece) have been used but most modern switchboards are made of synthetic resin bonded compounds. The necessary apparatus is mounted on the front of this and the connexions made at the back.

The main items common to both charge-discharge and float systems are as follows:

One side of the d.c. supply is connected to the exchange earth. All switching of batteries or charging circuits is carried out in the connexions to the non-earthed side of the supply. The switches are thus in the negative lead, for example, of the main 50 V supply in a telephone exchange because this is a negative supply, and the positive side of the exchange battery is connected to earth.

The switches used are single-pole knife-switches. The advantage of this type of switch is that it is robust and reliable and provides a low resistance contact of high current capacity. Typical contact actions for this type of switch are shown in Fig. 9.14.

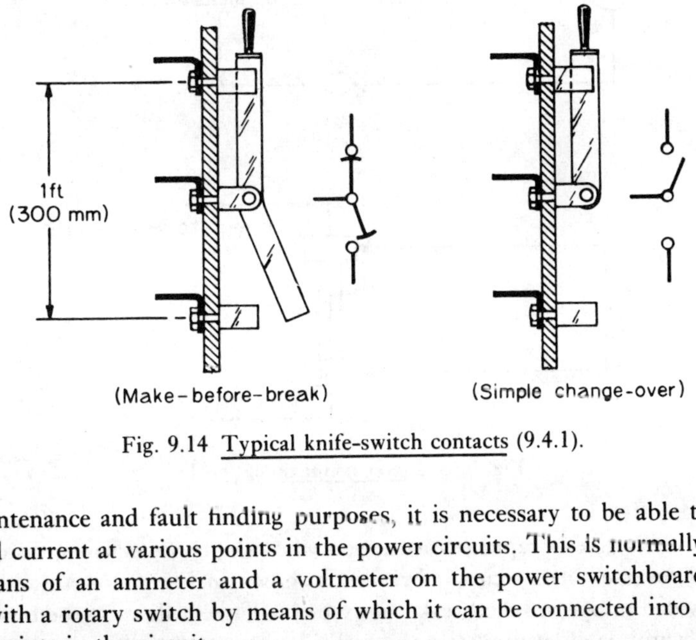

Fig. 9.14 Typical knife-switch contacts (9.4.1).

For maintenance and fault finding purposes, it is necessary to be able to measure voltage and current at various points in the power circuits. This is normally provided for by means of an ammeter and a voltmeter on the power switchboard. Each is provided with a rotary switch by means of which it can be connected into any of the necessary points in the circuit.

Measurement of voltage is easily arranged for, since the meter has only to be connected across the two ends of the component whose voltage is to be measured.

The measurement of current is not so easy because the ammeter has to be connected in series with the circuit it is measuring. Normally this would mean interrupting the circuit. To avoid this low resistance 'shunts' are connected permanently in those parts of the power circuit at which current measurements are necessary. When a measurement is required, the ammeter then has only to be connected across the appropriate shunt. The current divides between the meter and the shunt in a fixed proportion depending on their resistances. The meter, of course, has its scale marked to give a direct reading of the circuit current with this particular value of shunt resistance (see Fig. 9.15).

We saw earlier (**9.2.4**), that a d.c. generator would rotate as a motor if a reverse current were passed through it. This means that where motor-generators are used, any failure causing the motor to stop (e.g. mains failure) would cause the battery to discharge into the generator and drive it as a motor. This would be a waste of stored energy at the very time we need it most and could also damage the generator. To prevent this, a *circuit breaker* is connected in series with the generator. This disconnects the generator if current flows in the reverse direction.

The main circuit through the circuit breaker is made by a laminated bronze shoe pressing on a copper block giving a low resistance contact. An auxiliary carbon contact is provided in parallel with this which 'makes' first and 'breaks' last. Any arcing which occurs when the breaker operates takes place at the carbon contacts thus avoiding damage to the main contact and preserving its low resistance.

The breaker is closed by hand against a powerful restoring spring and toggle

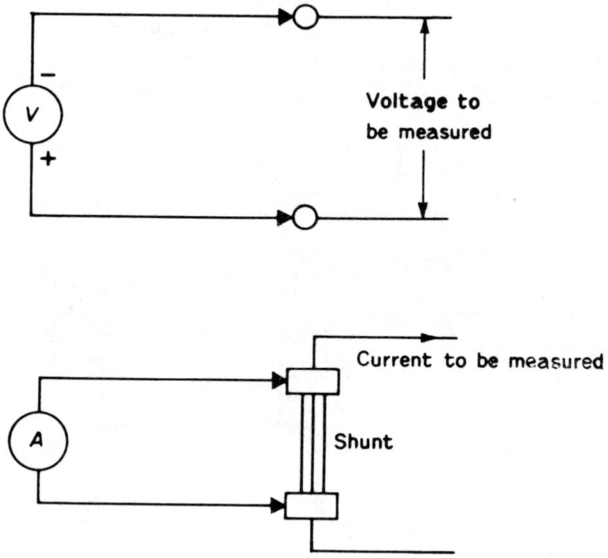

Fig. 9.15 Meter connexions (9.4.1).

mechanism. This gives a rapid release action when the trip mechanism is operated.

A typical reverse-current circuit breaker is shown in Fig. 9.16. An additional coil can be provided to operate the breaker if excessive current flows in the forward direction. This facility is often provided separately by means of an 'overload' relay.

One way of controlling the charging of a battery to avoid over charging it or discharging it too far is to keep a check on the quantity of electricity put into the battery and subtract from this the quantity taken out. This would indicate the state of charge at any time.

A device for monitoring the charging process in this way is called an *ampere-hour meter*, the principle of which is shown in Fig. 9.17. Current flows radially in the copper disk between the poles of the 'driving' magnet causing the disk to rotate. The other side of the disk, however, is between the poles of a second permanent magnet, the 'braking' magnet. This sets up eddy currents in the disk. The magnetic field of the eddy current reacts with that of the braking magnet to oppose the rotation of the disk. This opposing torque increases as the speed of rotation increases and the combined effect of the two magnets is that the speed of rotation of the disk is proportional to the current in the disk. Reversing the current reverses the direction of rotation.

9.4.2 Switchboard apparatus for float systems. The items described in the previous section may be used in charge-discharge or in float battery systems. The latter systems require additional apparatus in order to maintain a constant voltage.

It can be seen from Figs 9.6 and 9.11, that neither a rectifier circuit nor a d.c. generator produces a completely smooth output. The regular variations in output voltage are termed *ripple*. They constitute an a.c. component superimposed on the d.c. supply.

This is not important so far as charging a battery is concerned and no special steps need to be taken on a charge-discharge system.

In the case of a floated battery system, however, the ripple would be heard as an objectionable 'hum' on all telephone circuits through the exchange. It could also cause misoperation of signalling circuits in the same way as external a.c. interference.

The method employed to remove the 'ripple' is to connect the charging supply to the battery via a *smoothing circuit* or *filter* of the form shown in Fig. 9.18. The inductors *L*1 and *L*2 present a high impedance at the ripple frequency and the capacitor *C* presents a low impedance. *L*1 and *C* in series can be regarded as a potential divider across the source of the ripple voltage giving a low ripple voltage across *C*, (which has the lower impedance). This small voltage is still further reduced across the battery and the exchange load which constitute another low impedance in series with the high impedance of *L*2.

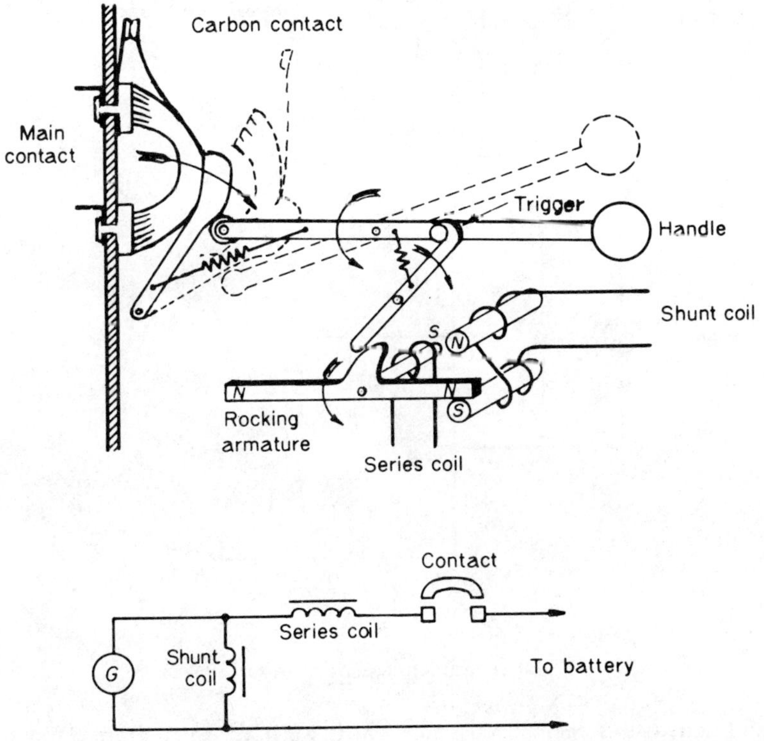

Fig. 9.16 Reverse-current circuit breaker. The magnet polarities *N* and *S* refer to current in the correct direction. Reversal of current in the series coil gives the rocking armature 'south' polarity so that its right-hand is attracted upwards by the shunt coil pole (9.4.1).

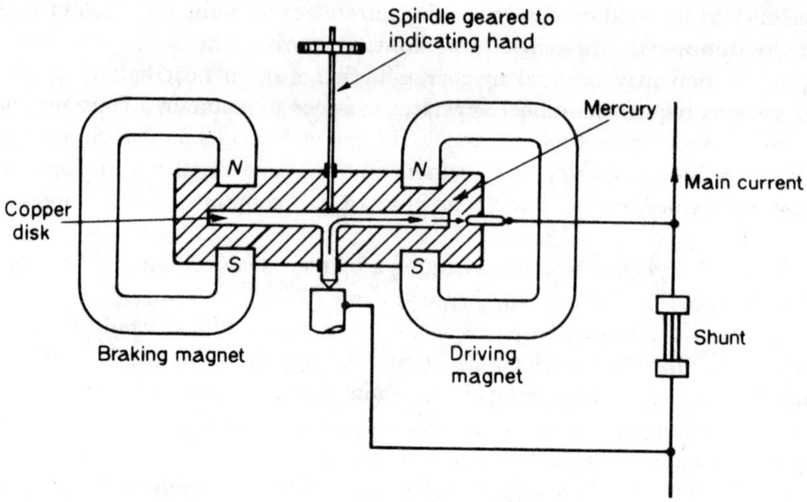

Fig. 9.17 Principle of ampere-hour meter (9.4.1).

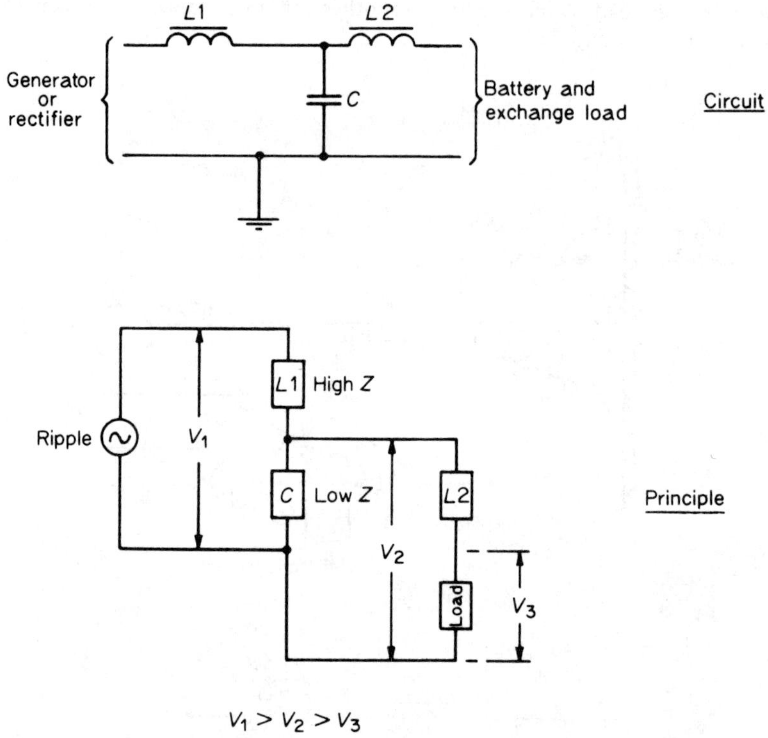

Fig. 9.18 Smoothing circuit (filter) (9.4.2).

In addition to this regular variation which we have called ripple, there are other variations in the output voltage of the rectifier or generator. These are caused by fluctuations in the load current taken by the exchange as the traffic varies and by changes occurring in the public mains supply voltage. Voltage control is necessary to keep the supply to the exchange within the required limits in spite of these variations.

In *full float* operation the whole of the power required by the exchange under normal conditions is obtained from the rectifier or generator. The battery is maintained in a fully charged condition. This is achieved by controlling the output voltage of the rectifier or generator directly under the control of a voltmeter device connected to the battery terminal.

In the case of a generator, a *regulator* is used to control the field current. A typical regulator is shown in Fig. 9.19. Several other types are also in use.

In the case of rectifier equipment, it is usual to control the input to the rectifier. One way of doing this is to use a *contact voltmeter* to switch the controls of motor-

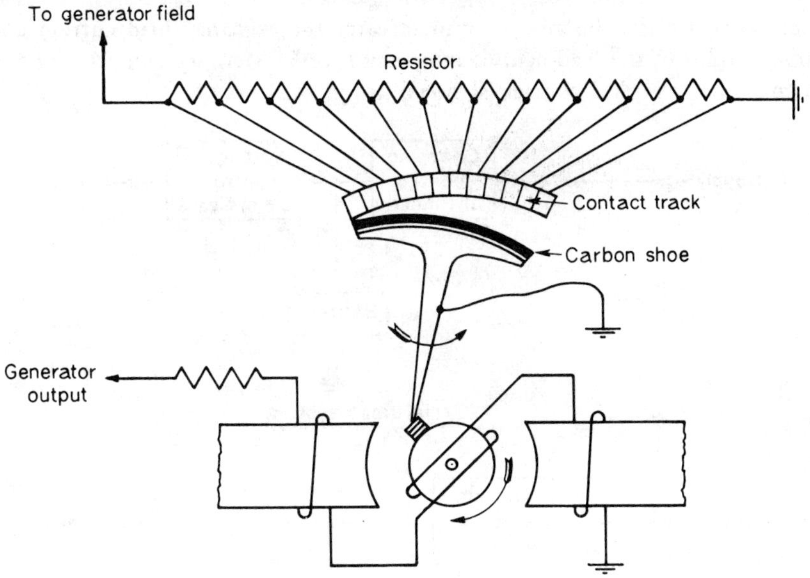

Fig. 9.19 Dynamometer regulator. An increase in the generator output voltage due to a reduction in the exchange load causes the dynamometer movement to rotate as shown increasing the resistance in series with the generator field winding (9.4.2).

driven variable inductors in series with the primary windings of the three-phase input transformer (Fig. 9.25).

In partial float systems, the battery is left connected to the exchange load but the charging supply is switched on and off as required under the control of an ampere-hour meter to maintain the battery charge within pre-determined limits. The battery is therefore partially discharged and then re-charged resulting in variations in the battery terminal voltage which are too great for supplying the exchange directly. *Counter-e.m.f. cells* (c.e.m.f. cells) are therefore switched in as required under the control of a *contact voltmeter* to compensate for these variations.

In all but the very small exchanges, the counter-e.m.f. cells are of the alkaline type consisting of nickel-plated iron electrodes immersed in potassium hydroxide solution (caustic potash). When a current is passing through the cell, electrolysis takes place with the production of hydrogen at the cathode and oxygen at the anode. These gases form a film over each of the nickel electrode surfaces so that the cell has effectively one electrode of hydrogen and the other of oxygen. This then generates an e.m.f. of 2 V. The action is similar to the effect of using dissimilar metals in an acid solution to form a primary cell.

The advantage of this type of cell is that it relies upon the passage of current through it to maintain the gas film on the electrodes and produces its e.m.f. in a direction which opposes this current. Consequently the e.m.f. falls away when the current is stopped. This allows the counter-e.m.f. cells to be short circuited without damage to the switching contacts used. If this were attempted with lead-acid cells, the stored energy would circulate a large current through the short-circuiting contacts.

A disadvantage of the alkaline type of counter-e.m.f. cell is that its e.m.f. becomes quite small if the current through it is reduced to a small value during periods of light traffic on the exchange. In small exchanges where this is likely, therefore, lead-acid cells are used. This means using change-over contacts instead of simple make contacts for switching the cells. These change-over contacts must have low resistance and high current rating because they must carry the exchange load current and the discharge current of the lead-acid counter-e.m.f. cells. Mercury contacts are suitable for this purpose; a typical contact is shown in Fig. 9.21.

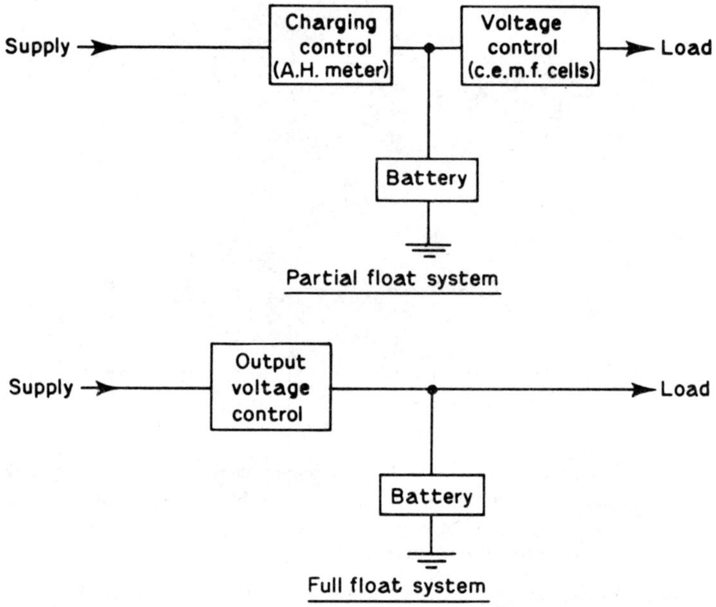

Fig. 9.20 Principle of partial-float and full-float operation. Compare this with Fig. 9.13 (9.4.2).

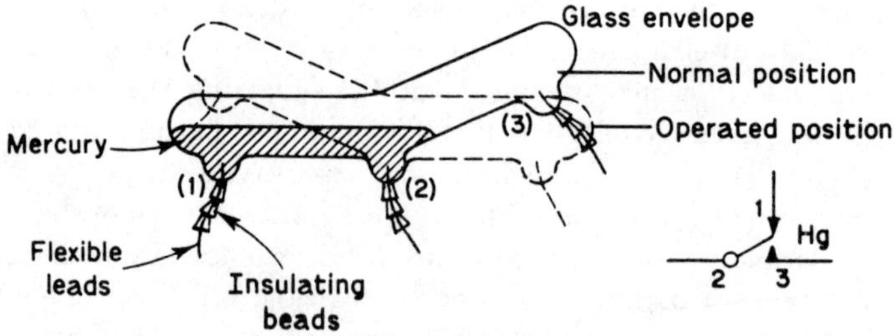

Fig. 9.21 Mercury contact. This is a change-over contact which would be mounted on a pivotted platform linked to a relay armature (9.4.2).

Now turn to page 247 and test yourself on questions 1–17 before reading the next part.

PART 3 TYPICAL PRACTICAL EXCHANGE POWER PLANTS

9.5 Telephone exchange power plants

9.5.1 Charge-discharge system. Although for most purposes charge-discharge operation has been superseded by floated battery operation, it is useful where there is no public mains supply because it does not require the prime mover to be kept running continuously.

A charge-discharge power plant suitable for small exchanges of up to 100 Ah power consumption per day is shown in Fig. 9.22.

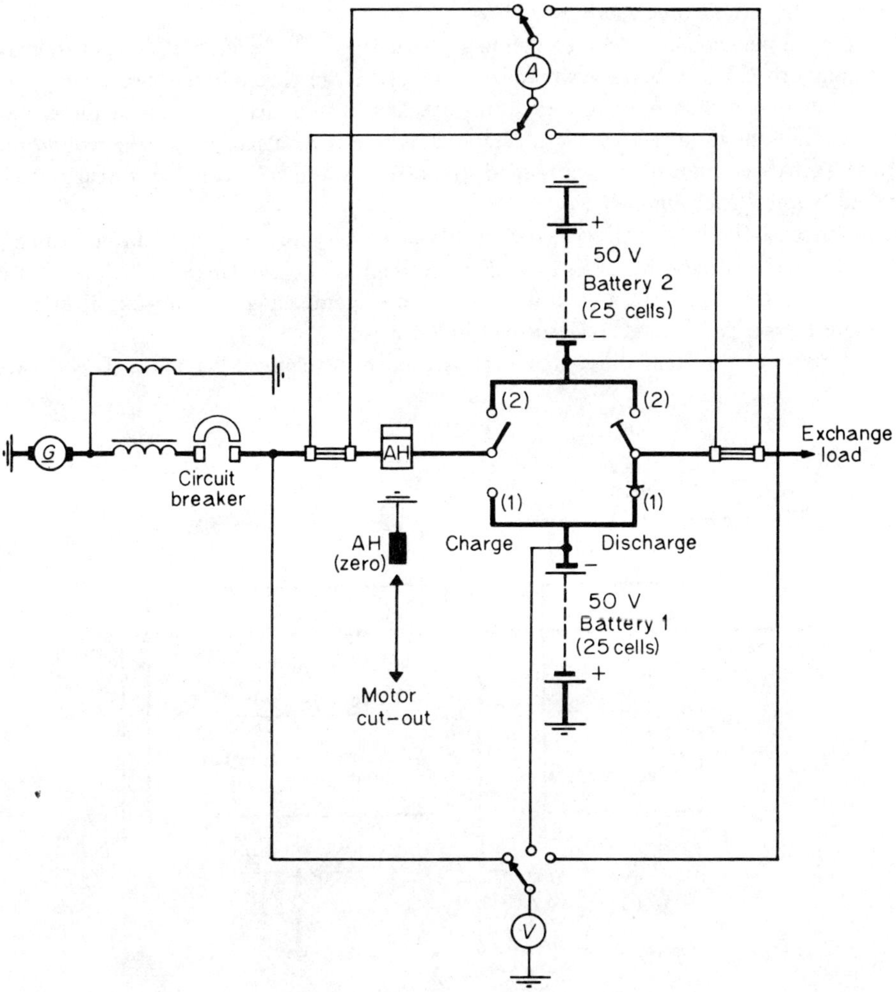

Fig. 9.22 Charge-discharge power plant. For small exchanges (up to 100 Ah per day) without mains power supply (9.5.1).

There are two batteries, each having a capacity of three times the ultimate daily power requirement of the exchange. Switches are provided so that either battery may be connected to supply the exchange load; the other battery is connected to the charging supply when required.

The charging supply is a petrol engine driven d.c. generator.

To commence charging, the engine is first started up and run up to speed. The ampere-hour meter is preset to the quantity of electricity required to bring the battery up to the fully charged state, and the circuit breaker set manually to complete the circuit, after first checking that the generator output voltage is slightly greater than the battery voltage.

Charging then proceeds without further attention, the ampere-hour meter turning back as the battery charges. When the required charge has been put into the battery, the ampere-hour meter operates a control which stops the engine and charging ceases. The battery starts to discharge into the generator and this momentary reverse current operates the circuit breaker (Fig. 9.16).

If the engine should run out of fuel or stop due to some mechanical fault during the charging process, the battery would start to discharge through the generator. This reverse current would operate the circuit breaker. An auxiliary contact of the circuit breaker, not shown in Fig. 9.23, is arranged to sound an alarm under this condition. If the exchange is normally unattended, the alarm would be located at a nearby fully staffed 'parent' exchange.

9.5.2 Partial float (parallel battery) system. A typical power plant in which a floated battery is partially charged and discharged as the exchange load fluctuates is shown in Fig. 9.22. This is a fully automatic system using counter-e.m.f. cells to regulate the output voltage as described in **9.4.2.**

The particular system illustrated here is used on exchanges having a daily power

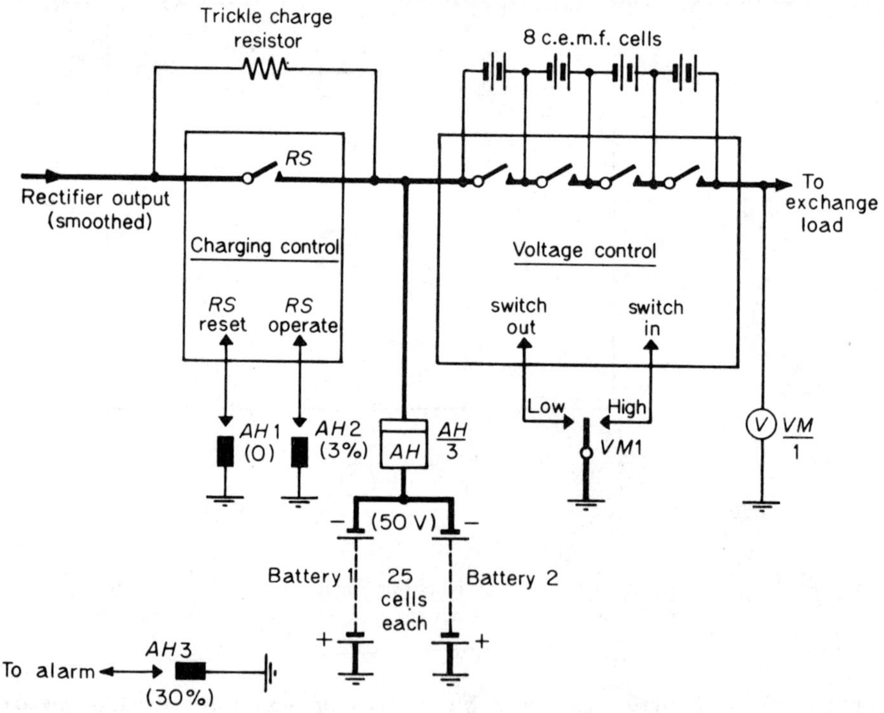

Fig. 9.23 Partial float systems (parallel battery power plant) for exchanges up to 2 000 Ah per day (9.5.2).

consumption of up to 2 000 Ah. Smaller exchanges of up to 120 Ah per day usually have only one battery of twenty-four cells and four counter-e.m.f. cells of the lead-acid type (**9.4.2**) switched in two stages by mercury tube contact relays (Fig. 9.21). This smaller type of power plant is known as the 'single battery' float system.

The parallel battery system of Fig. 9.23 gets its name from the use of two batteries which are connected in parallel under normal working conditions but can be separated when necessary so that one then carries the exchange load while maintenance work is carried out on the other, The total capacity of the two batteries is 10% more than the ultimate daily power requirement of the exchange.

During periods of heavy traffic such as the busy hour (**7.2.2**), the exchange load current exceeds the maximum rectifier output current. The battery therefore provides the difference, discharging as it does so. All c.e.m.f. cells would be short circuited at this time and the ampere-hour meter would register a discharge of between 4% and 30% of the battery capacity.

Later, as the exchange load falls, some of the rectifier output current re-charges the battery. The decreased load also causes the voltage of the supply to the exchange to rise. When the bus bar voltage rises to 52 V, voltmeter *VM* causes a pair of c.e.m.f. cells to be switched in so that the bus bar voltage drops back to 48 V. Two more cells are switched in each time the decreasing load causes *VM* to register 52 V. Not all battery losses are eliminated by floating the battery. Local action due to impurities in the plate materials causes a loss of charge equal to between 1% and 2%

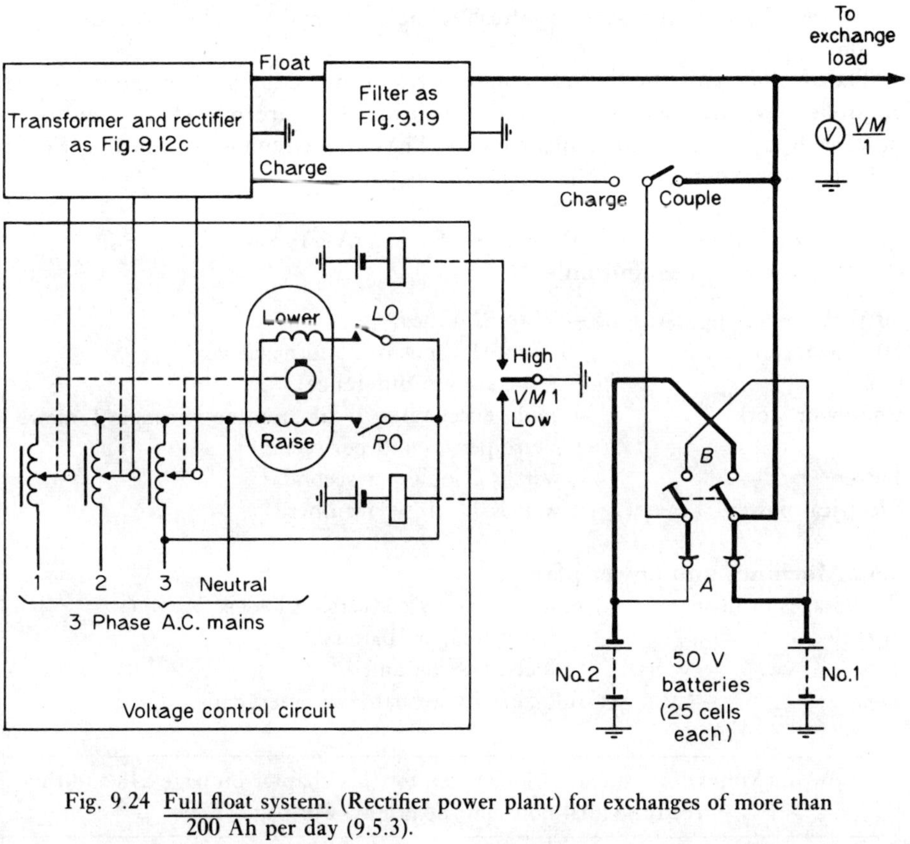

Fig. 9.24 Full float system. (Rectifier power plant) for exchanges of more than 200 Ah per day (9.5.3).

of the battery capacity in the course of a day. To compensate for this, a minimum charging current is maintained to a floated battery at all times. This is known as *trickle charging*. When the ampere-hour meter registers that the battery is fully charged, contact *AH*1 closes causing the charging control contact *RS* to open and the battery then receives only a trickle charge.

With the charging source thus disconnected, the battery now discharges to supply the exchange load under light traffic conditions.

When the traffic increases again, the exchange load current rises, causing a drop in the bus bar voltage. At 46 V, *VM* causes a pair of c.e.m.f. cells to be short circuited, raising the voltage to 50 V. Two more c.e.m.f. cells are switched out each time the increasing load causes *VM* to register 46 V.

Eventually the ampere-hour meter will register that the battery has discharged 4% of its capacity and *AH*2 closes to cause *RS* to close and re-connect the full rectifier output to supply the exchange load and re-charge the battery as before.

9.5.3 Full float system. A typical power plant operating as a full float system is shown in Fig. 9.24. This is a rectifier system suitable for large exchanges of more than 2 000 Ah daily requirement for which purpose it supersedes the older motor-generator method known as the 'divided battery' system. Voltage control is applied at the input to the rectifier so that the system is fully automatic.

The two batteries have a combined capacity equal to the ultimate daily requirement of the exchange. They are normally connected in parallel by means of knife switches (Fig. 9.14) so that they can be separated easily. As no trickle charge is provided for while the batteries are connected to the exchange load, each battery is disconnected once every 2 weeks and given a refresher charge to make up the losses due to local action.

The rectifier is a three-phase unit using silicon rectifiers. The input voltage is controlled by variable inductors, the sliders of which are moved by a chain drive powered by a reversible single-phase motor. The motor circuit is controlled by a contact voltmeter *VM* connected to the exchange bus bar.

PART 4 SUMMARY

9.6.1 Power supplies and units.

50 V exchange apparatus	:	46 V to 52 V negative.
80 V teleprinter	:	75 V to 92 V positive and negative (4 V maximum difference).
Energy or work	:	1 joule accelerates 1 kilogramme through 1 metre at 1 metre per second per second.
Power	:	1 watt is 1 joule per second.
Electrical power	:	1 watt is 1 volt × 1 ampere.

9.6.2 Machines and power plant

Advantages of float	:	reduced energy conversion losses, reduced wear on battery, reduced gassing and full capacity available in emergency.

Now test yourself on the questions given for this chapter on page 248 and then read the final notes on the last answer page (271).

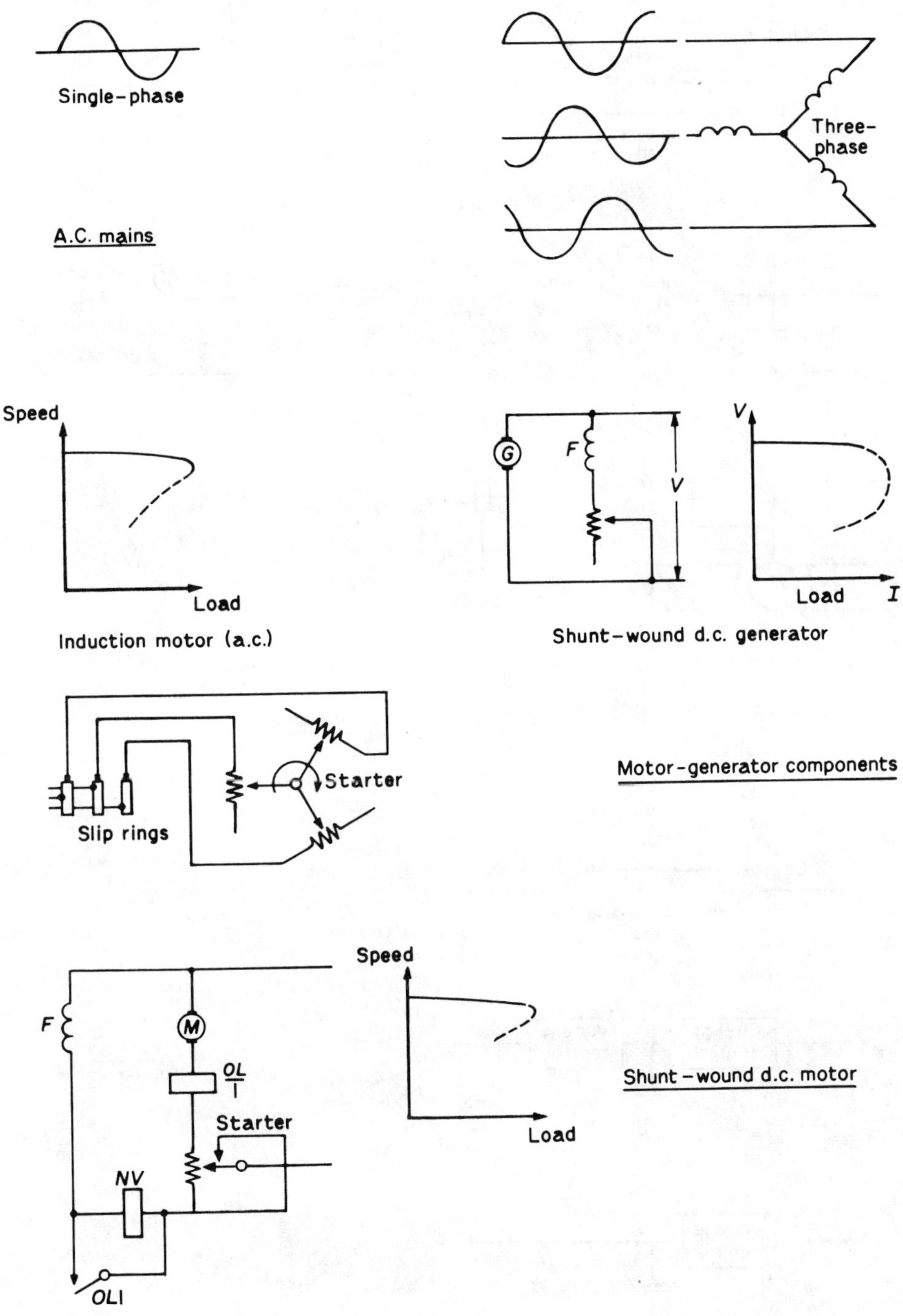

Fig. 9.25 Motors and generators—summary.

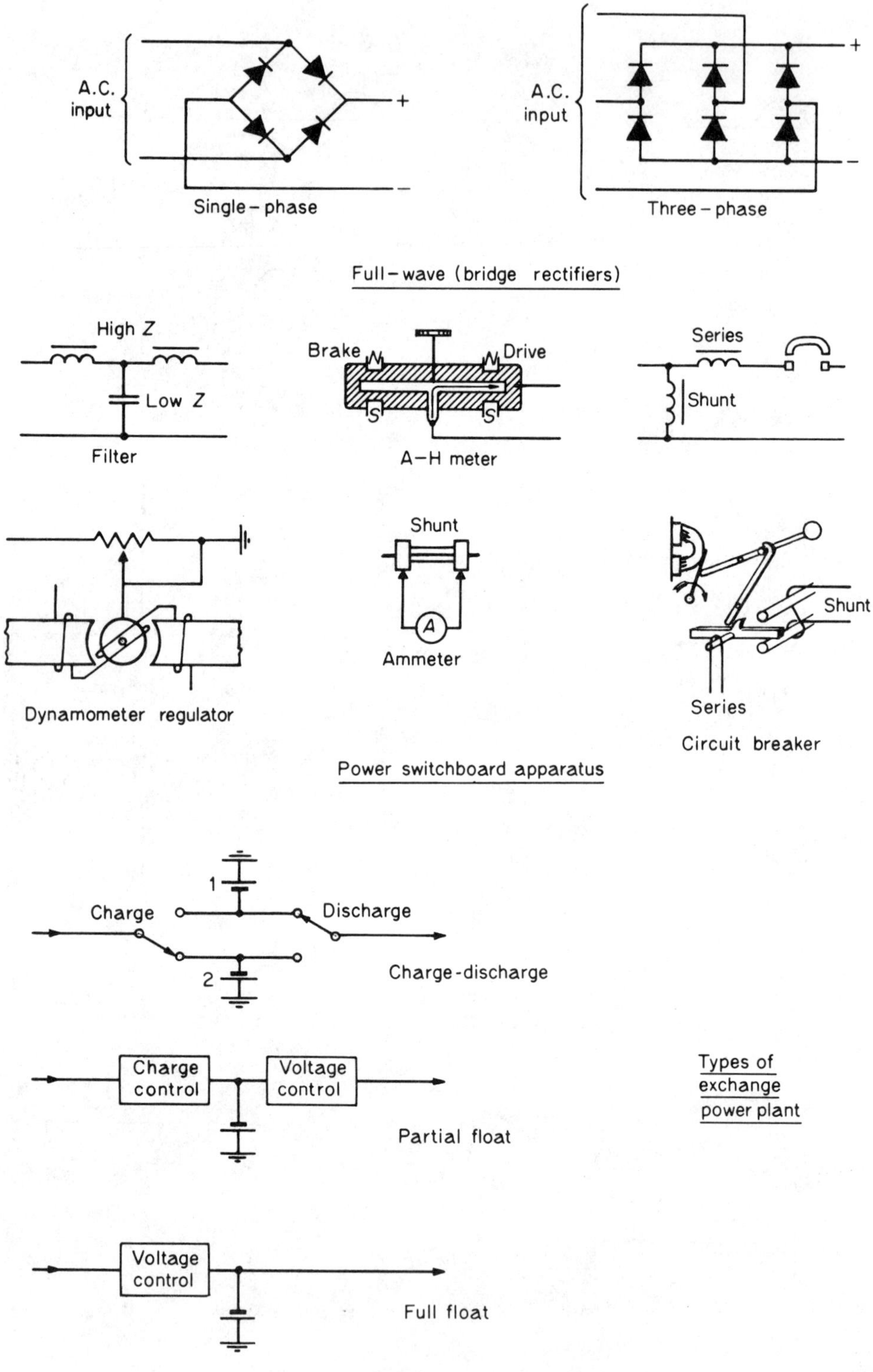

Fig. 9.26 Exchange power plant—summary.

PART 5 SELF-TEACHING NOTES

9.7.1 First read **9.1.1** to **9.4.2** inclusive.

9.7.2 After reading **9.1.1** to **9.4.2**, try the following seventeen questions. Write the answers to all the questions before turning to the answer page.

Q.1 Between what voltage limits must the d.c. supply be maintained for automatic exchanges?

Q.2 What is the SI unit of power?

Q.3 What power is dissipated by the current in a 100 Ω resistor connected to a 50 V supply?

Q.4 Why do motor-generator sets *not* use single-phase motors (when used for charging exchange batteries)?

Q.5 Under what conditions is the current in the rotor of an induction motor at its maximum?

Q.6 What does a d.c. generator have in place of the slip-rings of an alternator?

Q.7 Why is a shunt-wound generator used for exchange power plant in preference to a series-wound generator?

Q.8 Why is the starting handle of a d.c. motor starter provided with a restoring spring?

Q.9 How many rectifier elements are required for full-wave rectification of a three-phase supply?

Q.10 Under what conditions does 'gassing' occur in a secondary (lead-acid) cell?

Q.11 State two advantages of floated battery operation compared with charge-discharge.

Q.12 How is the ammeter used for measuring the current supplied to an exchange switched from one point in the circuit to another without interrupting the current?

Q.13 Why is a reverse-current circuit breaker required when using a motor-generator set to charge a battery?

Q.14 What instrument can be used to register the charge supplied to or withdrawn from a battery?

Q.15 What would be the effect on a telephone system of providing a floated battery power supply without a filter (smoothing circuit)?

Q.16 At what point is output voltage control applied to a motor-generator set?

Q.17 Why are alkaline counter-e.m.f. cells not suitable for use at small exchanges?

Now check your answers on page 269.

9.7.3 You should have answered most of Q.1 to Q.17 correctly. If you scored less than fifteen out of seventeen, see if the notes on the answer page help and then read **9.1.1** to **9.4.2** again and try Q.1 to Q.17 once more.

With fifteen or more correctly answered, read the rest of the chapter (**9.5.1** to **9.6.3**).

9.7.4 A few days after studying the whole of Chapter 9, try the following questions.

Check the answer to each question on page 270 before trying the next question.

Q.18 Explain with simple circuit diagrams the difference between telephone exchange plants working on the
(*a*) charge-discharge system, and
(*b*) floated battery system.
Name two items of equipment necessary for (*b*) but not for (*a*) and explain their functions. (C & G 1962)

Q.19 Briefly explain the principle of the d.c. generator, and describe the functions of protective equipment associated with such a generator when it is used to float an exchange battery.
Explain the precautions necessary when connecting an additional generator to augment the supply to the exchange. (C & G 1965)

APPENDIX 1

INTERNATIONAL MORSE CODE

A	•—	1	•———— (or •—)
B	—•••	2	••——— (or ••—)
C	—•—•	3	•••——
D	—••	4	••••—
E	•	5	••••• (or •)
F	••—•	6	—••••
G	——•	7	——•••
H	••••	8	———•• (or —••)
I	••	9	————• (or —•)
J	•———	0	————— (or —)
K	—•—		
L	•—••	.	•—•—•—
M	——	:	———•••
N	—•	?	••——••
O	———	,	——••—— (comma)
P	•——•	–	—••••—
Q	——•—	'	•————• (apostrophe)
R	•—•	/	—••—•
S	•••	( )	—•——•—
T	—	"	•—••—• (inverted commas)
U	••—	=	—•••—
V	•••—	+	•—•—• (end of message)
W	•——		
X	—••—	Error	••••••••
Y	—•——	End of schedule	•••—•—
Z	——••		

APPENDIX 2

INTERNATIONAL FIVE-UNIT TELEPRINTER CODE

Letters	5	4	3		2	1	Figures
A	○	○	○	•	●	●	—
B	●	●	○	•	○	●	?
C	○	●	●	•	●	○	:
D	○	●	○	•	○	●	Who are you
E	○	○	○	•	○	●	3
F	○	●	●	•	○	●	
G	●	●	○	•	●	○	
H	●	○	●	•	○	○	
I	○	○	●	•	●	○	8
J	○	●	○	•	●	●	Bell
K	○	●	●	•	●	●	(
L	●	○	○	•	●	○	)
M	●	●	●	•	○	○	.
N	○	●	●	•	○	○	, (comma)
O	●	●	○	•	○	○	9
P	●	○	●	•	●	○	0
Q	●	○	●	•	●	●	1
R	○	●	○	•	●	○	4
S	○	○	●	•	○	●	' (apostrophe)
T	●	○	○	•	○	○	5
U	○	○	●	•	●	●	7
V	●	●	●	•	●	○	=
W	●	○	○	•	●	●	2
X	●	●	●	•	○	●	/
Y	●	○	●	•	○	●	6
Z	●	○	○	•	○	●	+
·	○	●	○	•	○	○	Carriage return
·	○	○	○	•	●	○	Line feed
·	●	●	●	•	●	●	Letter shift
·	●	●	○	•	●	●	Figure shift
·	○	○	●	•	○	○	Space
·	○	○	○	•	○	○	

○ Spacing

● Marking (hole in tape)

• Feed holes in tape

APPENDIX 3

EFFECT OF POLARIZATION

1 General

For any magnet:

$$F\text{ (newtons)} = \frac{B^2A}{2\mu_0}, \quad \text{where } B = \text{flux density (T)},$$
$$A = \text{pole area (m}^2\text{)},$$
$$\mu_0 = 4\pi \times 10^{-7} \text{ SI units.}$$

2 Sensitivity

At one air-gap in a polarized device, the armature is acted on simultaneously by a flux density b (T) produced by the operating coil and polarizing flux density B (T) produced, for example, by a permanent magnet. If the fluxes are assumed to be in the same direction, their combined effect on the armature is a force:

$$F_{B+b} = \frac{(B+b)^2A}{2\mu_0} = \frac{(B^2 + 2Bb + b^2)A}{2\mu_0}$$
$$= \frac{B^2A}{2\mu_0} + \frac{BbA}{\mu_0} + \frac{b^2A}{2\mu_0}. \quad (1)$$

For the armature to have been in equilibrium under the action of B alone before current flowed in the operating coil, however, there must have been a restoring force (or reaction) F_R equal to the force due to B

$$\text{i.e. } F_R = \frac{B^2A}{2\mu_0}. \quad (2)$$

The resultant force F on the armature with current in the coil, if this restoring force is assumed to remain constant at the instant of operation, is $F_{B+b} - F_R$

$$\text{i.e. } F = \frac{B^2A}{2\mu_0} + \frac{BbA}{\mu_0} + \frac{b^2A}{2\mu_0} - \frac{B^2A}{2\mu_0} \quad \text{[from (1) and (2)]}$$
$$\text{or } F = \frac{BbA}{\mu_0} + \frac{b^2A}{2\mu_0}. \quad (3)$$

A non-polarized device, however, would have produced a force due to flux density b alone of: $F_b = \dfrac{b^2A}{2\mu_0}$ (4)

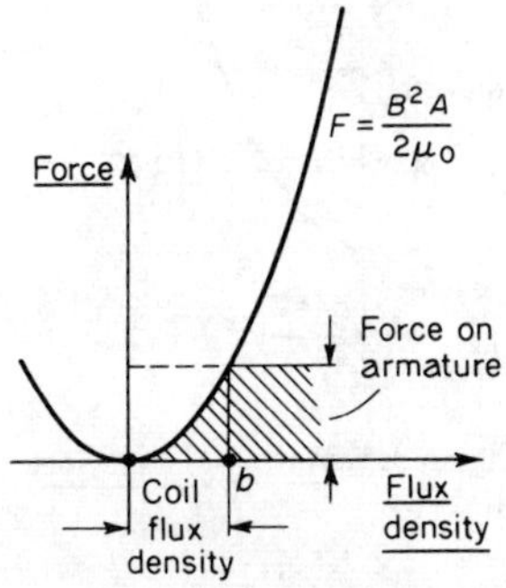

Fig. X3.1 Without polarization.

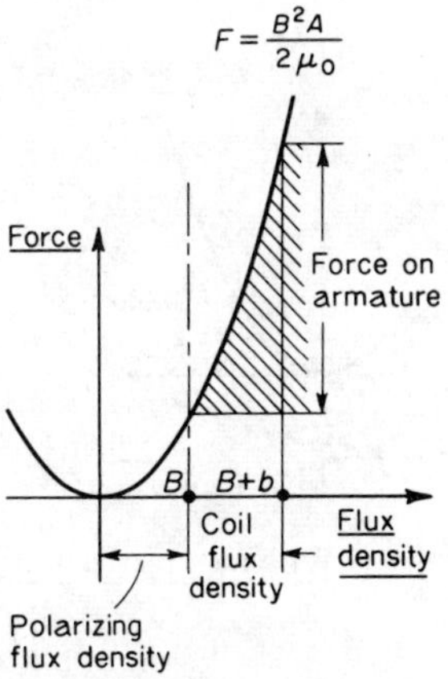

Fig. X3.2 With polarization.

It can be seen from (3) that the effect of polarization is therefore to increase the force on the armature for a given current by an amount $\dfrac{BbA}{\mu_0}$.

Polarization has thus increased the sensitivity.

This can also be shown graphically (Figs. X.3.1 and X.3.2).

In most practical arrangements, the armature is acted on by magnetic flux at more than one air-gap. The term $\dfrac{b^2A}{2\mu_0}$ in equation (3) may then be cancelled by a similar force at another air-gap leaving a resultant force proportional to the product Bb only. The increased sensitivity then depends on B being greater than b (i.e. permanent magnet stronger than electromagnet) making $Bb > b^2$.

3 Frequency doubling

For an a.c. input, the flux produced by the coil will be of the form $b \sin \omega$ where ω is the angular frequency.
For a non-polarized device, substituting $b \sin \omega$ in equation (4) gives:

$$F = \frac{(b \sin \omega t)^2 A}{2\mu_0} = \frac{b^2A}{2\mu_0} \sin^2 \omega t$$

$$= \frac{b^2A}{2\mu_0} \times \frac{1}{2}(1 - \cos 2\, \omega t). \qquad (5)$$

The effect on the armature (or diaphragm) is thus a force at frequency $2\,\omega$, twice the frequency of the input current.

For a polarized device, we can substitute $b \sin \omega t$ for b in equation (3), giving:

$$F = \frac{Bb \sin \omega t.\ A}{\mu_0} + \frac{b^2 \sin^2 \omega t.\ A}{2\mu_0} = \frac{BbA}{\mu_0} \sin \omega t + \frac{b^2A.}{2\mu_0}.\ \tfrac{1}{2}(1 - \cos 2\omega t).$$

The output thus consists of one component at the correct frequency ωt and one at twice this frequency $2\,\omega t$. Providing B is large compared with b, this second component is much the smaller of the two, so that the output (force) is substantially *at the same frequency as the input current.*

In the case of a rocking armature, the $\dfrac{b^2A}{2\mu_0}$ term provides an opposite and nearly equal force at each of two air-gaps so that its effect is further reduced.

Polarization thus effectively eliminates frequency doubling, an effect which can also be shown graphically (Figs. X.3.3 and X.3.4).

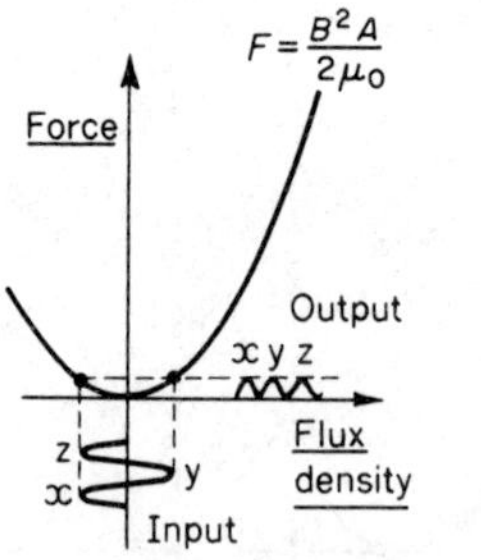

Fig. X3.3 Without polarization.

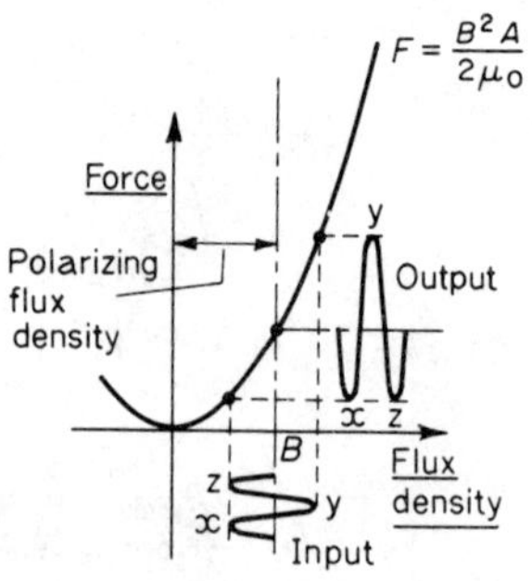

Fig. X3.4 With polarization.

APPENDIX 4

MAXIMUM POWER TRANSFER (MATCHING)

1 D.C. circuit

Consider a battery of e.m.f. E volts and internal resistance R_S Ω connected directly to a load of resistance R_L Ω. The current flowing will be:

$$I = \frac{\mathrm{E}}{R_S + R_L} \tag{1}$$

and the power P delivered to the load R_L will be

$$P = I^2R_L = \frac{E^2R_L}{(R_S + R_L)^2}\,. \tag{2}$$

It can be seen that the power P delivered is zero when R_L is either infinite or zero. At some intermediate value of R_L, therefore, the power must be a maximum. This can be found by plotting a curve of equation (2) or by differentiating P with respect to R_L and equating $\frac{dP}{dR_L}$ to zero.

By either method it is found that *maximum power is transferred when the load resistance equals the internal resistance of the battery.*

2 A.C. circuit

Consider now a generator of e.m.f. E volts (r.m.s.) and internal impedance Z_S connected to a load of impedance Z_L. Then the current is:

$$I = \frac{E}{Z_S + Z_L} = \frac{E}{(R_S + \mathrm{j}X_S) + (R_L + \mathrm{j}X_L)}$$

$$= \frac{E}{(R_S + R_L) + \mathrm{j}(X_S + X_L)}\,. \tag{3}$$

If the load reactance X_L is equal in magnitude but opposite in sign to the generator reactance X_S, equation (3) becomes $I = \frac{E}{(R_S + R_L)}$ which is the same as the d.c. circuit, maximum power transfer occurring when $R_S = R_L$.

Impedances which have equal resistance components and have reactance components equal in magnitude but opposite in sign are called *conjugate impedances.*

Maximum power transfer is obtained in an a.c. circuit when the generator and load have conjugate impedances.

In practice, impedances are matched by means of transformers which permit the magnitude of the load to be varied but not its angle so that a *conjugate match cannot be obtained.* In these circumstances, *maximum power is transferred when the magnitude of Z_L equals the magnitude of Z_S*

$$\text{i.e. when } \sqrt{(R_L{}^2 + X_L{}^2)} = \sqrt{(R_S{}^2 + X_S{}^2)}.$$

This can be shown by substituting the value of I from equation (3) into the equation for the power transferred, $P = I^2R_L$

$$\text{i.e. } P = \frac{E^2R_L}{(R_S + R_L)^2 + (X_S + X_L)^2}\,. \tag{4}$$

If the load impedance angle θ is constant we can now substitute $Z_L \cos\theta$ for R_L and $Z_L \sin\theta$ for X_L in equation (4) giving:

$$P = \frac{E^2 Z_L \cos\theta}{(R_S + Z_L \cos\theta)^2 + (X_S + Z_L \sin\theta)^2} \cdot \quad (5)$$

Equation (5) can then be differentiated with respect to Z_L and $\frac{dP}{dZ_L}$ equated to zero, giving $Z_L = \sqrt{(R_S^2 + X_S^2)}$ as the condition for maximum power transfer.

APPENDIX 5

CIRCUIT DIAGRAM SYMBOLS

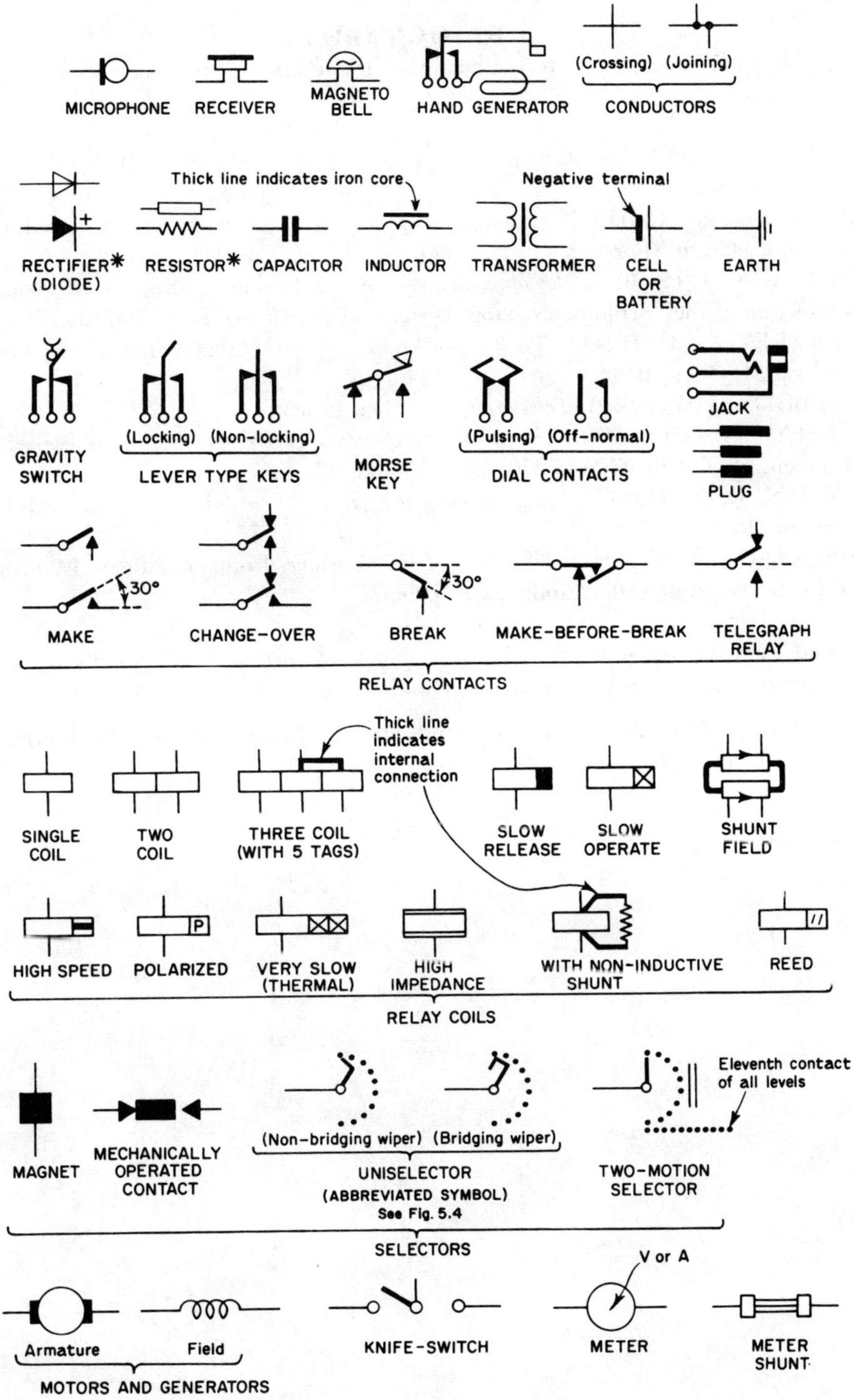

* The upper symbols shown for rectifier and resistors are now the preferred alternatives.

APPENDIX 6

BIBLIOGRAPHY

The following books are recommended for additional reading or additional information:

AGGER, L. T. (1971) *Introduction to Electricity*, Oxford, OUP. (Edited by Sydney F. Smith)

ARMITAGE, M. D. (1961) *Basic principles of telecommunications and electronics*, London, Harrap. (For C & G Principles A.)

ATKINSON, J. (1950) *Telephony*, Vols 1 & 2, London, Pitman. (A detailed exposition of the telephone exchange systems of the British Post Office.)

BERKELEY, G. C. (1949) *Traffic and trunking principles in automatic telephony*, 2nd Ed., London, Benn.

FREEBODY, J. W. (1959) *Telegraphy*, London, Pitman.

HALSEY, S. P. (1968) *A reference book for electrical and telecommunications technicians*, London, OUP. (For C & G 270, 271, 280 and 281 students.)

RENTON, R. N. (1965) *Telecommunication principles in M.K.S. units*, 3rd Ed., London, Pitman.

ROBERTSON, J. H. (1947) *The story of the telephone*, London, Pitman. (A history of the telecommunications industry of Britain.)

In addition, for the purpose of keeping abreast of current developments, students are recommended to read the quarterly *Post Office Electrical Engineers' Journal.*

For drawing symbols, see British Standards Specification: BSS 3939: *Graphical symbols for telecommunications.*

ANSWERS

These are the answers to the questions given in the self-teaching notes at the end of each chapter. The numbers in parentheses like this . . . (**1.1.2**) refer to the section in the text in which the information for this answer is given. You can therefore turn back to the appropriate section if you have the wrong answer or if you are uncertain whether your answer means the same as that given here because obviously you will not always choose exactly the same words as are used here.

Chapter 1 answers (Telephones and teleprinters)

Q.1 Written record, automatic reception or narrow bandwidth. (1.2.1)
Q.2 Easy to use, direct conversation or recognition of talker's voice. (1.2.1)
Q.3 A method of party line operation in which two telephone subscribers are connected to one line. (1.2.3)
Q.4 Telegram in writing at post office, phonogram by telephone to post office. (1.2.4)
Q.5 The a.c. component of the varying d.c. output from a carbon microphone. (1.3.1)
Q.6 To prevent frequency doubling. (1.3.2)
Q.7 Load resistance equal to source resistance. (1.3.4)
Q.8 Induction coil (transformer). (1.3.4)
Q.9 Talker lowers voice, impaired listening ability, or listener hears own room noise louder. (1.3.5)
Q.10 The circuit shown in Fig. 1.8 is the simplest. The diamond-shaped symbol is a good memory aid but does not explain anything by itself. You could otherwise use Fig. 1.18b or 1.20b. In either case show the currents or voltages on transmitted speech. (1.3.5)

Questions 1 to 4 are fairly general. If you have difficulty with them, you may find that Fig. 1.31 helps you to visualize the main features of the systems.

Questions 5 to 10 are concerned with telephone instruments. They are especially important to you if your work is concerned with the telephone system rather than the telegraph system. If you had difficulty with these, you probably do not know enough about a.c. theory and electromagnetism and should refer to a book on **telecommunications principles** and make sure you properly understand such terms as *frequency*, *power*, *induced e.m.f.*, and *transformer*.

Q.11 Start pulse (spacing, positive). (1.4.3)
Q.12 Combination bar. (1.4.4)
Q.13 To restore synchronization in readiness for next character (allow receiver to catch up). (1.4.3)
Q.14 Finger setting pin. (1.4.6)
Q.15 Bellcrank. (1.4.6)

Questions 11 to 15 are concerned with teleprinters. They are especially important to you if your work is concerned with the telegraph rather than with the telephone system.

Now turn back to page 28 (**1.8.4**).

Q.16 First of all, read your answer out loud. No cheating. Read just what's there. Does it make sense?
You should have stated separately and clearly those things which are the same and those things which are different. Similarities include: electrical operation, long distance communication, public network with exchange-switching and connexions by line or radio transmission. Differences include: ease of operation, written record, bandwidth, method of handling messages and automatic reception.
These things are described in **1.1.1** to **1.2.4**.
The main fields in which each is used can be deduced from the relative advantages of each (**1.2.1**). Telegraphy is used in business transactions which require writing; telephony for social and business purposes where rapid personal contact is required.
The advantages of switching are given in **1.2.2**.

Now read the rest of Chapter 1 (**1.5** to **1.7.3**).

Q.17 Figure 1.17a with the dial omitted is perhaps the simplest. Automatic exchanges employ central battery operation, however, so you could have drawn Fig. 1.17a or 1.19a complete with dial.
The best answer to the explanation asked for would be to draw Fig. 1.17b or 1.19b depending on which instrument you have chosen and then describe it as in **1.5.1**. An easier but less satisfactory answer is to draw the balanced transformer of Fig. 1.8 and give the explanation in **1.3.5** with an additional reference to the use of an auto transformer in practical circuits.
For what is gained, see **1.3.5** or the answer to Q.9 above.

Q.18 See Figs. 1.21 and 1.22 and the description in **1.6.1** which also gives limiting values of speed, and reasons.

Q.19 The selecting levers are shown in Fig. 1.14c and their operation described in **1.4.4**.

Now read Chapter 2 (**2.1.1** to **2.4**).

Chapter 2 answers (Relays)

Q.1 Armature, yoke and core. (**2.2.1**)

Q.2 A 'make' contact. (**2.2.2**)

Q.3 10 mA. $\left(\dfrac{100 \text{ ampere-turns}}{10\,000 \text{ turns}} = \dfrac{1}{100}\text{A} = \dfrac{1\,000}{100}\text{ mA} = 10\text{ mA}\right)$ (**2.2.3**)

Q.4 Relay *B* has the higher value of operate current. (**2.2.3**)

Q.5 Relay *B* has the longer operate lag. (**2.2.4**)

Q.6 To make the relay slow to operate and slow to release. (**2.2.5**)

Q.7 Connect a resistor or a rectifier in parallel with the relay as shown in Fig. 2.9 or release it on short circuit instead of disconnexion. (**2.2.6**)

Q.8 For pulsing, usually for inverting dial pulses in a selector circuit. (**2.2.7**)

Q.9 Use nickel-iron sleeves. (2.2.8)

Q.10 It still does not operate, the two are still opposing one another. The relay can only be made to operate by reversing one current without reversing the other current. (2.2.9)

Q.11 Because pulsing speeds are higher in telegraphy than in telephony. (2.3.1)

Q.12 Any two of the following:
Polarization for double current signals (reduced distortion).
Increased sensitivity (therefore faster operation).
Reduced power consumption (compared with using a polarizing winding as in a shunt-field relay). (2.3.2)

Q.13 The tongue moves to the other contact. (2.3.2)

Q.14 The tongue does not move. It stays on the contact to which the current last operated it, but with reduced contact pressure. (2.3.2)

Q.15 Because nickel-iron has a higher resistivity than soft-iron and therefore reduces eddy currents giving faster operation. (2.2.4 and 2.3.1)

If you had difficulty with questions 1 to 10, it is likely that you found most of this chapter hard to follow because you do not know enough about electro-magnetism. The general theory of such things has been dealt with very briefly here and is really only intended for those who already have some idea of it. For a more detailed treatment you must refer to a book on **telecommunications principles** or **electrical principles**. Make sure that you find out about *ampere-turns, flux, flux density, permeability and reluctance.*

The same difficulties would also probably apply to questions 11 to 15. For these, however, some understanding of *permanent magnets* is also necessary.

A brief study of the practical relays described in Part 3 of the chapter may help with the questions relating to the operation of the relays, particularly those described in **2.5.1** and **2.6.1**.

Q.16 There are two contacts on relay *A* so we write 2, making the designation $\frac{A}{2}$. (2.4)

Q.17 Relays *A*, *B* and *Z* are slow release. *A* is shunted by a rectifier, *B* has a heel-end slug and *Z* has a front-end slug. (2.2.5 and 2.2.6)

Q.18 Contact *A*1 operates *B*. (2.4)

Q.19 Contact *Y*1 releases *B*. (*Z*1 operates *Y*, *Z* having been operated by *A*2 . (2.4)

Q.20 None. The switch opening releases *A*, *A*2 releases *Z*, *Z*1 releases *Y* and *B* has already been released by *Y*1 as in Q.19. (2.4)

The purpose of questions 16 to 20 was to give you some practice in recognizing circuit symbols and in working out the sequence of operations in a circuit using relays. It doesn't really matter if you found this difficult: it is the sort of thing which comes with practice and you will meet other examples later. Chapter 3, for example, is almost entirely concerned with relay circuits.

Now turn back to page 57 (2.9.3).

Q.21 Any telegraph relay would do for part (*a*).
Part (*b*) should be answered with a B.P.O. 3000 type relay having an isthmus armature and usually also having nickel-iron sleeves. High speed relays are also used for this purpose in some circuits.

Your diagrams should not include irrelevant details of practical construction and contacts, but should indicate clearly the path of the magnetic flux as well as the components of the iron circuit.

Suitable diagrams are the telephone relay in Fig. 2.27 with a sketch of an isthmus armature in addition as in Fig. 2.11. Diagrams of the telegraph relay are shown either in Fig. 2.22 or 2.24.

Q.22 In order to leave enough time for the rest of the question, neither your sketch nor your description should be too detailed. Your description should mention the soft-iron armature, core and yoke, the contact springs and the buffer block and the fact that the slug is made of copper. Your sketch could be like Fig. 2.18 with the addition of the slug as shown in Figs 2.8b and 2.19. It should be clear from either your description or sketch that the slug is round the core not round the outside of the coil.

The action of the slug is given in **2.2.5** and the various methods of making relays slow to release without slugs in **2.2.6**.

Q.23 Sketch as Fig. 2.27.

Residual gap as **2.2.1**.

Enlarged pole face reduces air-gap reluctance and increases the tractive effort (2.2.3).

Now read Chapter 3 (**3.1.1** to **3.5.3**).

Chapter 3 answers (Signalling)

Q.1 Subscriber calling; exchange calling; subscriber clearing. (3.2.1)

Q.2 Dialling; answer. (3.2.1)

Q.3 Calling; clearing. (3.2.2)

Q.4 No operator to give information verbally. (3.2.1)

Q.5 Groups of character signals. (3.2.3)

Q.6 Dialling—control of selectors. (3.3.4)

If you had difficulty with Q.1 to Q.6, you may find it helps to have a brief look through some of the practical systems which provide these signals. (3.5.1 to 3.6.2)

Q.7 Single-wire. (3.3.1)

Q.8 Earth potential difference. (3.3.1)

Q.9 Unaffected by e.p.d.; induced e.m.f.s cancel; interference currents cancel if circuit balanced. (3.3.3)

Q.10 Polarized relay has equal lags; line capacitance charges and discharges at the same rate. (3.3.6)

Q.11 Double-current. (3.3.6)

If Q.7 to Q.11 gave you any trouble, your difficulty may be that you are not familiar enough with d.c. theory. In this case you should refer to a book on **telecommunications principles** and make sure you properly understand such terms as *potential*, *potential difference*, *e.m.f.*, *capacitance*, *resistance*, *charge*, and *discharge*.

Q.12 Inductance. (1.3.6 and 2.2.4)

Q.13 Relay coil. (3.3.7)

Q.14 Between elements —one dot;
between letters —three dots (one dash);
between words —seven dots. (3.4.1)

Q.15 Seven (including stop and start). **(3.4.2)**

If necessary, you should also make sure you are familiar with simple a.c. theory and understand such terms as *impedance*, *transformer*, *inductance* and *induced*.

Now read the rest of Chapter 3 (**3.5.1** to **3.7.2**).

Q.16 For d.c. signalling, see **3.6.1** and Fig. 3.17. Your description should include signals for calling, dialling, answering, clear back, and clear forward. If the exchanges were 300 miles apart, either the circuit would be routed via carrier or radio equipment providing no d.c. path, or it would be via an amplified audio circuit of very high resistance over which relay *A* (Fig. 3.16) would not receive enough current to operate. In these circumstances voice frequency (VF) Signalling would be used. **(3.3.9, 3.6.1, and 3.6.2)**

Q.17 (*a*) Change *S* wire from spacing to marking;
(*b*) double-current pulses: 60 ms spacing/40 ms marking;
(*c*) five-unit code: start—20 ms spacing,
code elements—20 ms spacing/marking;
stop—30 ms marking;
(d) change from marking to spacing for at least 375 ms.
(Spacing = +80 V, marking = −80 V).
For timing chart, see Fig. A3.1.

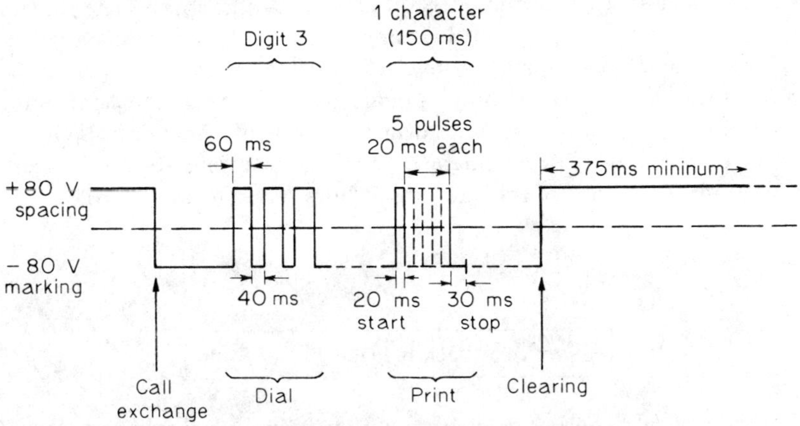

Fig. A3.1 Timing chart for Q.23. **(3.5.2)**

Q.18 See **3.4.2** for description. Although the question does not actually ask for a diagram, your description will be clearer if you include a suitable illustration in your answer. This should be similar to Fig. 3.12 but need only show one character (i.e. one letter). With a 30 ms 'stop' element, each character is 150 ms in length so the number per min is $60 \times \frac{1\,000}{150} = 400$.

(If a 20 ms stop is used, there are 428 characters/min).

Now read Chapter 4 (**4.1.1** to **4.3**).

Chapter 4 answers (Switching systems).

Q.1 Thirty-three crosspoints.
10 subs × 3 links = 30,
add 3 crosspoints for operator = 33. (4.2.2)

Q.2 Three simultaneous calls (one via each link). (4.2.2)

Q.3 Six plugs (two for each of the cord circuits which replace the three links of the co-ordinate switchboard). (4.2.3)

Q.4 The size of the jacks (operators' fingers) and the size of the jackfield (operators' arms). (4.2.3)

Q.5 Any three of these: reduced cost per call, faster setting up, immediate release, uniform service, privacy, fewer staff. (4.3.1)

Q.6 To provide the appropriate instruction signals to the switches to operate the crosspoint in the required path. (4.3.5)

Q.7 So that a finger at an operated crosspoint can bend to permit the selecting bar to operate another finger without the first one becoming dislodged. (4.3.6)

Q.8 Nine (see Fig. 4.13b). (4.3.7)

Q.9 Eight (4 + 4). (4.3.7)

Q.10 100 (ten outlets of each of ten switches to twenty inlets of each of five switches). See Fig. 4.11. (4.3.7)

Q.11 If two free trunks cannot be connected together because all suitable links are engaged, the call is said to be blocked. (4.3.8)

Q.12 Because they are taken into use on each call only while this is being set up. (4.3.9)

If you had difficulty with the numerical questions, read the appropriate paragraphs of the chapter again with this in mind. It is important that you appreciate the relationships between the various quantities involved and the choice of switching methods. Remember that an exchange is only complicated because of the large number of different connexions it has to be able to make. You will find much of the detail of any exchange system, whether automatic or manual, easier to understand once you can see the need for it in terms of the numbers of lines or circuits involved.

Now turn back to page 114 (4.8.3).

Q.13 A *switch matrix* consists of an array of switching devices in which these devices are provided at the intersections of the 'vertical' and 'horizontal' circuits (co-ordinates). (4.2.2)
This part of the question does not ask for sketches but a portion of Fig. 4.2 or Fig. 4.9a could be used to illustrate your explanation especially if you have difficulty finding the right words.
To explain how to provide a complete exchange you need to explain link trunking applied to *group selection* (Fig. 4.5b or Fig. 4.6) and *concentration* (Fig. 4.7) and then show how to combine these (Fig. 4.8). (4.3.7)

Q.14 The simplest approach would be to use the simple sketches of Fig. 4.11 or Fig. 4.12 and the associated description. (4.3.6, 4.4)

Now read Chapter 5 (5.1.1 to 5.3.4).

Chapter 5 answers (Strowger mechanisms)

Q.1 Ten levels. (4.3.3)
Q.2 The pawl. (4.3.3)
Q.3 To prevent the ratchet from stepping backwards when the armature *operates*. (4.3.3)
Q.4 In series with the magnet (Fig. 5.1). (5.2.1)
Q.5 Three ranks, first, second and final. (4.3.4)
Q.6 10 000 numbers.
(If you answered 9999, remember 0000 is also a possible four-digit number.) (4.3.4)
Q.7 Hunter: call*ed* circuits on bank, call*ing* circuits connected at wipers.
Finder: call*ing* on bank, call*ed* on wiper. (4.3.4)
Q.8 This would mean giving the selector ten selectors in the next rank all to itself. It could only use one of them at a time so there would always be at least nine standing idle.
(It would not do to multiple this one selector on each level because then it would not be possible to switch more than one call at a time to a given group in the next rank). (4.3.4)
Q.9 The last contact carries the fewest calls—because it is only used on those occasions when all the other contacts are also in use. (4.3.4)
Q.10 Can use a simple type of interrupter contact. (5.2.1)
Q.11 A bridging wiper. (5.2.2)
Q.12 Thirty teeth. The pitch of the teeth is the same as the pitch of the bank contacts but they must fill 360° instead of only 120°. (5.2.2)
Q.13 The pawl of a reverse-acting mechanism is in the ratchet tooth when at normal so that it (*a*) causes friction on one ratchet to interfere with the action of the other ratchet, and (*b*) must be disengaged from the ratchet on release. (5.3.1)
Q.14 Vertical pawl guide. (5.3.1)
Q.15 To hold the pawl in the ratchet tooth during stepping. (5.2.1 and 5.3.1)
Q.16 A toggle type of interrupter contact. (5.3.2)
Q.17 Two vertical rows of eleven contacts each, a total of twenty-two contacts. (5.3.3)
Q.18 Because the bank contacts are in two rows on each level with a separate wiper for each row. The two wipers are therefore connected to different points in the circuit. (5.3.3)
(In a uniselector, the two wipers make contact with opposite sides of one row of contacts as shown in Fig. 5.11 and have no clearance.)
Q.19 The *NR* springs (rotary off-normal contacts). (5.3.4)

If you got more than one question wrong out of the five questions concerned with trunking (i.e. Q.6 to Q.9), read **4.3.4** again.

The other fifteen questions (Q.1 to Q.4 and Q.10 to Q.19) are concerned with mechanisms. Any difficulties here will probably be cleared up by reference to the practical selectors in the next part of the chapter. (5.4)

Now turn back to page 138 (5.7.3)

Q.20 As the question asks for a description of a mechanism rather than for an explanation of a principle, your answer should be related to specific practical mechanisms. For (*a*) either the vertical or the rotary mechanism of any type of two-motion selector would do, e.g. Fig. 5.9, 5.16 or 5.20. The description should name and state the functions of all the components shown in Fig. 5.9 and should include a description of the operation (**4.3.3** and **5.3.1**).

For (*b*), describe a uniselector mechanism (Fig. 5.2; **4.3.3** and **5.2.1**).

Make sure in each case that your sketch has the right shape ratchet teeth, that there is room for the reverse-acting pawl to move without hitting the detent and that the position of the stops and guides is such that your pawls would move in the right direction and for the right distance to step the ratchet one tooth.

Q.21 The springsets asked for are the *N*, *NP* and *NR* springs. The 4000 type selector has the easiest ones to draw but you could do any other type of selector if you know it better.

You could either draw each separately as in Fig. 5.12 or together as in Fig. 5.21 (or Fig. 5.18).

Descriptions are given in **5.3.4**, **5.4.3** and **5.4.4**.

Other mechanical springsets are *S* (eleventh contact) and interrupter contacts.

Q.22 Use Fig. 5.11 and **5.3.3**.

The 220 points referred to are the eleven contacts per row arranged in two rows for each of the ten levels.

The pitch of the teeth and the contacts are the same, i.e. the *distance* between vertical teeth is equal to the distance between levels and the *angle* between rotary teeth is equal to the angle between contacts round the level.

Now read Chapter 6 (**6.1.1** to **6.5**).

Chapter 6 answers (Strowger circuits)

Q.1 Loop. (**3.5.1** and **6.2**)

Q.2 Disconnexion. (**3.5.1** and **6.2**)

Q.3 *LR* is not connected during speech or tones, *A* is used as a tone transformer. (**3.3.3** and **6.2**)

Q.4 Final selector. (**4.3.4** and **6.2**)

If you had difficulty with Q.1 to Q.4, you need to revise the earlier work (see **6.8.1**).

Q.5 Because it has a release lag much longer than a break pulse. (**6.3.1**)

Q.6 Any two of the following: more winding space, faster flux rise, reduced sparking. (**6.3.1** and **6.3.3**)

Q.7 The selector *N* contact or a relay operated by *N*. (**6.3.3**)

Q.8 At the end of the dial pulse train (digit). (**6.3.2**)

Q.9 Relay *E* must switch the pulsing circuit from the vertical to the rotary magnet and also re-operate relay *CD*.

If you had difficulty with Q.5 to Q.9, you may need to revise either relays, especially slow releasing relays, or signalling principles.

Q.10 Outlet testing in which the busy outlets are indicated by an earth condition (on the *P*-wire) and the free outlets are indicated by an absence of this earth marking. (**6.4.1**)

Q.11 Because it is short circuited, having earth at one coil tag from the *LR* contact and at the other tag from the bank contact via the *P* wiper. (**6.4.2**)

Q.12 Because they use forward-acting mechanisms, operating the magnet via the wiper would cause the wiper to step and disconnect the magnet. (**6.4.2** and **6.4.3**)

Q.13 It can be made to release faster on a free outlet than the alternative type of testing relay could be made to operate on a free outlet. (**6.4.3**)

Q.14 Double connexions can be prevented by suitable design of the testing relay. Circuit need not be unguarded during release either. (**6.4.5**)

Q.15 The final selector wipers have finished stepping before outlet testing commences. In the group selector, the testing relay has to stop the wipers stepping on when it detects a free outlet. (**6.4.6**)

You cannot expect to commit all the circuit elements in this chapter to memory. You will find it useful, however, to be able to recognize certain of them when you see them as part of complete circuits.

The more important ones in this respect are Figs 6.5b, 6.10, 6.11 and 6.13. Learn to draw these from memory.

Now turn back to page 171 (**6.8.3**).

Q.16 You could use either Fig. 6.11 or the equivalent pre-2000 type circuit but unless you are particularly familiar with the older circuit through your work, there seems to be nothing to recommend answering this question with it in preference to the modern circuit (Fig. 6.11), the circuit operation of which is described in **6.4.3**.

Battery testing is dealt with in **6.4.5**.

Q.17 The circuit element is as Fig. 6.5 (preferably b). State that the relays must be slow to release in order to remain operated under pulsing conditions. State the pulsing conditions for each relay and explain the function of each relay (as **6.3.3**, second paragraph) in order to justify the need to hold during pulsing which is the ultimate reason for the lags.

Q.18 Suitable elements are those of Figs 6.10 and 6.13, or if you can remember that much detail, Fig. 6.14. These are described in **6.4.2** and **6.4.5**. Notice that 'describe' as used in this question can only mean 'describe the operation of'. No bridging wiper (**6.4.2**) is required in the battery testing case because the disconnexions as the wiper moves from one contact to the next cannot be misinterpreted as 'free' markings as they can on an absence-of-earth testing system.

Q.19 A good answer would give the whole of the circuit element shown in Fig. 6.7. The question can be answered more simply, however, if you notice that it asks only for rotary stepping not for vertical *and* rotary stepping. This means that your circuit element could be confined to the conditions existing in the circuit *after* relay *E* has operated. This would be effectively the vertical stepping circuit of Fig. 6.5b but with an *NR* contact in place of *N*1, *RM* in place of *VM* the addition of *MR*1 in series with *R*1 and with *R*1 connected directly to battery. You would not be far out of course if you only substituted *NR* for *N* and *RM* for *VM* in Fig. 6.5b.

A group selector has a different element because it must hunt for a free outlet (see **4.3.4** and Chapter 5 Q.8 answer). (The different element referred to here is, of course, the outlet testing element Fig. 6.11.)

Now read Chapter 7 (**7.1.1** to **7.3.3**).

Chapter 7 answers (Traffic and trunking)

Q.1 The traffic flow or intensity in erlangs for a specified period, usually the busy hour, is numerically equal to the average number of simultaneous calls in that period (or circuits simultaneously engaged, if your definition refers to a group of circuits).
Or:
1 erlang is a traffic flow (or intensity) equivalent to one circuit continuously occupied. (7.2.1 and 7.2.3)
Or:
1 erlang is a traffic flow of 1 call-hour per hour (or 1 call-minute per minute).

Q.2 13 erlang. (Don't forget to state the units in your answer.)
$$\frac{10 + 15 + 16 + 11}{4} = 13.$$ (7.2.1)

Q.3 5 erlang.
Using $A = Ct$, $A = 100 \times 30/60 = 5$,
(notice that t must be divided by 60 to express it as a fraction of the busy hour).
Otherwise, use $TA = Ct$, so that $60 \times A = 100 \times 30$
and again we have $A = 100 \times 30/60 = 5$. (7.2.1)

Q.4 Lunch (people do not usually make telephone calls while eating meals!). (7.2.2)

Q.5 45 min (0·75 h). (7.2.3)

Q.6 100 erlang.
Number of calls = 2 000 × 1·5 = 3 000 calls.
$A = Ct = 3\,000 \times 2/60 = 100$ erlang. (7.2.3 and 7.3.3)

Q.7 0·75. (7.2.3)

Q.8 Two calls lost.
400 × 0·005 = 2. (7.3.1)

Q.9 4 erlang.
Calls carried = calls offered − calls lost = 80 calls.
$\therefore A = 80 \times 3/60 = 4$ erlang. (7.3.1)

Q.10 Reduced capacity for overload (greater deterioration in g.o.s. for a given percentage overload). (7.3.2)

You may have made mistakes in your arithmetic in some of your answers and so arrived at the wrong result. Providing you are quite sure that the method you used was otherwise correct, count any such answers as being right: the questions here are a test of telephony or telegraphy, not mathematics.

If you found it difficult to extract the information for the calculations from the wording of the questions, go back to Chapter 5 and make sure you understand terms like *selector* and *group selector*, and to Chapters 5 and 6 for the general idea of selectors connected in groups with outlet testing.

Now turn back to page 203 (7.9.3).

Q.11 When the availability of the selector level (number of outlets) is less than the number of trunks provided. (7.4.1)

Q.12 Levels 2 and 5.
These are the levels with limited availability. (7.4.1, 7.4.2 and 7.4.3)

Q.13 As Fig. 7.14b (three individuals, four pairs). (7.4.3)

Q.14 Four groups.
From $g = 2n/a = 38/10 = 3{\cdot}8$ or from the table of modified values for small numbers of groups in Fig. 7.25. (7.5.1)

Q.15 (*c*) is the best grading.
The sums of successive differences for the three gradings are 17, 8 and 1 respectively. **(7.5.3** example, also **7.5.2)**

Q.16 Three commons.
$a + b + c =$ availability, $c = 10 - (3 + 4) = 3.$ **(7.5.3)**

Q.17 Thirty-five trunks.
$6a + 3b + 2c + d = 18 + 12 + 4 + 1 = 35.$ **(7.5.4)**

Q.18 Four pairs.
$a = 3, \ a + b + c = 10$ 7.24.
$4a + 2b + \quad = 23$
$3a + b \quad = 13, \ b = 13 - 3a = 13 - 9 = 4$ trunks. **(7.5.3)**

Disregard arithmetical errors as before. (If you have a large number of such errors you must do something about it. The usual cause is untidy work; even the simplest calculation should be written out properly.)

If you had difficulty with the wording of Q.11 to Q.18, go back to Chapters 5 and 6 and make sure you understand terms like *availability*, *level* and *multiple*.

Now turn back to page 204 (7.9.5).

Q.19 For definitions, see **7.2.2**, **7.2.3** and **7.3.1** respectively.
Traffic offered $= 100 \times 3/60 = 5$ erlang.
Calls carried $= 100 - 1 = 99.$
Traffic carried $= 99 \times 3/60 = 4{\cdot}95$ erlang.
(NOTE: 5 erlang is *not* correct here).

Q.20 First design a four-group, ten contact, twenty trunk grading:

$$4a + 2b + c = 20$$
$$a + b + c = 10 \quad \therefore c = 10 - (a + b),$$

Subtract: $3a + b = 10 \quad \therefore b = 10 - 3a.$

	a		$b = 10 - 3a$		$c = 10 - (a + b)$		S
	1	6	7	5	2	+ =	11
Best grading	2	2	4	0	4	+ =	2
	3	2	1	5	6	+ =	7

Since the question does not ask for the best grading, you could use any of these three gradings.

Your sketch would then be based on Fig. 7.18 as shown in Fig. A7.1.

Q.21 Graph as Fig. 7.4.
Explanation of busy hour as **7.1.1** and **7.2.2**. This does not need to be very long or there will not be enough time for the rest of the question. Say what the busy hour is and that equipment sufficient for this traffic will be more than adequate for the rest of the time.
Number of calls $= 2\,000 \times 0{\cdot}5 = 1\,000$ calls in busy hour.
Traffic $A = Ct = 1\,000 \times 3/60 = 50$ erlang.
Calls lost $=$ calls offered $\times 0{\cdot}002$
$= 1\,000 \times 0{\cdot}002 = 2$ calls lost.

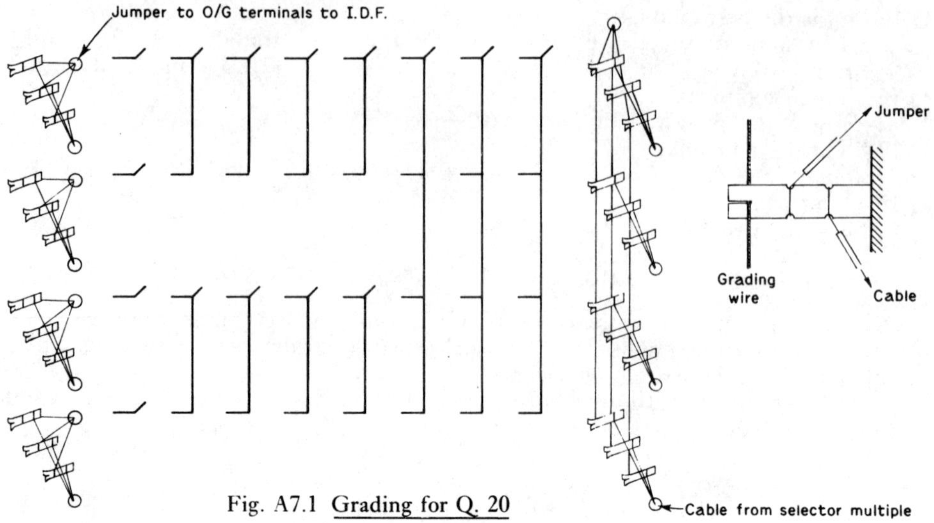

Fig. A7.1 Grading for Q. 20

Q.22

$$4a + 2b + c = 18$$
$$a + b + c = 10 \quad \therefore c = 10 - (a + b),$$

Subtract: $3a + b = 8 \quad \therefore b = 8 - 3a.$

	a	$b = 8 - 3a$	$c = 10 - (a + b)$	S
Best grading	1	5	4	
	4	+	1	= 5
	2	2	6	
	0	+	4	= 4

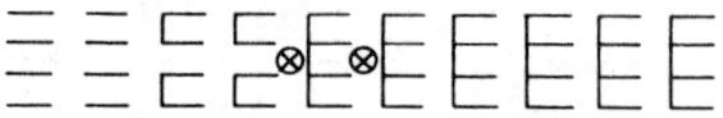

To convert this to a twenty trunk grading, cut the vertical commons at the points marked ⊗ and connect two additional jumpers at the new 'pairs' so formed. This twenty trunk grading is actually the one we designed for the answer to question Q.20.

Now read Chapter 8 (**8.1.1** to **8.4**).

Chapter 8 answers (Distribution frames)

Q.1 Local side and multiple side. (8.2)

Q.2 M.D.F. exchange side to I.D.F. multiple side. (8.2)

Q.3 At least one extra wire per line (*P*- or *S*-wire) would be required between M.D.F. and I.D.F. if the multiple and meters were connected to the M.D.F. (8.2)

Q.4 Change the jumpering at the I.D.F. to distribute the high calling rate lines on to other selector groups by re-allocation of the line circuits. The high calling rate lines can be identified by observation of the subscribers' meters. Lower calling rate lines would be jumpered to the line circuits freed by this change and having access to the previously overloaded selectors.

Q.5 Local side (via grading terminals on the racks). (8.2)

Q.6 Line side.
(This can be remembered as HORIZON:LINE.) (8.3)

Q.7 M.D.F. line side. (8.4)
Q.8 Overhead lines (more than three spans). (8.4)
Q.9 M.D.F. exchange side. (8.4)

Now turn back to page 219 (8.9.2).

Q.10 Both are shown in Fig. 8.1.
For details of protective devices, see **8.4**.
It is not necessary to draw these devices or describe their construction. It *is* necessary to state their function in order to answer the last part of the question.
Q.11 M.D.F. —for terminating all lines (8.1.1)
cross-connecting into numerical order (8.2)
mounting protective apparatus (8.4)
and providing access to lines for testing (8.4)
I.D.F. —for terminating the final selector multiple and meters, and cross-connecting the lines to the line relays to provide an even distribution of traffic.

The fittings referred to are the jumper rings (Fig. 8.11) and the orderly system of jumpering is the use of vertical and horizontal jumpering on the two sides of a frame as described in **8.3**.

An I.D.F. jumper has additional wires for the meter and *P*-wire connexions. The M.D.F. connects lines only and therefore does not have to provide jumper conductors for any of the control wires needed between selectors (Fig. 8.1).

Now read Chapter 9 (**9.1.1** to **9.4.2**).

Chapter 9 answers (Power plant)

Q.1 46 V to 52 V negative. (9.1.1)
Q.2 The watt (1 joule per second). (9.2.1)
Q.3 25 W.
From Ohm's law, $I = V/R = 50/100 = \frac{1}{2}$ A.
Power $= VI = 50 \times \frac{1}{2} = 25$ W.
Or, from power $= V^2/R = 50^2/100 = 25$. (9.2.1)
Q.4 The heavy load it would put on one phase of the three-phase mains would unbalance the supply. (9.2.3)
Q.5 When stationary (i.e. when starting). (9.2.3)
Q.6 A commutator. (9.2.4)
Q.7 Because its output voltage is more or less constnt under varying load conditions. (9.2.4)
Q.8 So that it will restore when the electromagnet in series with the field winding is de-energized. This puts the starting resistance back into circuit in readiness for starting the motor next time. (9.2.4)
Q.9 Six rectifiers: see Fig. 9.11. (9.2.5)
Q.10 If the charging current is maintained after the cell is fully charged. (9.3.1)
Q.11 Any two of the following:
reduced energy conversion losses, reduced wear on the battery, reduced gassing or full battery capacity available in emergency. (9.3.2)

If you had difficulty with these first eleven questions it may be that your knowledge of some basic electrical theory is inadequate. If this is so, you should consult a book on **telecommunications principles** or **electrical principles** for such things as *SI units, d.c. circuit calculations, induced e.m.f.s, motor principle* and *secondary cells.*

Q.12 By providing low resistance ammeter shunts wired permanently in the circuit so that the ammeter can be connected in parallel with any of these shunts when required. **(9.4.1)**

Q.13 To prevent the battery from discharging through the generator if the prime-mover stops (or the mains fail). **(9.4.1)**

Q.14 An ampere-hour meter. **(9.4.1)**

Q.15 An objectionable 'hum' on all the telephone circuits. **(9.4.2)**

Q.16 At the field winding—by means of a regulator providing a variable series resistance. **(9.4.2)**

Q.17 Because the e.m.f. depends on having sufficient current through the cell. **(9.4.2)**

Now turn back to page 247 (**9.7.3**).

Q.18 A good answer would give circuit diagrams like Fig. 9.22 (without the ammeter or voltmeter connexions) and Fig. 9.24. The question does not specify whether full or partial float is required, so you could have used Fig. 9.23 in place of Fig. 9.24.

The essential information to support an explanation is given more simply by Fig. 9.13, however. The explanation is given in **9.2.3**. You should refer to the advantage of a float system because this is an essential part of the difference between the two systems.

The items necessary could refer specifically to the examples you have given if your circuits show this much detail and could include such things as c.e.m.f. cells, contact voltmeter, trickle charge resistor and contactors if a partial float system. For the full float system there are no c.e.m.f. cells but there is a trickle charge supply from the rectifier, variable inductors in the mains supply connexions etc. Other components could be selected if you have chosen some system not described here in detail.

About the only components common to all float battery systems and not used on any charge-discharge systems are the filter and contact voltmeter. Some form of voltage control and trickle charge is also provided, but it varies from one float system to another.

Q.19 For d.c. generator, see **9.2.4**.

For protective equipment, see circuit breaker and overload relay, **9.4.1**.

First check that the generator output voltage is slightly more than the voltage of the point to which it is to be connected. Adjust its voltage if necessary. Too low a voltage would cause the reverse-current circuit breaker to operate; too high a voltage would cause the overload relay to operate when the generator is connected. **(9.5.1)**

Now that you have finished the whole book, you are certain to have forgotten much of what you learned in the earlier chapters at least.

If you are taking an examination, you will need to do some revision.

Whatever your next step, examination or not, you will want at some time in the future to refer to specific items you recall having seen in this book.

In either of these two cases, the notes entitled 'How to use this book' have been designed to help you. You will find these notes on page xiii.

INDEX

absence-of-earth testing, 150
ABS signal, 72
a.c. machines, 223
a.c. signalling, 61
AC9 signalling system, 75
air-gap, 30, 31
alarms, 136, 242
algebraic method, grading, 191
alkaline cell, 239
alternator, 223
ampere-hour meter, 236
ampere turns, 33, 34
answer jack, 84
  plug and cord, 83
  answer signal, 58, 70, 74, 75
anti-chatter, 50, 52
anti-sidetone induction coil (A.S.T.I.C.), 9, 18, 20
A relay, 70, 141
armature, 30
  end slug, 37
  isthmus, 39, 46
  rocking, 19, 20
  stops, 117, 122
  travel, 31
auto-manual, 80, 84
automatic (auto.) exchange, 3, 70, 80, 85
  sub's line signalling, 70
automatic
  power plant, 242
  telegraph message switching, 17
  Telex system, 72
  transmitter, 15
auto-transformer, 9
auxiliary armature, 131
availability, 86, 181, 183
A-wire, 62

backward holding, 150
backward signal, 60
bakelite (plastic) front check, 36, 47
balanced
  circuit, 62
  tone, 65
  transformer, 9
balance network, 9, 20
ballast resistor (barretter), 168
bandwidth, 2
bank contact, 86, 117, 123
bank multiple, 90, 91, 194
battery
  storage, 232
  testing (outlets), 149, 155
baud, 50, 59
bell, 6
bell-crank, 18
best grading, 189
biassed relay, 43
bimotional, 86
blocking, 97
bothway junction, 59
bounce, 47, 124
braking magnet, 236
break
  contact, 31
  pulses, 63, 70
B relay, 37, 69, 143, 146
bridge control, 85
bridge, crossbar, 104
bridging wiper, 118, 125, 152
buffer block, 32
buffer spring, 32
busy hour, 177
  calling rate, 182
busy tone, 58, 65, 70, 141
B-wire, 62

cabling, 195, 205
call-connected signal, 72
called subscriber held (C.S.H.), 168
calling rate, 90, 182
call packing, 98
call tracing, 199
cam, 12
  plate, 131
capacity of battery, 241
carbon granule microphone, 5
carpenter relay, 51
case-shift, 68
CD relay, 80, 148, 156
central battery (C.B.), 10, 18
centralized trunk distribution, 194
central office, 3
centre-stable relay, 43
change-over contact, 31
character, 59
charge, 232
charge-discharge, 234
circuit breaker, 236
clear back, 59, 75
clear forward, 59, 75
clutch, 12, 13
code selectors, 100
coil retardation, 47
comb, 18
  plate, 121, 131
  relay, 32
  setting fingers, 18
combination bars, 14
common battery, 10
common control, 92, 98

concentration, 89, 96
concentrator, 81
connexion strip, 197, 215
conservation of energy, 223
contact, 11, 31
  material, 32, 47, 50
  pressure, 32, 125
  tongue, 14, 50
  voltmeter, 239
  wetting, 33
co-ordinate switching, 81, 92
cordless switchboard, 82
counter-e.m.f. (c.e.m.f.) cell, 239
crossbar, 80, 92, 101, 104, 105
crosspoint, 81, 92

day calling rate, 182
day to busy hour calling rate ratio, 183
d.c. machines, 226
d.c. signalling, 61
de-centralized trunk distribution, 194
delayed action fuse, 210
DER signal, 72
detached contact diagram, 44
detent, 87
dial, 12, 21
  pulse, 13, 63
  speed and ratio, 23, 63
  tone, 141, 163
diaphragm, 5, 20
differential, 66
digit absorption, 121
diode, 42, 228
direct control, 99
director, 100
discharge, 232
distortion, 39, 63
distribution frames, 205
distribution, traffic, 95
district switching centre (D.S.C.), 4
divided battery, 244
double
  connexion, 155
  current signals, 65
  -ended wipers, 119
D relay, 71, 74
dry reed relay, 53, 82
dual testing, 154
dynamometer regulator, 239

earth
  potential difference (e.p.d.), 61
  return, 61
  testing, 150
eddy current, 36, 38, 40, 130, 236
efficiency, traffic, 181, 185
electromagnet of teleprinter, 17
electronic exchange, 80, 82, 98, 107, 108
electronic teleprinter, 13
eleventh step, 126
Ericsson crossbar, 102
energy, 221
equipment I.D.F., 208
E relay, 150, 170
erlang, 174, 178
exchange side (M.D.F.), 205, 209

expansion, 97

fanning strip, 195, 215
feeder spring, 119
field winding, 224
figure case and shift, 68
filter, 237
final selector, 88
  circuit, 147, 165
  outlet testing, 158
finder, 89
finger
  plate, 13, 21
  setting pin, 18
  stop, 13, 21
first selector, 88, 165
five-unit code, 14, 68, 250
fleeting test, 159
flexible finger, 92, 103
floated battery, 234, 237, 242, 244
flux in relay, 30, 34, 143
folded network, 97
forward
  -acting mechanism, 87, 88, 122
  holding, 150
  signal, 60
four-wire trunk, 75
F relay, 69, 167
frequency, 2, 6
frequency doubling, 6, 256
front (fore)-end slug, 37
full
  availability, 181, 183
  float, 239, 244
  wave rectifier, 231
fuses, line (M.D.F.), 210, 212

gassing, 232
go and return, 75
graded groups, 188
grade of service, 180
grading, 91, 173, 186
  chart, 188, 199
  design of, 188
  terminals, 194
gravity switch, 12, 19
G relay, 123, 152, 167
group
  selector, 91
    circuit, 143, 145, 163
    outlet testing, 152, 164
    rack and grading, 136, 197
    switching centre (G.S.C.), 4

HA and *HB* relays, 154
half-wave rectifier, 231
heat coil, 210
heel-end slug, 37
hertz (Hz), 2
hold
  bar, 93
  current, 34
holding, *P*-wire, 150
holding time, 176
home contact, 161
homing arc, 126, 161

horizontal
  bar, 93
  side, 202
H relay, 141, 150, 152, 158, 161
hum, 237
hunters, hunting, 89, 141
hybrid coil, 9, 75
hysteresis, 40

immediate release (*IR*), 72
impedance, 7, 40
index wheel, 127
indirect control, 100
inductance of relay, 34, 40
induction coil, 6, 9, 18
induction motor, 225
interacting relay, 123, 152
interaid, 98
interception, 210
interconnecting, 185
inter digital pause (I.D.P.), 23, 63
interference, 61
intermediate distribution frame (I.D.F.), 205, 215
interrupter contact, 12, 116, 122, 131
IR signal, 72
isthmus armature, 39, 46

jack, 82
joule, 221
J relay, 160, 167
jumper, 195, 202, 217
  ring, 195, 202
junction, 4
  signalling, 59, 74
junctor, 105

K-action contact, 31
key, 81
keybar, 15
keyboard perforator, 17
knife-edge, 46, 117
knife-switch, 235
K relay, 141, 151, 161

La Porte (exchange), 80, 140
large groups, 181
lead-acid cell, 232, 240
leakage current, 60
leakage flux, 30, 46
letter case and shift, 68
level, 86
lever, 12
  spring, 32
lightning, 209
limited availability, 183
line, 2
  capacitance, 63
  disengaged signal, 72
  fuses (M.D.F.), 210, 212
  resistance, 64, 74
  side (M.D.F.), 205, 210
  splitting, 76
  uniselector, 88
linefinder, 90
link, 82, 95, 110
linked numbering, 100

local
  battery telephone, 10
  exchange, 3
  side (I.D.F.), 206
loop
  -disconnect pulsing, 63, 74
  resistance, 74
  signalling, 62, 69
    junction, 74
lost calls, 180
lost motion period, 23
lower case, 68
LR relay, 141, 161
l.s.i., 13

magnet, force, 251
magnetic circuit, 30, 41, 50, 54
magneto, 223
  bell, 6
magnetomotive force (m.m.f.), 33
main distribution frame (M.D.F.), 205, 210
mains, 223
  rectifier, 228
main switching centre (M.S.C.), 4
make-before-break contact, 31
make contact, 31
manual exchange, 3, 80
marker, 92, 98, 104
marking, 14, 52, 67, 82
  bias, 43
matrix, 81, 98
maximum power (matching), 6, 7, 253
mechanically-operated springs, 125, 132, 135
mercury contact, 240
message switching, 15, 17
mesh connexion, 232
metering, 58, 71, 75, 99, 159
microphone feed current, 5
minor exchange, 4
Morse code, 2, 59, 67, 249
Morse key, 2
motor, 227
  characteristics, 228
  -generator, 225
  starters, 225, 229
multi-frequency (m.f.), 67, 71
multiple, 83, 90
  side (I.D.F.), 205, 219
  terminals, 197

N and *NR* springs, 125
NC signal, 59, 72
negative wire, 67
neutral signalling (see single-current)
new exchange, traffic, 182
nickel-alkaline cell, 239
nickel-iron (NiFe), 36
  core, 36, 46
  sleeves, 40, 46
non-blocking array, 98
non-bridging wiper, 118
non-director, 99
non-inductive winding, 46
normal post (*NP*, *NPA* and *NPB*) springs, 125
NP signal, 72
n-type semiconductor, 229

numerical digits, 101
number unobtainable tone (N.U.), 70, 159, 210
NV magnet, 228

off-normal contacts, 13, 125
operate current, 34
operate lag, 34, 37
operating, ampere-turns, 34
outlet testing, 149
overload relay, 236

paper tape, 16
parallel battery, 242
partial float, 234, 239, 242
party line, 3
pawl, 12, 88, 116, 122, 130, 134
  abutment, 12
  arm, 122
  guide, 122, 130, 134
  spring, 117, 122, 130, 134
  stops, 117, 122, 130, 134
paystation, 3
peak traffic, 177
Pentaconta, 103, 105
Pentex, 107
perforator, 15
permanent magnet
  machine, 223
  polarization, 5, 6, 20, 42, 251
  reluctance, 20
permeability, 30, 40
P.G. alarm, 165
phonograms, 4
pitch of ratchet teeth, 117
planar transfer, 98
plane, 109
plant provision, 180
plastic (bakelite) front check, 36, 47
plug, 82
point-to-point, 3
polarization, 5, 251
polarized relay, 29, 40, 42, 49
polarizing current, 5
polar signalling (see double-current)
positive battery, 159, 221
positive wire, 62
power plant, 221
power switchboard, 234
pre-operated *CD* relay, 145, 163, 167
pre-operated testing relay (*H*), 152, 164
pre-2000 type selector, 121, 128
primary
  cell, 232
  character, 68
  winding, 7
prime mover, 223
printing re-perforator, 15
private manual branch exchange (P.M.B.X.), 82
private wire (*P*-wire), 149, 159
probability, 178
proceed-to-select signal, 72
protection and protectors, 209
p-type semiconductor, 229
pulse
  -controlled stepping, 87, 115, 143
  distortion, 39, 63
  speed and ratio, 23, 63
pure chance traffic, 176
push-button dialling, 13, 71

quantity of electricity, 231

racks, 135, 197
rack type M.D.F., 210
radio, 4, 19
random traffic, 176
rank, 91
ratchet
  for two-motion selectors, 121
  mechanism, 87
  wheel, 12
ratio, 23, 63
rear-end slug, 37
receiver, telephone, 5, 6
receiver, teleprinter, 14, 18
receiving cam, 14
receiving comb, 18
rectifier
  full-wave and half-wave, 231
  polarization of relay, 42
  power plant, 228, 242
  semiconductor (silicon), 42, 228
  shunt, relay, 38
  thermionic, 228
  three-phase, 231
reed relay, 53, 82
reflected impedance, 8
refresher charge, 244
register, 92, 98, 104
regulator, telephone, 21
regulator, voltage, 240
relay, 29
  high speed, 36, 48
  inductance, 34, 40
  mounting plate, 47
  practical types, 46
  slow, 36, 38
  wiring tags, 46
release
  current, 34
  lag, 34, 137
  magnet (*Z* or *ZM*), 120, 133
  of selector, 120, 160
reluctance, 33
residual
  air-gap, 31
  magnetism, 31
  screw, 47
  stud (pin), 31, 47, 126
resistance shunt, relay, 38
retard coil, 47
reverse
  -acting mechanism, 87, 115, 126
  battery supervision, 74
  current circuit breaker, 236
ring, 83
ringer, 6
ringing, 58
  frequency, 141
  key, 81
  machine, 69, 228
ripple, 237

rocking armature receiver, 20
room noise and sidetone, 9
rotary
  disk, 121, 131
  hunting, 149
  off-normal contact (*NR*), 125
  stepping circuit, 147
  stepping mechanism, 120, 130, 133
routing digits, 101

sandwich winding, 40
secondary
  cell, 232
  character, 68
  winding, 8
second selector, 88
sectionalization, 109
select bar, 93
selecting lever, 14
selector
  cradle, 124
  crossbar, 103
  mechanism, 86, 115
  multiple, 90
  racks, 135, 197
self-drive, 87, 115
semiconductor (silicon) rectifier, 42, 228
series-wound machines, 227
service signals, 72
SE50 (4000 type) selector, 132
shared service, 3
shelf jack, 124
shelf multiple, 135, 194
shift signal, 68
short-circuited relay, 38, 143
shunt
  ammeter, 235
  -field relay, 41
  -wound machines, 227
side-stable relay, 43
sidetone, 9
signal element, 59
signalling, 58
  codes, 59
single
  battery float, 244
  -current signals, 63
  -ended wipers, 119
  frequency, 67, 75
  -phase, 224
  wire signals, 60
siting of exchange, 4
SI units, 33, 221
sleeve, 83
  control, 84
sleeves, nickel-iron, 40, 46
slip ring, 227
slow relays, 36, 38
slugged relays, 36, 143
slug, size, 47
smooth grading, 190
smoothing, 237
solid segment (arc), 117, 126
sounder, 2, 3, 4
sound waves, 1, 5
spacing, 14, 52, 67
  bias, 43
spare cable pairs, 213
spark quench, 13, 19, 163, 165, 168
speak key, 82
speech current, 5
speech frequency, 2
speed, dial, 23, 63
speed, stepping, 128, 132
split, 76
spring load, 30
springset, 11, 12
squirrel cage motor, 225
S springs, 125
standard operator's load, 174
star connexion, 231
starting resistance, 226
start pulse, 14, 68
start-stop telegraphy, 13, 14
stehlic *B* relay, 144
step-by-step, 86
stepping, 87
  circuit, 143
  mechanism, 87, 115, 120, 130, 132
  speed, 128, 132
stop pulse (signal), 14, 68
storage battery, 232
straps, 195
striker, 123
Strowger, Almon B., 80
Strowger system, 80, 86
  trunking, 88, 173
Strowger, Walter, 140
subscribers'
  I.D.F., 208
  line, 3
    circuit, 48, 161
    signals, 58, 72, 141
    uniselector, 88, 161
suite of racks, 135
sum of successive differences, 190
supervisory (supy.)
  alarm, 164, 166
  relay set, 107
  signal, 59
swift, 217
switchboard, 80, 82
switched network, 3
symbols, 255

tandem, 101
telecommunication, 1
telegram, 4
telegraph, 1
  instrument (see teleprinter)
  message switching, 15, 17
  network, 1, 4
  relays, 29, 42, 49
  signals, 2
  speed, 51
telephone, 1
  instruments, 18
  network, 1
  receiver, 5, 6
  relays, 29, 30, 46
  -telegram, 4

transmitter, 5
telephonist, 80
teleprinter, 2, 13, 227
electronic, 13
keyboard, 2
receiver, 14, 17
transmitter, 14, 15, 16
Telex, 4
selector circuits, 156
signalling, 59, 72
terminating set, 75
testing of lines, 209
testing of outlets, 149
test jack, 136, 211
frame (T.J.F.), 209
thermal relay, 52
thermionic rectifier, 228
three
-magnet selector, 120, 129, 132
-phase rectifier, 231
-phase supply, 224
tie, 197
time constant (inductive), 36
timing,
relays, 34
selector magnets, 128, 132
tip, 83
toggle interrupter, 123, 134
tone calling, 6
tones, 50, 65, 67, 70, 141
torn-tape, 15
total traffic, 174
tractive effort, 30, 33
traffic, 173
intensity, 174
on selector level, 91
recorder, 176
unit (T.U.), 174
variation, 176
transformer, 7, 65
transit switching centre (T.S.C.), 4
transmission bridge, 11, 40, 65, 71, 84, 99, 104, 150
transmitter
feed current, 5
telephone, 5
teleprinter, 14
transmitting cam, 14
transposed links, 95
transposition method (grading), 193
travel, 31
trickle charge, 244
trigger dial, 22
trunk, 96, 173
distribution frame (T.D.F.), 194
exchange and switchboard, 4, 84
line, 4
trunking, 173
diagram, 89, 100
twin contact, 33
two-motion selector, 86
circuits, 152, 163
magnet times, 132
mechanism, 120
practical types, 128
release, 120, 133, 160
wiper and bank, 86, 123, 129

TXE2, 48, 107
TXE4, 98, 108
TXK1, 104
TXK3, 105
typebar, 18

unguarded interval, 152, 155
unidirectional junction, 59
uniselector, 86, 88
line circuit, 88, 161
magnet times, 128
mechanism, 115
outlet testing, 150
practical types, 126
racks and grading, 135, 197
unit call, 174
units, 33, 173, 221
U-points, 124
upper case, 68

variable inductor voltage control, 244
variation of traffic, 176
vertical, 210
and rotary ratchet, 121
and rotary stepping circuit, 147, 167
jumper, 202
marking bank (V.M.B.), 124, 129
off-normal contacts, 125
side (M.D.F. and I.D.F.), 208
stepping circuit, 145
stepping mechanism, 120, 130, 132
swift, 217
voice frequency (V.F.) signalling, 67, 75
voice immunity, 67, 76
voltage regulator, 239
voltage of supply, 221

watt, 221
wiper, 86
carriage, 121, 129
cords, 124
two-motion selector, 123, 129
uniselector, 119, 126
work, 221
wrapped joint, 215

x action contact, 47

y action contact, 47
yoke, 30, 126, 133

Z or *ZM* magnet, 120, 133

2B polarized relay, 51
4-group grading, 187, 191, 196
6-group grading, 192
12 type relay, 40
25-point uniselector, 126
100 line final selector, 165
100 output group selector, 152, 163
200 line final selector, 165
200 outlet group selector, 154
299 type relay, 49
328 type relay, 50
332 type telephone, 18
600 type relay, 48
706 type telephone, 19
2000 type selector, 129
3000 type relay, 46
4000 type selector, 132
5005 crossbar, 104